Lecture Notes in Computational Vision and Biomechanics

Volume 7

For further volumes:
http://www.springer.com/series/8910

Lecture Notes in Computational Vision and Biomechanics

The research related to the analysis of living structures (Biomechanics) has been a source of recent research in several distinct areas of science, for example, Mathematics, Mechanical Engineering, Physics, Informatics, Medicine and Sport. However, for its successful achievement, numerous research topics should be considered, such as image processing and analysis, geometric and numerical modelling, biomechanics, experimental analysis, mechanobiology and enhanced visualization, and their application to real cases must be developed and more investigation is needed. Additionally, enhanced hardware solutions and less invasive devices are demanded.

On the other hand, Image Analysis (Computational Vision) is used for the extraction of high level information from static images or dynamic image sequences. Examples of applications involving image analysis can be the study of motion of structures from image sequences, shape reconstruction from images and medical diagnosis. As a multidisciplinary area, Computational Vision considers techniques and methods from other disciplines, such as Artificial Intelligence, Signal Processing, Mathematics, Physics and Informatics. Despite the many research projects in this area, more robust and efficient methods of Computational Imaging are still demanded in many application domains in Medicine, and their validation in real scenarios is matter of urgency.

These two important and predominant branches of Science are increasingly considered to be strongly connected and related. Hence, the main goal of the LNCV&B book series consists of the provision of a comprehensive forum for discussion on the current state-of-the-art in these fields by emphasizing their connection. The book series covers (but is not limited to):

- Applications of Computational Vision and Biomechanics
- Biometrics and Biomedical Pattern Analysis
- Cellular Imaging and Cellular Mechanics
- Clinical Biomechanics
- Computational Bioimaging and Visualization
- Computational Biology in Biomedical Imaging
- Development of Biomechanical Devices
- Device and Technique Development for Biomedical Imaging
- Digital Geometry Algorithms for Computational Vision and Visualization
- Experimental Biomechanics
- Gait & Posture Mechanics
- Multiscale Analysis in Biomechanics
- Neuromuscular Biomechanics
- Numerical Methods for Living Tissues
- Numerical Simulation
- Software Development on Computational Vision and Biomechanics
- Grid and High Performance Computing for Computational Vision and Biomechanics
- Image-based Geometric Modeling and Mesh Generation
- Image Processing and Analysis
- Image Processing and Visualization in Biofluids
- Image Understanding
- Material Models
- Mechanobiology
- Medical Image Analysis
- Molecular Mechanics
- Multi-modal Image Systems
- Multiscale Biosensors in Biomedical Imaging
- Multiscale Devices and Biomems for Biomedical Imaging
- Musculoskeletal Biomechanics
- Sport Biomechanics
- Virtual Reality in Biomechanics
- Vision Systems

Manuel González Hidalgo
Arnau Mir Torres · Javier Varona Gómez
Editors

Deformation Models

Tracking, Animation and Applications

Springer

Editors
Manuel González Hidalgo
Department of Mathematics
and Computer Science
University of the Balearic Islands
Palma de Mallorca
Spain

Javier Varona Gómez
Department of Mathematics
and Computer Science
University of the Balearic Islands
Palma de Mallorca
Spain

Arnau Mir Torres
Department of Mathematics
and Computer Science
University of the Balearic Islands
Palma de Mallorca
Spain

ISSN 2212-9391 ISSN 2212-9413 (electronic)
ISBN 978-94-007-5445-4 ISBN 978-94-007-5446-1 (eBook)
DOI 10.1007/978-94-007-5446-1
Springer Dordrecht Heidelberg New York London

Library of Congress Control Number: 2012949703

Printed on acid-free paper

Springer is part of Springer Science+Business Media (www.springer.com)

Preface

The computational modeling of deformations has been actively studied for the last 30 years. This is mainly due to its large range of applications that include computer animation, medical imaging, shape estimation, face deformation as well as other parts of the human body, and object tracking. In addition, these advances have been supported by the evolution of computer processing capabilities, enabling realism in a more sophisticated way.

This book encompasses relevant works of expert researchers in the field of deformation models and their applications. The primary audience for this work are researchers from different multidisciplinary fields, such as those related with Computer Graphics, Computer Vision, Computer Imaging, Biomedicine, Bioengineering, Mathematics, Physics, Medical Imaging, and Medicine.

This book is divided into two main parts. Part I presents recent object deformation techniques from the point of view of computer graphics and computer animation. First, Palmer et al. present a survey of all the most modern techniques in the representation of deformable objects such as NURBS representation, free form representation, level sets, mass-spring models, algebraic surfaces, modal decomposition, and so on. Next, in Chap. 2, Raffin presents the free-form deformation techniques and its application to deform objects maintaining their topology and geometry.

Chapter 3 introduces us to the cage deformation techniques. Nieto and Susin show how these techniques are useful in modeling, texturing, and animation. In addition, the advantages and drawbacks of these techniques are shown, as well as how they are used in order to deform more complex systems. In the following chapter, Xie shows how to deform an image gradient using level set. The 2D and 3D deformable model segmentation is used to capture complex geometries and to deal with difficult initializations, weak edges, and broken boundaries.

In Chap. 5, Buades et al. present a new methodology to design shoes based on the biomechanical anatomical structure of the foot and of the deformable shape. The use of a deformable model is introduced in order to design shoes that are perfectly adapted to the foot's shape. The last chapter of this part introduces a deformable model, which allows the interactive simulation of objects with heterogeneous material properties and complex geometries. The model presented by Gilles et al. combines the realism of physically based continuum mechanics models and the usability of frame-based skinning methods.

The work presented thus far covers the geometric- and physics-based deformation models. In computer vision, however, modeling deformations is necessary in order to study the variations of natural shapes so as to extract useful real information from image or video sequences. Thus, deformation models are important for the accurate localization of complex structures in applications ranging from facial feature detection to tracking of anatomical structures in medical images.

Part II of this book presents six works that study deformations from a computer vision point of view with a common characteristic: deformations are applied in real-world applications.

First, the work of Marques et al. presents four methods to initialize deformable models in order to properly extract 2D and 3D shape estimations. These algorithms show an improved performance when compared to the classical techniques. This is proved experimentally in the estimation of facial features in 2D face images and in the detection of deformations of the left ventricle in an ultrasound 3D volume. The following chapters are mainly devoted to these two fields of application of deformations models. Particularly, one of the most recent impressive advances in facial feature tracking is explained in Chap. 8, where Saragih reviews the constrained local models (CLM). This approach for deformable face alignment leverages the generalization properties of local appearance representations of parts and the strong global constraints imposed by the geometrical relationships between part locations. Specifically, this work places CLM in the general context of deformable face alignment, highlighting its similarities and differences with other approaches and justifying its benefits.

In the field of medical imaging, the work of the Siemens research team has achieved a great interest from the scientific community by using a robust approach to deformation models in medical applications. Chapter 9 presents their robust learning-based fusion framework, demonstrating it by means of various medical image analysis applications. The framework combines the prior information with traditional tracking approaches based on template matching and registration, in order to maintain an anatomically consistent representation of target appearance. This representation has proved it can cope with inherent changes due to target movement, imaging device movement, varying imaging conditions, and is consistent with the domain expert clinical knowledge. A different approach in medical imaging applications is presented in the next chapter, where Igual et al. describe an automatic segmentation method for brain medical images. Its approximation defines a deformable model by combining an atlas-based segmentation strategy with a well-known computer vision technique, the Graph-cut model, which is adapted to make it suitable for segmenting small and low contrast structures. The obtained results show improved performance in terms of segmentation accuracy compared to current approaches. The last two chapters present two particular applications of deformation models in medical applications. In Chap. 11, Fürtinger et al. map a Purkinje fiber network from a real heart. The elastic deformations guarantee that despite even large differences in the endocardial geometries of both models, the artificial Purkinje fiber network is mapped sufficiently close to the real endocardium. In the last chapter, Bilgen applies deformation models in order to obtain an accurate and precise motion estimate in elastography.

Finally, we want to thank the chapter authors for providing work of great quality and for their collaboration in all the book edition process. We believe that thanks to the achieved contributions, this book gives an excellent overview of the state of the art of the applications of deformation models in animation and computer vision.

School of Engineering
University of Balearic Islands
March 2011

Manuel González-Hidalgo
Arnau Mir
Javier Varona

Contents

Contributors

Mehmet Bilgen Health and Translational Medicine, University of Malaya, 50603 Kuala Lumpur, Malaysia, e-mail: mehmet.bilgen@yahoo.com

Guillaume Bousquet INRIA LJK-CNRS, University of Grenoble, Grenoble, France, e-mail: guillaume.bousquet@inria.fr

J. M. Buades Computer Graphics Vision and Artificial Intelligence Group, Department of Mathematics and Computer Science, University of Balearic Island, Anselm Turmeda Building, Crta. Valldemossa Km 7.5, 07122 Palma de Mallorca, Spain, e-mail: josemaria.buades@uib.es

Terrence Chen Siemens Corporate Research, Princeton, NJ 08540, USA, e-mail: terrence.chen@siemens.com

Dorin Comaniciu Siemens Corporate Research, Princeton, NJ 08540, USA, e-mail: dorin.comaniciu@siemens.com

Sergio Escalera University of Barcelona, Gran Via de les Corts Catalanes 585, 08007 Barcelona, Spain; Computer Vision Center (CVC), Edifici 0, Campus UAB, Bellaterra, Barcelona, Spain, e-mail: sergio@maia.ub.es

François Faure INRIA LJK-CNRS, University of Grenoble, Grenoble, France, e-mail: francois.faure@inria.fr

Stefan Fürtinger Institute for Mathematics and Scientific Computing, Karl Franzens University, Graz, Austria, e-mail: stefan.fuertinger@uni-graz.at

Bogdan Georgescu Siemens Corporate Research, Princeton, NJ 08540, USA, e-mail: bogdan.georgescu@siemens.com

Benjamin Gilles INRIA LIRMM-CNRS, University of Montpellier, Montpellier, France, e-mail: benjamin.gilles@inria.fr

Manuel González-Hidalgo Department of Mathematics and Computer Science, University of Balearic Islands, Anselm Turmeda Building, Crta. Valldemossa Km 7.5, 07122 Palma de Mallorca, Spain, e-mail: manuel.gonzalez@uib.es

Antonio Hernández-Vela University of Barcelona, Gran Via de les Corts Catalanes 585, 08007 Barcelona, Spain; Computer Vision Center (CVC), Edifici 0, Campus UAB, Bellaterra, Barcelona, Spain, e-mail: ahernandez@cvc.uab.es

Laura Igual University of Barcelona, Gran Via de les Corts Catalanes 585, 08007 Barcelona, Spain; Computer Vision Center (CVC), Edifici 0, Campus UAB, Bellaterra, Barcelona, Spain, e-mail: ligual@ub.edu

Razvan Ionasec Siemens Corporate Research, Princeton, NJ 08540, USA, e-mail: razvan.ionasec@siemens.com

Stephen Keeling Institute for Mathematics and Scientific Computing, Karl Franzens University, Graz, Austria, e-mail: stephen.keeling@uni-graz.at

Xiaoguang Lu Siemens Corporate Research, Princeton, NJ 08540, USA, e-mail: Lvxiaogu@gmail.com

Jorge S. Marques Institute for Systems and Robotics, Instituto Superior Tecnico av. Rovisco Pais, 1049-001 Lisboa, Portugal, e-mail: jsr@isr.ist.utl.pt

Arnau Mir Department of Mathematics and Computer Science, University of Balearic Islands, Anselm Turmeda Building, Crta. Valldemossa Km 7.5, 07122 Palma de Mallorca, Spain, e-mail: arnau.mir@uib.es

Majid Mirmehdi Department of Computer Science, Bristol University, Merchant Ventures Building, Bristol BS8 1UB, UK, e-mail: majid@cs.bris.ac.uk

E. Montiel INESCOP, Footwear Technological Institute, Polig. Ind. Campo Alto, C/. Alemania 102, Aptdo. Correos 253, 03600 Elda-Alicante, Spain, e-mail: emontiel@inescop.es

Jacinto C. Nascimento Institute for Systems and Robotics, Instituto Superior Tecnico av. Rovisco Pais, 1049-001 Lisboa, Portugal, e-mail: jan@isr.ist.utl.pt

Jesús R. Nieto Universitat Politecnica de Catalunya, Barcelona, Portugal, e-mail: jrodrigueznieto@gmail.com

Perumal Nithiarasu College of Engineering, Talbot Building, Swansea University, Singleton Park, Swansea SA2 8PP, UK, e-mail: P.Nithiarasu@swansea.ac.uk

A. Oliver Computer Graphics Vision and Artificial Intelligence Group, Department of Mathematics and Computer Science, University of Balearic Island, Anselm Turmeda Building, Crta. Valldemossa Km 7.5, 07122 Palma de Mallorca, Spain, e-mail: antoni.oliver.tomas@gmail.com

Dinesh K. Pai University of British Columbia, British Columbia, Canada, e-mail: pai@cs.ubc.ca

Pere Palmer Department of Mathematics and Computer Science, University of Balearic Islands, Anselm Turmeda Building, Crta. Valldemossa Km 7.5, 07122 Palma de Mallorca, Spain, e-mail: pere.palmer@uib.es

Francisco J. Perales Computer Graphics Vision and Artificial Intelligence Group, Department of Mathematics and Computer Science, University of Balearic Island, Anselm Turmeda Building, Crta. Valldemossa Km 7.5, 07122 Palma de Mallorca, Spain, e-mail: paco.perales@uib.es

Gernot Plank Institute for Biophysics, Medical University, Graz, Austria, e-mail: gernot.plank@medunigraz.at

Anton J. Prassl Institute for Biophysics, Medical University, Graz, Austria, e-mail: anton.prassl@medunigraz.at

Petia Radeva University of Barcelona, Gran Via de les Corts Catalanes 585, 08007 Barcelona, Spain; Computer Vision Center (CVC), Edifici 0, Campus UAB, Bellaterra, Barcelona, Spain, e-mail: petia@cvc.uab.es

Romain Raffin LSIS UMR 7296, Aix-Marseille University, Domaine Universitaire de Saint Jerome, Av. Escadrille Normandie-Niemen, 13397 Marseille cedex 20, France, e-mail: romain.raffin@lsis.org

S. Ramis-Guarinos Computer Graphics Vision and Artificial Intelligence Group, Department of Mathematics and Computer Science, University of Balearic Island, Anselm Turmeda Building, Crta. Valldemossa Km 7.5, 07122 Palma de Mallorca, Spain, e-mail: silvi.ragu@gmail.com

Carlos Santiago Institute for Systems and Robotics, Instituto Superior Tecnico av. Rovisco Pais, 1049-001 Lisboa, Portugal

Jason M. Saragih CSIRO, 1 Technology Crt, Pullenvale, QLD 4096, Australia, e-mail: jason.saragih@csiro.au

Igor Sazonov College of Engineering, Talbot Building, Swansea University, Singleton Park, Swansea SA2 8PP, UK, e-mail: i.sazonov@swansea.ac.uk

Joan Carles Soliva Department of Psychiatry, IAPS Hospital del Mar., Universitat Autònoma de Barcelona (UAB) Unitat de Recerca en Neurociència Cognitiva (URNC), Passeig Marítim 25-29, 08003 Barcelona, Spain, e-mail: 24744jsv@comb.es

Antonio Susín Universitat Politecnica de Catalunya, Barcelona, Portugal, e-mail: toni.susin@upc.edu

Oscar Vilarroya Department of Psychiatry, IAPS Hospital del Mar., Universitat Autònoma de Barcelona (UAB) Unitat de Recerca en Neurociència Cognitiva (URNC), Passeig Marítim 25-29, 08003 Barcelona, Spain, e-mail: oscar.vilarroya@gmail.com

Peng Wang Siemens Corporate Research, Princeton, NJ 08540, USA, e-mail: WangPengAt-Work@gmail.com

Yang Wang Siemens Corporate Research, Princeton, NJ 08540, USA, e-mail: yangwang@siemens.com

Wen Wu Siemens Corporate Research, Princeton, NJ 08540, USA, e-mail: wenwu@cs.cmu.edu

Xianghua Xie Department of Computer Science, Swansea University, Faraday Tower, Singleton Park, Swansea SA2 8PP, UK, e-mail: x.xie@swansea.ac.uk

Si Yong Yeo College of Engineering, Talbot Building, Swansea University, Singleton Park, Swansea SA2 8PP, UK, e-mail: yeosy@ihpc.a-star.edu.sg

Yefeng Zheng Siemens Corporate Research, Princeton, NJ 08540, USA, e-mail: yefeng.zheng@siemens.com

Part I
Fundamentals and Animation Applications

Deformable Objects Representation

Pere Palmer, Arnau Mir and Manuel González-Hidalgo

Abstract In order to have a good representation of deformable objects, is clever to have an adequate model to represent them. Since the origins of computer graphics, a large number of representation models have been presented. Not all of them are adequate to represent deformable objects. The way in which the objects can be handled, allowing local changes in their shape, or the possibility of the creation of objects matching with data gathered from one or more datasources, are some of the desirable characteristics in those models. This chapter represents a study of the way in which the deformable objects can be represented. It is possible to classify them into some categories allowing to select the most adequate depending on the use given to the model.

1 Introduction

In order to simulate the deformable objects behaviour in a computer system, this must be able to represent the shape of those objects. To do this, it will be necessary to describe the geometry of these objects and also determine the evolution of this geometry along the deformation process [88].

In the literature related to deformation models many surveys has been appeared some of them are classics [54, 85], and other are more recent [51, 94], and some covered in depth some topics more specific [17, 27, 48, 102, 125].

P. Palmer (✉) · A. Mir · M. González-Hidalgo
Department of Mathematics and Computer Science, University of Balearic Islands, Anselm Turmeda Building. Crta. Valldemossa Km 7.5, E-07122 Palma de Mallorca, Spain
e-mail: pere.palmer@uib.es

A. Mir
e-mail: arnau.mir@uib.es

M. González-Hidalgo
e-mail: manuel.gonzalez@uib.es

M. González Hidalgo et al. (eds.), *Deformation Models*, Lecture Notes in Computational Vision and Biomechanics 7, DOI: 10.1007/978-94-007-5446-1_1,

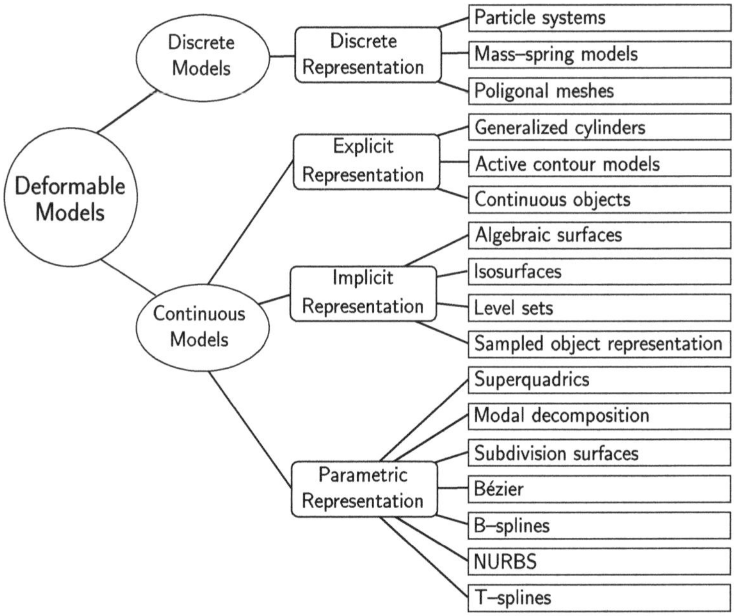

Fig. 1 A classification of the geometric representation of deformable objects

In this chapter we will study the approaches, the representation of the geometry of deformable objects, that has been raised over the time. In each case, an indication of the aspects for which they are more appropriate will be given.

A first classification of object representation models states that such models can be continuous or discretes. In the case of the discrete representation the shape is inferred from a finite set of known points on the object surface. The knowledge is incomplete, so that numerical problems can arise when performing certain differential calculations (normal at a given point, higher derivatives, ...). It is quite usual to consider as a discrete representations those having only C^0 continuity on the surface. By contrast, in a continuous representation the whole surface is well known, so in principle, all geometric parameters are well defined and are known at any point on the surface. This second approach has its own drawbacks, because if a computational treatment is required, it will be necessary to perform a spatial discretization of the model.

Within the continuous representation models is possible to make a distinction between parametric models and the explicit and implicit models, see Fig. 1.

2 Discrete Models

This kind of object representation has been used long time before the development of computers. During the Renaissance, from the hand of Leonardo da Vinci, Luca

(a) (b) (c)

Fig. 2 Discrete models are, in fact, prior to computing. Their use is clearly documented since the fifteenth century. **a** Paolo Uccello, *Perspective study of a chalice*, around 1450, pencil on paper, the Uffizi Gallery, Florence, Italy. **b** An example of *intarsia*, mosaic of engraved wood, made by *Fra* Giovanni Pacioli for the Cathedral of Santa Maria in Organo, Verona, Italy, in the early sixteenth century. In it, it can be seen some shapes obtained from drawings by Leonardo da Vinci, as the high rombicuboctahedron, **c**, which appears in *De Divina Proportione* of the same Pacioli printed in 1509

Pacioli, o Paolo *Uccello*,[1] there is already geometric representations of three dimensional objects in a way in which nowadays are so-called polygonal meshes, see Fig. 2.

The aim of these models is to eliminate the parameters present in the other models and, thereby, eliminate the disadvantages that these parameters imply. However, not everything is profit. On one hand most of the difficulties faced by other models are solved but on the other hand these models lead to an explicit numerical scheme which can lead to numerical problems because the equations of motion obtained do not allow the regularization of the surface and may be necessary to impose restrictions.

2.1 Polygonal Meshes

A polygonal mesh, $\mathcal{M}$, allows to describe an object from a discrete set of points and a list of connectivity between them, so that each point has a relation of neighboring points. Therefore, a mesh is defined by a couple, $\mathcal{M} = \langle \mathcal{V}, \mathcal{N} \rangle$, where

- $\mathcal{V}$ is the vertex set, defined in a space $\mathbb{R}^d$:

$$\mathcal{V} = \{P_i\}_{i \in I} ,$$

---

1 one of whose works appears on the cover of *Computer Aided Geometric Design.*

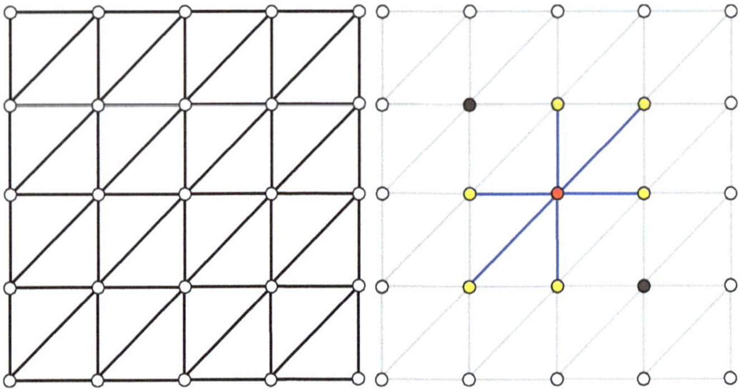

Fig. 3 In **a** a polygonal mesh is shown, and in **b** a representation of the connectivity function, showing, for the *red vertex*, their neighbours (the *yellow* ones) and the edges connecting them (in *blue*). Note that only the connected vertexes by an edge are considered neighbours. In the figure, the *gray vertexes* are not considered as neighbours

- and $\mathcal{N}$ is the connectivity function of the mesh:

$$\begin{aligned} \mathcal{N} : \mathcal{V} &\longrightarrow \mathcal{V}^* \\ P_i &\mapsto \left\{P_{i,1}, P_{i,2}, \ldots, P_{i,m_i}\right\}. \end{aligned}$$

  Where $\mathcal{V}^*$ is the set of finite parts of $\mathcal{V}$ and m_i $(m_i \geq 2)$ indicates number of neighbours of the vertex P_i. The Fig. 3 clarifies the concept.
  The function $\mathcal{N}$ and the elements set $\mathcal{V}^*$ satisfies:
- If two vertices are connected, they do so by a single edge:

$$\forall i \in I, \quad P_{i,j} \neq P_{i,k} \forall j,k : 1 \leq j,k \leq m_i \wedge j \neq k.$$

- A vertex is not joined to itself:

$$\forall i \in I, \quad P_i \notin \mathcal{N}(P_i).$$

The edge set, $\mathcal{E}$, of a polygonal mesh is defined as:

$$\mathcal{E} = \left\{(P_i, P_{i,j}) \,\forall P_i \in \mathcal{V}, \; \forall P_{i,j} \in \mathcal{N}(P_i)\right\},$$

from $\mathcal{E}$ the set of polygons P is defined as:

$$\mathcal{P} = \left\{f = (e_{i_1,i_2}, e_{i_2,i_3}, \ldots, e_{i_l,i_1}), e_{i_k,i_{k+1}} \in \mathcal{E}\right\} \subseteq \mathcal{E}^*, \tag{1}$$

being l the number of edges of each polygon and $\mathcal{E}^*$ is the set of finite parts of $\mathcal{E}$.

There are multiple techniques to generate polygonal meshes from objects defined by some other method of representation. In [90] the objects are defined with cubic meshes, that is: the space is divided into regular cubes, in an analogous way as in a surface; and the mesh is generated automatically from a polygonal mesh of the surfaces defining the object.

If the value of l in (1) is always 3, then all the polygons defining the mesh are triangles, and the mesh is usually called triangulation. The use of triangles offers several advantages, for example: the connectivity of each polygon and the coplanarity of the three vertex related, makes the surfaces representation algorithms defined by triangulation very simple and efficient. In [133], a method of representing triangular and tetrahedral meshes is presented: triangular meshes defining objects, not only the surface of the objects. The particularity of this proposal is that the model allows to define complex objects without a high computational cost.

Other approaches based on this model are, among others: [1] which proposes an image segmentation based on triangular meshes which adapt their topology to fit the objects in the image.

The explicit models are difficult to use when the topology of the objects represented can change along the deformation process. Some works in this direction are dealing with this problem. In [122] is presented a model that uses triangular meshes to represent objects capable to be divided or cracked. Also in [138] the objects can also be split but also they can be merged or blended.

Another way in which the deformable object are represented is presented in [69], where the deformable objects are represented by a skeleton and over it a mesh which determines the volume and the shape.

Another interesting application of this model can be seen in [95], which describes a graphical tool, *FiberMesh*, with which a user can easily design complex free-form objects by the interpolation of a set of 3D curves and using some basic operations (curve creation, cutting, extrusion and tunneling).

Although the advantage is its simplicity, this kind of representation has some drawbacks. Among them, being dependent on the scale or level of approximation of the object.

2.1.1 Delaunay Triangulation

One of the ways in which a mesh can be established, and also a very effective way to do this, is known as Delaunay triangulation and its dual problem, the Voronoi diagrams:

Given a set of plane points, $\mathscr{P}$, a triangulation of the set is a subdivision of its convex hull in triangular regions by using edges whose vertexes are points of the set. Of all these, a Delaunay triangulation verifies [118, 119]:

- All the triangles are such that the circumference which passes for its vertexes does not contain any other point of $\mathscr{P}$.
- All the edges of the triangulation are such that there exists a circumference passing by its extremes and without any other point of the set.
- The minimal angle between edges is maximal. That is, any other possible triangulation has edges forming minor angles.

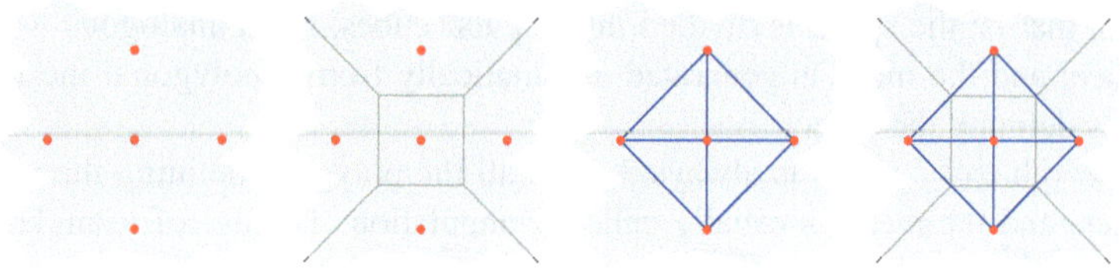

Fig. 4 For a set of points in a 2D space $\mathscr{P}$ (**a**), the space can be divided into a Voronoi diagram (**b**), that is, it is divided in regions such that the points of each region are nearer to the corresponding point of the set than to the others. Also the points can be interpreted as a vertexes of a Delaunay triangulation (**c**). The Voronoi diagram and the Delaunay triangulation are related, as it can be seen in (**d**)

Furthermore, it is also fulfilled that the Delaunay triangulation of a set of points is the dual graph of the Voronoi diagram of that set, see Fig. 4.

A Voronoi diagram is a tessellation of the plane based on the euclidean distance of the plane points to the points of the set $\mathscr{P}$, because it represents a plane partition into regions such that every point of the set is mapped into the geometrical place of the points of the plane nearest to it more than any other point of $\mathscr{P}$.

2.1.2 Simplex Meshes

Simplex meshes are discrete representations of objects which are characterized by the constant connectivity between vertexes [38, 39]. For the surfaces representation, 2-simple meshes are used, see Fig. 5. In these meshes, every point is joined through edges to, exactly, three different points.[2] The geometry is very simple because of the constant connectivity between vertexes.

A k-simple mesh, (M), defined in $\mathbb{R}^d$ represents a meshing of connectivity $k+1$. The main difference with respect to the others polygonal meshes is that the connectivity function is defined as follows:

$$\begin{array}{rcl} N: \quad V & \longrightarrow & \{V\}^{k+1} = \overbrace{V \times \cdots \times V}^{k+1} \\ P_i \in V & \longrightarrow & (N_1(i), N_2(i), \ldots N_{k+1}(i)). \end{array}$$

These meshes are a particular kind of meshes in which all vertexes have exactly the same number of neighbours. What makes interesting this approach is:

- Its generality: it is possible to represent any kind of orientable surfaces, being able to represent deformations of k-dimensional surfaces with $k+1$-simples meshes.
- Its easy and efficient implementation.

[2] There exists a duality between the representation based on triangulations and the one based on simplex meshes.

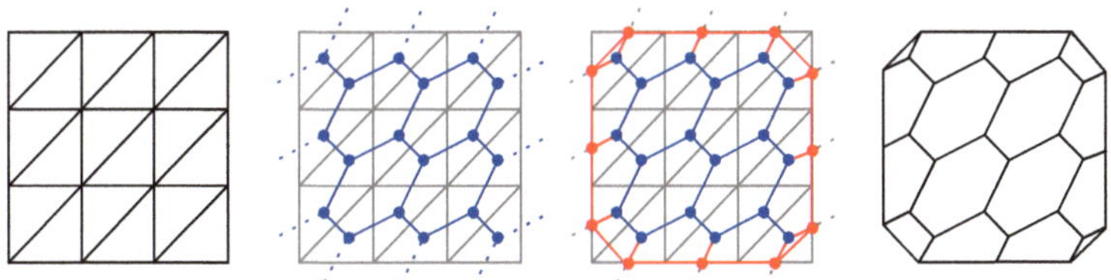

Fig. 5 Duality between a triangulation and a 2-simplex mesh. **a** Surface representation with a regular triangulation. **b** 2-simplex mesh, in *blue*, over the same surface, there exists a problem at the contour. **c** The meshing is *corrected* with the addition of the dual of the triangulation contour, in *red*. **d** The 2-simplex mesh finished with the contour

- Its adaptability: it is possible to refine the meshing to increase the number of vertexes in areas with high curvature. If the surface to build is based on a given series of points, it is also possible to refine the meshing to correct values or to adapt them to create contours where data is incomplete.

Furthermore, it can be shown [38], that a triangulation is equivalent to a simplex mesh in which the formed faces by the edges are flat. It is also possible to define a model with both approaches, this way it is possible to have the best features of each one combined [52].

This kind of representation is used in medical image segmentation [53, 88]. This model can also be used in image segmentation. Delingette [40] proposes a general tridimensional reconstruction algorithm of range and volumetric images, based on deformable simplex meshes, the objects recovered can be of any topology, it is also possible divide a single object into multiple parts.

Simplex Meshes are valuable models thanks to their good propensity to handle a large variety of shape alterations altogether with a fine resolution and stability. However, despite all these great characteristics, Simplex Meshes are lacking to cope satisfyingly with other related tasks, as rendering, mechanical simulation or reconstruction from iso-surfaces [52].

2.2 *Particle Systems*

The particle systems are based on the concept that every object to represent is formed by a set of *particles* that can be characterized by a set of physical parameters (mass, position, velocity, acceleration, ...). The behaviour of the particles is determined by the laws of Newtonian mechanics. The interaction of the particles is performed by attraction/repulsion forces with which it is possible to obtain a certain organization.

Due to its design, particle systems are well suited for representing viscous objects or fluids [33]. Although they are also used to represent the behaviour of fabrics [19–21, 46].

In [124] a model is proposed, the oriented particles, based on particles which have potential functions added that favour the organization of these particles in certain forms. Each particle has associated a matrix of rotation with respect to a reference system. With this matrix, together with a normal vector associated with each particle, it is possible to achieve potentials that induce particles to organize themselves according to certain restrictions of *co-planarity*, *co-normality* and *co-circularity*. This way, it is possible to get easily a representation of the surfaces.

The use of particle systems combined with other techniques has been used to represent objects from data obtained empirically. In [64], parametric curves has been used first to define the contours of organs obtained from medical images. Once the contour is obtained, the interior is created using particles of different sizes and weights arranged in layers. In [18] particles are used together with implicit surfaces to represent clouds. Each cloud is defined by successive particle layers. The final cloud shape is obtained by an explicit formulation. Heïgéas et al. [61] presents a modeling process in order to produce a realistic simulation of crowds based on particle systems. In particular the non-deliberative emergent crowd phenomena, that is: the self-organization of a group of people. Finally, in [82] the particle systems are proposed as a model to represent 1D and 2D shapes, as hair and cloth respectively.

The main drawback of the models based on particles is the difficulty that represents to obtain accurate measures of the way the objects are represented (for example: the surface curvature). Another drawback is the higher computational cost of the model.

2.3 Mass–Spring Models

Mass–spring models are one of the most used physical techniques. An object represented with this model consists on a set of punctual masses connected by a series of springs, with null mass and a certain length, forming a deformable structure.

The springs are often linears, based on de Hooke's law, but it is possible to use any other behavioral model for them. The object behaviour along the deformation process will be established by the application of Newton's second law to each particle, the forces acting are those fixed externally by the user and also those generated by the stretching of the different springs as a result of the motion of the particles [59, 101].

The particles can be connected with the springs in a way more or less complex, see Fig. 6. For example, in [105] three different kind of spring are stated: the *structural springs*, the *shear springs* and the *flexural springs*; each kind of spring acts as a reaction to a determined kind of stress (traction–compression, shearing and bending respectively).

These models are frequently used to represent cloth because the fabric structure, made by weft and warp, simplifies their description, representing the surface by a set of points, each one joined with its neighbours by several links giving to the set some properties of elasticity and flexibility [42, 86].

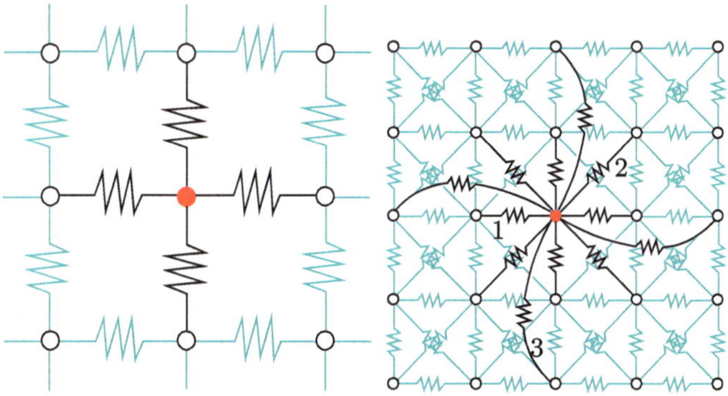

Fig. 6 Different mass and spring configurations. **a** Simple configuration: each mass is connected by springs with its immediate neighbours. **b** Configuration stated in [105]: each mass is connected by *structural springs* (1), *shear springs* (2) and *flexural springs* (3)

This is not the only way in which this model is used. In [129, 134] masses and springs, the so-called *adaptative meshes*, are used to reconstruct objects from images. In the same way, in [92] a mass–spring model is used to perform the segmentation of images and also to track the objects detected in them.

Far from being abandoned, this model remains as a topic of great interest. Thus, in [84] this model is used, paradoxically, to represent rigid object with high degree of realism in its movement. In [66], starting from oriented particles [124], a system in which the particles maintain a reference vectors is proposed. These vectors allow to represent the object with great stability without being necessary to add a lot of springs to ensure such stability.

In general it is very difficult to determine the correct elasticity parameters to represent the real behaviour of an object, although in [93] an automatic method is proposed.

In [109] the deformable objects are defined as a simplex meshes, but they really represent lumped masses (that is: a mass–spring model). The object evolution is driven by an image, the result is the segmentation of this image.

In a different approach, [113] uses a mass–spring model to represent hundred thousands hairs in an human head. This model takes into account the hair-to-hair interactions using physical techniques.

3 Explicit Models

Explicit models are based on the premise that objects can be represented in a way in which every of their point can be well known. In order to do this, it is usual describe them mathematically by using several parameters that determine their shape.

Thus, one object, $S_{\mathbf{q}}$, can be described from a vector of shape parameters[3] of the form, $\mathbf{q} = (q_1, \dots, q_{n_{\mathbf{q}}})^{\top}$, so that if Ω is the parametric space where the object is defined, then the representation of the object will be:

$$\begin{aligned} S_{\mathbf{q}} : \mathbb{R}^{n_{\mathbf{q}}} \times \Omega &\longrightarrow \mathbb{R}^3 \\ (q_1, \dots, q_{n_{\mathbf{q}}}, \mathbf{u}) &\longrightarrow (\mathbf{x}(\mathbf{q}, \mathbf{u}), \mathbf{y}(\mathbf{q}, \mathbf{u}), \mathbf{z}(\mathbf{q}, \mathbf{u})). \end{aligned} \tag{2}$$

The characteristics of the shape obtained, and the deformations it may receive, will be one or another depending on the type of the chosen parameters. For example, if the parameters allow controlling local deformations, that is, the deformations which affect a small part of objects, then the objects obtained will have very elaborated shapes, but with a significant computational effort. In contrast, if the parameters are controlling the whole object, then the computational cost will be lower and also the complexity of the objects obtained with them will be lower too.

This model has several drawbacks:

- They are continuous models, so they need a discretization in the parametric space. This discretization necessarily lead to some degree of error.
- The decomposition of the parametric space at regular intervals doesn't lead to a regular decomposition of the surface.
- To define the topology of the object, the explicit representation requires the introduction of boundary conditions.

However, this way the objects representation tends to be simple. In the first approaches to deformable objects, it was usual that the representation model was an approach that can be classified as an explicit model.

3.1 Generalized Cylinders

Among the earliest representations of deformable objects, it highlights those which were based on the way in which an outline evolved along a curve (Fig. 7).

Generalized cylinders consist of a space curve or axis and a set of cross sections described on this axis. They describe in a natural and intuitive way pieces which possess elongation, or which have axial symmetry. The generalized cylinders may be linked together in various ways to form Complex Objects [7]. The way in which the solid is obtained is by sweeping a curve along the axis. The cross sections define how the solid object is obtained from this swept. The curve can be deformed by a scaling function during its evolution along the axis. This scaling function can be described as a curve secant and perpendicular to the curve used as a model for the cross sections see Fig. 7.

[3] Several authors consider explicit models within the group of parametric models. However, some specific features also allow them to be considered as an independent group.

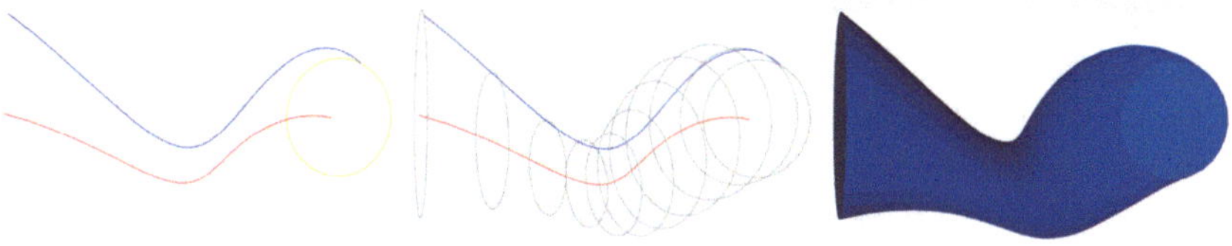

Fig. 7 A generalized cylinder is formed from a curve that serves as the axis, see **a**, on which another curve, the contour, can be used to define cross sections (**b**), to generate a shape (**c**)

The first references to jobs using this model date back in the early years of the 1970s of the twentieth century [7]. In [8], the generalized cylinders are used to recover the shape of real objects obtained from a primitive 3D laser-based scanner. The 3D object recovery from data gathered from images of real objects using generalized cylinders has been used in some other works afterwards, for example in [34, 132].

It is possible to build complex shapes through a combination of generalized cylinders. However, it is not always easy to represent a shape by using an axis and an outline, nor is it the most intuitive way of describe objects. Thus it does not appear to be an appropriate model to be used regularly.

The concept of generalized cylinders has been adopted as the basis for other models, such as the so-called *swung surfaces* [106], and the *swept surfaces* [103], which, in fact, are actually parametric surfaces although essentially they follow the approach of generalized cylinders.

Recently some new applications of generalized cylinders has been appeared. For example, [50] uses generalized cylinders to model soil microstructures, specifically the generalized cylinders are used to represent the pores which represent a complex volume into the soil space. The scope of the generalized cylinder has been extended to include fields such as the representation of medical images. In [87] generalized cylinders are used to segment and reconstruct 3D vascular trees. In the same field of application, in [137] the generalized cylinders are used to represent tendons, ligaments, and surgical sutures (in the same article, other biological tissues are modeled using triangular and quadrangular meshes).

An equivalent approach to generalized cylinders is presented in [75], where is presented a model capable to represent hollow tubes with deformable cross-sections.

3.2 Active Contour Models

Stated first in [68], the active contour models, also known as *snakes*, are curves which change their form in order to match a given shape into an image. The adjustment mechanism depends upon a number of parameters which impose restrictions. The *snakes* evolution allows to find the contours and lines, and also other characteristics of the objects represented on the images. In principle, the *snakes* are one-dimensional curves that can adapt to contours, track movements, etc. responding to internal tensions opposing the stretching and bending, to the forces the user can impose and those that can be inferred from the images. Some authors [36], have been extended the active contour model so it can be applied also in 3D contours.

The *snakes* are used regularly for the image enhancement and detection of organs on medical imaging [22–24]. In [107] their are used to gather the information needed to reconstruct, with a parametric model, the left ventricle of the heart. The natural context in which this model can be applied is in the image segmentation.

In this line, [76] propose an alternative method to reconstruct cranial defects. The authors combine medical information, low-resolution images and the active contour model to suppress partial volume defects, the result is a 3D skull with the defective part reconstructed.

In general, the active contour models are computationally expensive. This drawback is partially reduced in [145], its proposal is to analyze the curve gradient flow in order to study the evolution of active contour. When the curve is far from the object, the curve suffers a global translation, and when it is near to the object, the curve suffers a local deformation. With this approach the evolution of the active contour is more efficient and the computational effort less expensive. Another approach is proposed in [41], consists in discretize the space, replacing the continuous curve by a polygon, and defining the internal energy as a function based on the minimum length polygon.

Another problem of the active contour models, in the field of the images segmentation, is that it is not easy to separate adjacent objects. In [26] a graph-based method of active contour models, called network snakes, is presented and investigated. The main idea is to identify the objects into the image and also the relations between them, allowing differentiate adjacent objects correctly.

3.3 Continuous Objects

The models based on a continuous representation consider objects defined on a subset, Ω, of the 2D or 3D space. This approach is valid both for curves as for surfaces. Some of the papers with the greatest significance in the deformable objects representation context have proposed models following this approach [127, 130, 131].

The objects represented according to this model are in an euclidean space, 3D, under a framework, Φ. At a given time, t, the position of each point $u \in \Omega$, $x(u,t)$, is defined in several ways.

In a first formulation, [131], so-called *primal formulation* in [126], the placement of each material point is defined directly as:

$$r(u,t) = (r_1(u,t), r_2(u,t), r_3(u,t))^\top$$

This formulation leads to getting objects that can deform and move freely, but there is always some risk of getting an ill-conditioned system as revealed in [99]. This means that for the representation of objects with a degree of stiffness this formulation is not completely adequate.

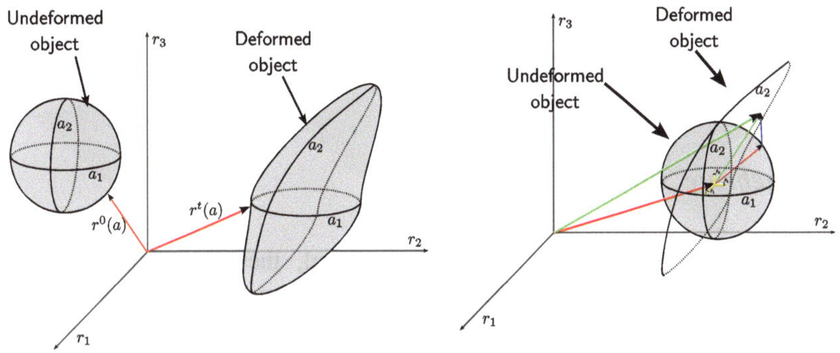

(a) **(b)**

Fig. 8 An explicit object representation: **a** defined upon a reference system, and **b** defined as a rigid displacement with respect to a reference system and a deformation

There exists another formulation, the *hybrid formulation*, described on [130], which is a combination of the concepts of the rigid body dynamics and the deformable objects. In this second formulation, the position $\mathbf{q}$, of each point of an object will be defined from a reference system local to the object ϕ, therefore these positions are represented as:

$$\mathbf{q}(u, t) = \mathbf{r}(u, t) + \mathbf{e}(u, t),$$

where $\mathbf{r}(u, t)$ is a reference component, defining the shape of the object without any deformation, and $\mathbf{e}(u, t)$ is the deformation component, adding to the reference component the relative displacement of each point at every time. Both $\mathbf{r}$ and $\mathbf{e}$ are defined with respect to ϕ, which is placed in the center of mass of the object, $c(u, t)$, whose position in turn is defined with respect to the inertial reference system Φ (see Fig. 8). The center of mass moves according to the laws of rigid body dynamics.

This *hybrid formulation* has no ill-conditioning problems that appear on the *primal formulation*, so it is possible to represent objects with a high degree of rigidity. The greater the rigidity of objects, the system will be better conditioned. Moreover, this system does not get good results when simulating flexible objects (fabrics, elastic objects, ...). However, in [127], the authors represent the behaviour of viscoelastic and plastics materials as well as fractures by using *generalized splines* functionals.

4 Implicit Models

Another way to represent the shape of an object, S_f, is taking null values in a given real-valued function, f:

$$\begin{aligned} f &: \mathbb{R}^3 \longrightarrow \mathbb{R} \\ S_f &= \{\mathbf{p} \in \mathbb{R}^3 \mid f(\mathbf{p}) = 0\}, \end{aligned} \tag{3}$$

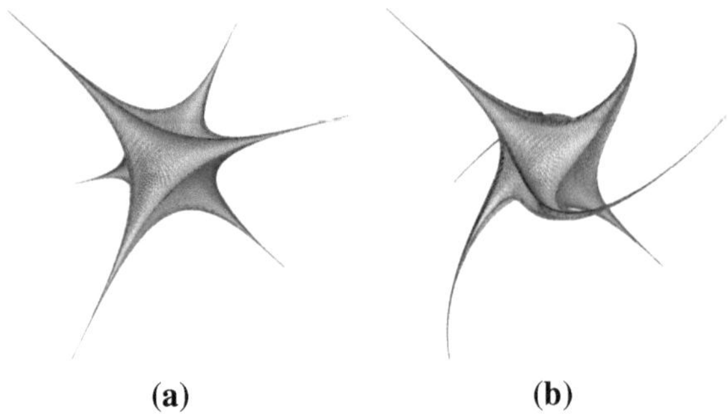

Fig. 10 A superquadric surface example: a superellipsoid with parameters $s = a_1 = a_2 = a_3 = 1.0$, $\varepsilon_1 = \varepsilon_2 = 5.0$. **a** Normal shape, **b** with a deformation

3. Two sheet super hyperboloids:

$$\begin{aligned} C_1^{\varepsilon 1}(\phi) &= \sec \phi^{\varepsilon 1},\ C_2^{\varepsilon 2}(\theta) = \sec \theta^{\varepsilon 2}, \\ S_1^{\varepsilon 1}(\phi) &= \tan \phi^{\varepsilon 1},\ \ S_2^{\varepsilon 2}(\theta) = \tan \theta^{\varepsilon 2}, \\ -\tfrac{\pi}{2} &\le \phi \le \tfrac{\pi}{2}, \\ -\tfrac{\pi}{2} &\le \theta \le \tfrac{\pi}{2} \ \text{(sheet 1)}, \\ \tfrac{\pi}{2} &\le \theta \le \tfrac{3\pi}{2} \ \text{(sheet 2)}. \end{aligned} \tag{7}$$

4. Supertoroids:

$$\begin{aligned} C_1^{\varepsilon 1}(\phi) &= a_4 + \cos \phi^{\varepsilon 1},\ C_2^{\varepsilon 2}(\phi) = \cos \phi^{\varepsilon 1}, \\ S_1^{\varepsilon 1}(\theta) &= a_4 + \sin \theta^{\varepsilon 2},\ \ S_2^{\varepsilon 2}(\theta) = \sin \theta^{\varepsilon 2}, \\ -\pi &\le \phi \le \pi, \\ -\pi &\le \theta \le \pi, \\ a_4 &= \tfrac{R}{\sqrt{a_1^2+a_2^2}}. \end{aligned} \tag{8}$$

Being the exponentiation a signed operation defined as:

$$\begin{aligned} a^b &= sgn(a)|a|^b, \\ sgn(a) &= \begin{cases} -1 & \text{if a} < 0, \\ 1 & \text{if a} \ge 0. \end{cases} \end{aligned}$$

However, a surface defined by a superquadrics is regular and symmetrical, *a priori* it does not seem adequate to represent a deformable object. However, some studies [128] propose superquadrics to which a local disturbance component is added, so that an object will be expressed by:

$$\tilde{Q}(\phi, \theta) = \mathbf{c} + \mathbf{R}(Q(\phi, \theta) + \mathbf{d}(\phi, \theta)) \tag{9}$$

Being **c** the inertial center of the superquadric $Q(\phi, \theta)$, **R** is a rotation matrix, and $\mathbf{d}(\phi, \theta)$ is a displacements field over the surface. The result obtained is a regular shape except at the point where the displacements field modifies the shape of the superquadric.

The superquadric representation can be also defined from implicit functions[5] [9, 10]:

1. Superellipsoids:

$$F(x, y, z) = \left(\left(\frac{x}{a_1} \right)^{\frac{2}{\varepsilon_2}} + \left(\frac{y}{a_2} \right)^{\frac{2}{\varepsilon_2}} \right)^{\frac{\varepsilon_2}{\varepsilon_1}} + \left(\frac{z}{a_3} \right)^{\frac{2}{\varepsilon_1}}. \tag{10}$$

2. One sheet superhyperboloids:

$$F(x, y, z) = \left(\left(\frac{x}{a_1} \right)^{\frac{2}{\varepsilon_2}} + \left(\frac{y}{a_2} \right)^{\frac{2}{\varepsilon_2}} \right)^{\frac{\varepsilon_2}{\varepsilon_1}} - \left(\frac{z}{a_3} \right)^{\frac{2}{\varepsilon_1}}. \tag{11}$$

3. Two sheet superhyperboloids:

$$F(x, y, z) = \left(\left(\frac{x}{a_1} \right)^{\frac{2}{\varepsilon_2}} - \left(\frac{y}{a_2} \right)^{\frac{2}{\varepsilon_2}} \right)^{\frac{\varepsilon_2}{\varepsilon_1}} - \left(\frac{z}{a_3} \right)^{\frac{2}{\varepsilon_1}}. \tag{12}$$

4. Supertoroids:

$$F(x, y, z) = \left(\left(\left(\frac{x}{a_1} \right)^{\frac{2}{\varepsilon_2}} - \left(\frac{y}{a_2} \right)^{\frac{2}{\varepsilon_2}} \right)^{\frac{\varepsilon_2}{2}} - a_4 \right)^{\frac{2}{\varepsilon_1}} - \left(\frac{z}{a_3} \right)^{\frac{2}{\varepsilon_1}}. \tag{13}$$

In [141] this model is used together with the *geons* model (a volumetric model that allows to extract qualitative features) to provide *superquadric geons*, which can be used to define parts of 3D objects, and also describe the topological relations between different parts. Also, [104] gives a solution to the problem of detecting contact between convex superquadric surfaces using implicit equations.

Although the superquadric surfaces are a good global approximation, the shapes that can be represented with this family of surfaces are too limited to represent complex forms with accuracy [9, 10].

The superquadrics model is also used to extract information from images [12]. See also [144], where is presented an automatic 3D hybrid segmentation approach based on free-form deformation and superquadrics.

---

[5] It is also possible to represent the superquadrics as a parametric function. So they are also among the parametric models.

4.1.2 Hyperquadrics

The hyperquadric surfaces are proposed as an extension of the superquadrics [60]. A hyperquadric surface is defined as:

$$f_{\mathbf{q}}(\mathbf{p}) = \sum_{j=1}^{n\geq 3} \left|a_j p_x + b_j p_y + c_j p_z + d_j\right|^{\varepsilon_j} = 1, \tag{14}$$

being $\varepsilon_j > 0 \; \forall \; j$ y $\mathbf{q} = (a_1, b_1, c_1, d_1, \varepsilon_1, \ldots, a_n, b_n, c_n, d_n, \varepsilon_n)^\top$. For a value of n greater than three, the Eq. (14) does not have an explicit representation. The form represented is inscribed in the convex hull defined by the planes $a_j p_x + b_j p_y + c_j p_z + d_j = \pm 1$. The coefficient ε_j is a fitting parameter of the surface to its convex hull. The number of shapes that can be represented with this kind of surfaces is greater than with the superellipses, but it is only possible to define shapes homeomorphics with a sphere [37].

4.2 Isosurfaces

Another approach is to define the surface by a potential function, that is, a function that maps each space point to a numerical value:

$$F : \mathbb{R}^3 \longrightarrow \mathbb{R}.$$

With this function, a surface can be defined as the set of the space points where the function has certain value, v.

$$S_v = \{\mathbf{p} \in \mathbb{R}^3 | F(\mathbf{p}) = v\}. \tag{15}$$

For a given function many surfaces can be obtained by simply changing the value v (Fig. 11).

The generated surfaces depend on the used function F. For example, if F is defined as $F(\mathbf{p}) = p_x^2 + p_y^2 + p_z^2$, then for each possible value $v > 0$, S_v will define the surface of a sphere of radius $\sqrt{v}$.

This kind of surfaces, depending on the potential function used, are called:

1. *Blobs*: surface model described initially in [14] with a potential function inspired in the behaviour of the atoms when joining together to create molecules.

$$F(\mathbf{p}) = \sum_i b_i e^{-a_i r_i^2}, \tag{16}$$

being r_i the distance from the point $\mathbf{p}$ to the center of the i molecule. For each molecule, a Gaussian centered at r_i, a height a_i and a standard deviation b_i are defined. The potential for each space point will be expressed by the contribution of every particle.

2. *Metaballs*: this approach is slightly later [97], it defines the potential function for every *metaball* as:

$$w_i(\mathbf{p}) = \begin{cases} d_i\left(1 - 3\left(\frac{r_i}{b_i}\right)^2\right) & 0 \leq r_i < \frac{b_i}{3}, \\ \frac{3d_i}{2}\left(1 - \frac{r_i}{b_i}\right)^2 & \frac{b_i}{3} \leq r_i < b_i, \\ 0 & b_i \leq r_i, \end{cases} \tag{17}$$

in which r_i is the distance from the point $\mathbf{p}$ to the center of the ith *metaball*, being b_i its radius and d_i a weighting coefficient.
If M is defined as a set of *metaballs* which are combined, a *fusion cluster*, then the potential function is:

$$F(\mathbf{p}) = \sum_{m_i \in M} w_i(\mathbf{p}). \tag{18}$$

3. *Soft objects*: it is an evolution of the *blob* molecules [139, 140] where the potential function is approximated with a polynomial whose value is null from a certain distance from a *key point*, the influence radius:

$$C_i(\mathbf{p}) = \begin{cases} -\frac{4}{9}\left(\frac{r_i}{R_i}\right)^6 + \frac{17}{9}\left(\frac{r_i}{R_i}\right)^4 - \frac{22}{9}\left(\frac{r_i}{R_i}\right)^2 & r_i \leq R, \\ 0 & r_i > R. \end{cases} \tag{19}$$

In this case r_i is the distance from the point $\mathbf{p}$ to each key point and R_i is the influence radius of the ith key point. Therefore if it is desired to calculate the potential field corresponding to a set of key points, the potential function is:

$$F(\mathbf{p}) = \sum_i C_i(\mathbf{p}) \tag{20}$$

The three models described are very similar, although there are notable differences between them. Thus, in the *blobs* model, to calculate the intensity of the field in each space point, it is necessary to calculate the contribution of every molecule, and this can make the calculation too expensive. Originally, the problem was solved with the setting of an influence sphere for each molecule, neglecting its effect in the areas outside this sphere. In the *metaballs* model there is no such problem due to the form of the functions, because they convert the potential of each element farther from some distance in a null value. However, a polynomial function, such as the case of Eqs. (17) and (19), is less expensive than an exponential function, which is the case of (16). Moreover, the *soft objects* have as a main advantage the fact that their potential

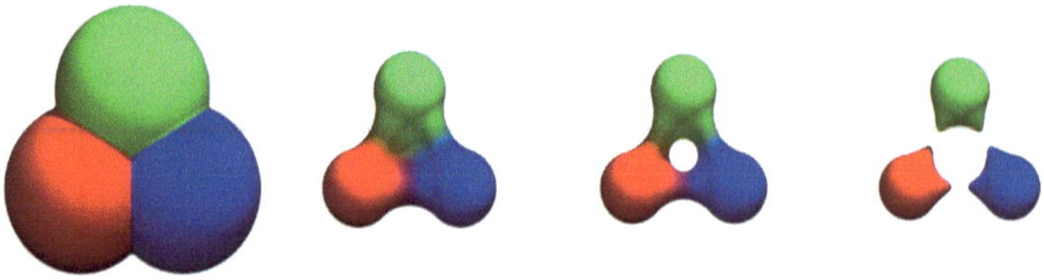

Fig. 11 Isosurface defined from three equally spaced spheres with an influence radius expressed as a percentage of their respective radius: in (**a**) 99% , in (**b**) 45%, in (**c**) 30% and in (**d**) 25%

function is the simplest one to evaluate, even the square roots implicit in the distance calculations can be simplified thanks to the function exponents.

Later, models which a more complex shape than a sphere have been appeared. In fact, also in the seminal works, the possibility of extensions and variations were presented. *A posteriori*, isosurfaces has been build from a geometric *skeleton*, more or less elaborated.

There exists another aspect related to the isosurfaces: their representation. One of the most widely used algorithms is *marching cubes* [80]. Basically, the algorithm consists in dividing the space into small cubes or *voxels*; the smaller the size, the greater the *smoothness* of the surface generated. Then, it is performed a scanning through all the vertexes of the cubes to determine whether the value of the scalar field is greater than or less than the threshold indicated. After the scan is performed, each one of the eight vertex of the cube will have a boolean value, true or false, depending on the results of the comparison. In total there are 256 possible combinations of true and false values of the eight vertexes.[6] Each combination determines a possible form for the surface in this particular cube. Thus, what it must be done is to change each cube by the surface piece determined by its combination.

The result of the algorithm is the isosurface of a given value in the scalar field. The smoothness will depend on the chosen size of the cubes. In the Fig. 12 two *blobs* molecules are shown, they are represented with the described algorithm and with different cube sizes. Other approaches in the same line are proposed in [15, 16, 117].

This kind of surfaces can be used to define deformable contours. In this way, in [28] they use isosurfaces defined from a *skeleton*.

In more recent works [18] this technique is used to modeling clouds from a set of particles.

Other of the aspects in which the use of this method offers advantage in front of others is in the object representation from image data, or from measuring mechanisms that get clouds of points with some noise level. In [63], an hybrid mechanism is used,

[6] In practice it is enough with 16 different combinations, the rest can be obtained from affine transformations.

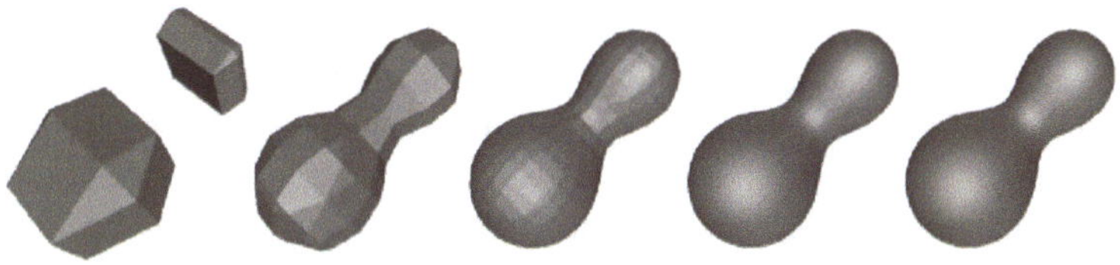

Fig. 12 Two *blobs* [14] represented, from *left* to *right*, with a higher level of detail

Fig. 13 The *level set* can be visually interpreted as the contours of the intersection of two surfaces. In the picture, from *left* to *right* can be seen in the *top* row, how a 3D surface generates a 2D shape (the *level set*) over a plane. Depending on the displacement of the 3D surface, the 2D shape is changed

in part an explicit model and in part an implicit one. With it, surfaces can be generated, which in turn can be deformed.

4.3 Level Sets

The level sets were originally proposed by Osher and Sethian [98, 114, 117]. Its function is to follow the contours of the objects which change their shape, this approach is very similar to the active contour models introduced in Sect. 3.2.

Basically their approach is to place the object to deform in a space with a larger number of dimensions. If the object changes shape, the function of movement must also be projected.

Thus, if $\Gamma(t)$ represents the object, seen as a closed hypersurface in motion:

$$\Gamma(t) : [0, \infty) \rightarrow \mathbb{R}^n.$$

The motion can be considered in a direction normal to the object. Let $\pm d$ the distance from the propagation front. Assuming that the propagation front is the 0 level-set of a function ϕ, defined on the higher dimensional space, then $\phi(x \in \mathbb{R}^n, t = 0) = \pm d$. Then the evolution of ϕ can be represented with the partial differential equation:

$$\phi_t + F|\nabla\phi| = 0,$$

for a given $\phi(x \in \mathbb{R}^n, t = 0)$, F is the speed of Γ, see Fig. 13.

One of the main advantages of this model is that the level set remains valid even if the hypersurface changes its topology. The other models described are based on the determination of the position of some points of the contour, and in the calculation of the shape of this contour from the calculated points. This approach is not generally applicable if there are changes in the objects topology, such as when a break occurs.

This approach is perfectly applicable to surfaces. But its natural scope is the images segmentation [83], and medical imaging. It is therefore an alternative to the active contours model. An early review on level set methods in 2D/3D medical images can be found in [123].

In the medical imaging field, [91] uses the level sets to identify and track biological cells using video microscopy techniques. Also, [120] proposes an automatic lungs segmentation from medical images.

Also [2–5] use level sets for image segmentation and edge detection. In those works is presented a new approach based on morphology operators implemented as fast algorithms that significantly reduce the high computational cost associated with this model. See also [74]. However, [143] the level sets are used to reconstruct complex 3D models by using a geometric external force field, which determines the evolution of the system.

This model can also be used to describe fluids and other complex objects (smoke, fire, ...) [116].

4.4 Sampled Object Representation

The *Sampled Object Representation*, SOR, defines a graphical model using data obtained from a sampling process, which takes a collection of samples at discrete positions in space in order to capture certain geometrical and physical properties of one or more objects of interest [32].

This technique does not allow to have the topological, geometric o semantic information that is required in order to handle the objects in a correct way. Such information must be obtained in some otherwise.

Some properties of this technique are:

- Limited geometric information. Although it need not always be like this, it is usual that the sampled data do not contain geometric information and this, though a certain limitation, can easily describe amorphous objects—smoke, fire, dust—that are particularly difficult to be represented with other models.
- Limited topological information. The only topological information available is that of spatial or temporal arrangement in which samples are obtained.
- Limited semantic information. The sampled information can hardly contain semantic information of source objects of the samples.
- Multiple data channels. The data obtained can be of very different types: temperatures, images, sounds, ...

- Multivalued data channels. The values of the samples can give rise to some uncertainty due to the type of data they describe.

Depending on the specific information to deal with this model, other models can be used to generate objects.

5 Parametric Models

A traditional way of posing the surface deformations is determining that these shapes should be adjusted according to some constraints. Thus, starting from an initial surface, more or less adjusted to the final shape desired, the appropriate modifications must be made to adjust the surface to the desired final result as much as it is possible. The constraints are posed on the basis of mathematic functions combined with the surface definition to achieve a system of equations. The surface definition must have enough degrees of freedom to allow the system of equations be solvable [58].

With this model the object representation is more restrictive than the one given by the explicit models, this is due to the constraints it must be imposed. On the other hand, parametric modeling has a much more simpler representation.

The parametric representation of an object, mathematically speaking, is that where each point is represented separately by an explicit function with a set of independent parameters. Thus, a parametric surface is represented as an application, S, defined in its domain $\Omega \subseteq \mathbb{R}^2$ so that each point, (u, v), of the domain has a corresponding point in $\mathbb{R}^3$, $(x(u, v), y(u, v), z(u, v))$:

$$\begin{aligned} S : \Omega \subseteq \mathbb{R}^2 &\longrightarrow \mathbb{R}^3 \\ S(u, v) &\longrightarrow (x(u, v), y(u, v), z(u, v)). \end{aligned} \tag{21}$$

The presentation cannot be unique, in fact there exists multiple parametric representations for a given object, from all of the possible representations, the interesting are those that [103]:

- Are capable to represent with accuracy all shapes the users want.
- Arc implcmcntablc cfficicntly in a computcr systcm, it is cspccially important that:
  - The calculation of the points and the derivatives can be performed efficiently.
  - The numeric operations are robust, without numeric errors.
  - They do not need too much storage space.
- Are simple and well understood from a mathematical point of view.

Among the different representations proposed, the simplest one is the based on polynomials, although there exist so many objects that cannot be represented with them.

5.1 Superquadrics

The superquadrics, described in Sect. 4.1.1, can be represented in a parametric way. In this case, the coefficients are the values characterizing the posed superquadric. In [128] they are used to represent 3D deformable objects.

5.2 Modal Decomposition

The essence of these models is to consider the object to be depicted as included in a space that can be expressed as a linear combination of an orthonormal basis of the space. The coefficients of the linear combination are the parameters defining the object. The advantage of this representation is that it is possible to achieve an approximation of the object by a finite subset of coefficients.

Thus, if an object, f, can be considered as an element of a Hilbert space H, that is, a space with a scalar product $\langle \cdot, \cdot \rangle$, then the object can be expressed as:

$$f = \sum_{n \in \mathbb{Z}^+} \alpha_n \phi_n ,$$

where $\{\phi_n\}_{n \in \mathbb{Z}^+}$ is an orthonormal basis of H and

$$\alpha_i = \langle f, \phi_i \rangle \quad \forall i \in \mathbb{Z}^+ .$$

So, for a certain value, $p \in \mathbb{Z}^+$, an approximation of f is:

$$f \approx \sum_{k=1}^{p} \alpha_k \phi_k ,$$

so that the vector $(\alpha_1, \alpha_2, \ldots, \alpha_p)$ forms the set of parameters defining the shape of the object with a certain degree of approximation. Obviously, the larger number of parameters, the larger the accuracy of the representation.

In [121] there is an approach using Fourier basis, so the Fourier coefficients determine the shape of objects. The essence of this paper is to put the object in the frequency space. Each frequency waves that describe an object consists of a set of harmonics. The parametric representation of objects following the modal decomposition approach requires that the parameters are the weights corresponding to these harmonics. Depending on the number of harmonics used, the object can be represented more or less accurately. In practice the interest is to minimize this number, provided that the model fits enough to the object to be represented.

In [100] this model is used to represent a simplified dynamic model, having a reduction of the representation complexity and a simplification of it thanks to the elimination of the vibration modes of higher frequency; as a counterpart, there are some limitations on the deformations that can be represented.

Using modal analysis it is possible to decompose a deformable shape into meaningful parts, requiring only a single input pose, see [62]. The modal analysis is able to determine which parts tend to move rigidly and it is possible to determine the deformation field associated to the shape. Another application of modal analysis can be seen in [43], where is presented a real-time method to animate complex scenes of thousands of trees under a user-controllable wind load, the main modes of deformation are pre-calculated and used later to deform the trees under wind load. This approach can be efficiently implemented on graphics hardware and requires a low computational effort.

5.3 Subdivision Surfaces

Subdivision is a technique to generate smooth curves and surfaces, which extends classical spline modeling approaches. Its algorithmic implementation is very simple and efficient. Subdivision defines a smooth curve or surface as the limit of a sequence of successive refinements.

The main idea of the subdivision of curves and surfaces is: given an initial set of control points $\mathbf{p}^0$, forming a grid, where $\mathbf{p}^0 = \left(\dots p^0_{-2}, p^0_{-1}, p^0_0, p^0_1, p^0_2, \dots\right)^\top$ (the values p_i^0 are points of $\mathbb{R}^2$ or $\mathbb{R}^3$ depending on if we are in the plane or in the space) and a matrix S, called *subdivision matrix*, the set of control points $\mathbf{p}^1$, can be found in this way: $\mathbf{p}^1 = S\mathbf{p}^0$. And, in general, the series of points $\mathbf{p}^k$ can be calculated as $\mathbf{p}^{k+1} = S\mathbf{p}^k$. The points $\mathbf{p}^k$ are considered a level k approximation to the curve or surface.

When a subdivision is defined, it must be taken into account:

- The eigenvalues of the matrix S. In order to have convergence, all the eigenvalues λ_i of S must verify: $|\lambda_i| \leq 1$, $\forall i$. In order to have affine invariance, S must have the eigenvalue 1 with eigenvector $(1, \dots, 1)^\top$.
- In the case of curves, if $\lambda_0 = 1 < \lambda_1 < \lambda_i$, and if $\mathbf{p}^0 = \sum_{i=0}^{n-1} a_i \mathbf{x}_i$ where $\mathbf{x}_i$ is the eigenvector corresponding to the λ_i eigenvalue; then the tangent vector to the curve in $\mathbf{p}^k$, for a k large enough, can be approximated by a_1, (a_i are in $\mathbb{R}^2$ or $\mathbb{R}^3$) vectors.
- In the case of surfaces, if $\lambda_0 = 1 < \lambda_1 = \lambda_2 < \lambda_i$ and $\mathbf{p}^0 = \sum_{i=0}^{n-1} a_i \mathbf{x}_i$, then the vectors which generate the tangent plane in $\mathbf{p}^k$ for a k large enough, can be approximated by a_1 and a_2.
- Regularity of the subdivision-generated surface. The control points $\mathbf{p}^k$ are forming a grid. The number of edges of the grid bearing this point are so-called *control point valence*. Then, a control point p_i^k is extraordinary if it has a different *valence* with respect to the rest of the supposed regular grid points. Where there is regularity, the subdivision-generated surface has continuity $\mathscr{C}^1$, but at the extraordinary point, the $\mathscr{C}^1$ continuity is not guaranteed.

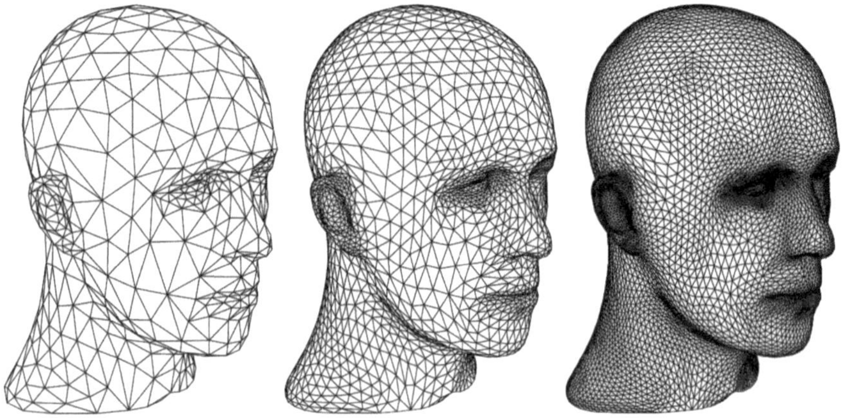

Fig. 14 An example of subdivision surface, showing three successive levels of refinement [146]. On the *left*, an initial *triangular* mesh as a first approximation to the surface. Each *triangle* is split into 4 according to a particular subdivision rule. On the *right* the mesh is subdivided a second time following the same scheme

- Characteristic map. From the eigenvalues $\mathbf{x}_1$ and $\mathbf{x}_2$ (which depend on the subdivision scheme) and the vectors a_1 and a_2 (which depend of the initial subdivision grid, $\mathbf{p}^0$), it is possible to make a local study of the subdivision generated surface in $\mathbf{p}^k$ for a k large enough. From this study, it is possible to define the characteristic map indicating the regularity of the surface in an environment of $\mathbf{p}^k$.

This model arose from [30] where an algorithm to represent smooth curves based on a control polygon were stated. The main idea is cut the corners of the control polygon to generate another polygon, for each vertex of the original polygon two new ones are created. This new polygon can be used this way again. This procedure can be applied recursively and the successive control polygons obtained converge to a smooth curve. The procedure stops when the distance between two consecutive polygons is less than some threshold (for example, the resolution of the display device where the shape is rendered).

The first references to this model are [29, 44], although the model has been widely developed afterwards. In fact examples of this method can be often viewed in any animation movie of today, as an example see Fig. 14.

The subdivision methods can be organized in categories based on four criteria [146]:

- Type of refinement rule: face split or vertex split.
- Type of grid generated: triangular or quadrilateral.
- Type of scheme: approximation or interpolation.
- Smoothness of the limit surfaces for regular meshes (C^1, C^2, ...).

Next, there is a classification of some of the most used methods:

An application of this model can be seen in [6], a meshless approach to represent deformable objects. The objects are represented as a set of coarse nodes which

Face split

	Triangular meshes	**Quadrilateral meshes**
Approximating	Loop [78]: C^2	Catmull–Clark [29]: C^2
Interpolating	Butterfly [45]: C^1	Kobbelt [70]: C^1

Vertex split

Doo–Sabin [44]: C^1

separation is described in a similar way as the fast marching method [115]. The deformation are solved by using modal decomposition techniques, see Sect. 5.2. The results illustrate that this model can handle complex shapes at interactive rates producing realistic deromable 3D shapes and their motion.

This model is commonly used in the context of free-form deformations, but usually a final conversion to other polygonal models is necessary due to its computational cost, specially when there is a real-time constraint as for example in the field of video-games. To solve this it is possible to use specific hardware, such as Graphics Processing Units (GPUs). But this kind of processors are specially designed to speedup the polygonal calculations, although something has been done in this orientation [79].

5.4 *Bézier Curves*

The Bézier curves were studied independently by Bézier and de Casteljau at the same time (endings of the 1950 decade in the case of de Casteljau and beginnings of the 1960 decade for the Bézier case).

Their solution, never before proposed, were based on the use of Bernstein polynomials, dating back the beginning of twentieth century, and the control polygons. The approach was revolutionary: the curve were not defined from points *on* the curve, but for some other points *near* the curve, the so-called control polygon. Instead of modifying directly the curve, the points of the control polygons could be modified, and doing this, the curve follows the change in an intuitive way. The work of both researchers was developed later and was shown to be equivalent.

A Bézier curve of degree n is defined as

$$\mathbf{C}(u) = \sum_{i=0}^{n} B_{i,n}(u)\mathbf{P}_i \quad 0 \leq u \leq 1, \tag{22}$$

being $B_{i,n}(u)$ the Bernstein polynomials of degree n:

$$B_{i,n}(u) = \binom{n}{i} u^i (1-u)^{n-i} = \frac{n!}{i!(n-i)!} u^i (1-u)^{n-i}. \tag{23}$$

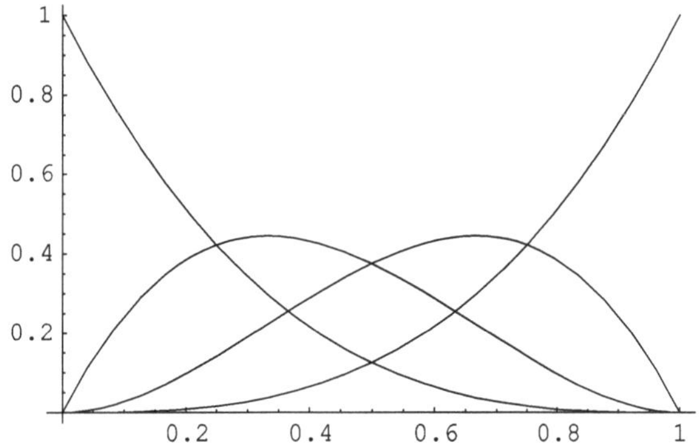

Fig. 15 The Bernstein polynomials set the influence of each control point on every point of the curve. In the image, four Bernstein polynomials are show, those corresponding to a degree three curve

All control points influence to a greater or lesser extent in the shape of the curve, the degree of influence is given by Bernstein polynomials, see Fig. 15.

The geometrical coefficients, $\{\mathbf{P}_i \in \mathbb{R}^3, 0 \leq i \leq n\}$, are the *control points*. The polygon formed by joining the control points by straight lines is called *control polygon* and it results to be an approximation of the curve. The control points also define a convex hull of the curve. Another interesting characteristic is that the curve passes through the first and last control points.

In the field of deformable objects representation, in [135] is proposed the use of triangular Bézier patches for the representation of surfaces constructed from restrictions as a variational approach.

Among all the existing ways for the representation of parametric surfaces one of the most commonly used is based on the tensor product. In this case, the bivariate basic functions, functions of variables u and v, are constructed as the product of the respective univariate basic functions. The geometric coefficients are arranged in a network, and the tensor product has the form

$$S(u, v) = \sum_{i=0}^{n} \sum_{j=0}^{m} f_i(u) \cdot g_j(v) \cdot \mathbf{b}_{ij}, \quad \begin{cases} \mathbf{b}_{ij} = (x_{ij}, y_{ij}, z_{ij}), \\ \quad a \leq u \leq b, \\ \quad c \leq v \leq d, \end{cases}$$

that can be expressed in matrix form as:

$$S(u, v) = [f_i(u)]^\top \left[\mathbf{b}_{ij}\right] \left[g_j(u)\right] = \mathbf{f}^\top \mathbf{b}\mathbf{g},$$

being $[f_i(u)]^\top$ a row vector, $\left[g_j(u)\right]$ a column vector, both representing the univariate basis. At the same time, $\left[\mathbf{b}_{ij}\right]$ is a $(n+1) \times (m+1)$ matrix of three dimensional points.

The domain of (u, v) is a rectangle $[a, b] \times [c, d] \subseteq \mathbb{R}^2$, although is quite usual the use of the rectangle $[0, 1] \times [0, 1]$ as a domain, due this does not suppose any restriction and, instead, it allows some simplifications.

In the case of Bézier surfaces the tensor product form is:

$$S(u, v) = \sum_{i=0}^{n}\sum_{j=0}^{m} B_{i,p}(u) \cdot B_{j,q}(v) \cdot \mathbf{P}_{ij} \quad 0 \leq u, v \leq 1 \tag{24}$$

The control points form a polyhedron, the *polyhedron of control*, which proves to be a linear approximation of the surface. The properties that satisfy Bézier surfaces are analogous to those described for curves, extending them for a two-dimensional parametric spaces.

Both curves as Bézier surfaces, being polynomial representations, have a number of drawbacks:

- A very high degree is required for the polynomial if the shape to represent has to adjust to a large number of constraints (in general, a $n-1$ degree polynomial is required to satisfy n constraints). The same occurs if the shape is complex. The representation algorithms are less efficient as higher is the degree; and also, the higher the degree, the higher the numerical instability.
- The control of the shape is no local, then the manipulation of any control point affects the entire shape in a not intuitive way. This fact makes this model be unattractive to their use in interactive environments.
- There is a great amount of shapes that can not be represented by a polynomial curve: the conics and quadrics (circles, cones, spheres, ...). An extension of the Bézier shapes are the rational Bézier shapes proposed in [47]. With the use of rational Bézier polynomials some of the cited limitations are solved, because it is possible to define conics with this kind of curves.

5.5 B-Splines

A solution to the problems posed by the Bézier schemes takes into account the piecewise polynomial functions.

The idea is to build a complex shape by the union of different sections with a polynomial structure and a low degree, without the drawbacks derived from the use of high degree polynomials. Therefore, a curve arises as a set of polynomial segments joint one after another. The unions between segments are called *break points*, also known as *knots*. Each segment is independent from the others, each one of the polynomial segments is defined between two consecutive knots. The junctions are made in a form that it is possible to guarantee some degree of continuity along the entire curve (see Fig. 16).

B-Splines are defined from the so-called B-Spline Basis functions. Given a set of real values $U = \{a = u_0, u_1, \ldots, u_m = b\}$, such that $u_i \leq u_{i+1} \forall i \in \{0 \ldots m-1\}$, the ith B-Spline basis function of degree p is defined recursively as[7]:

[7] There exists the possibility to obtain a quotient of the form $\frac{0}{0}$ or $\frac{\cdot}{0}$. To algorithmic development purposes these cases are considered as 0.

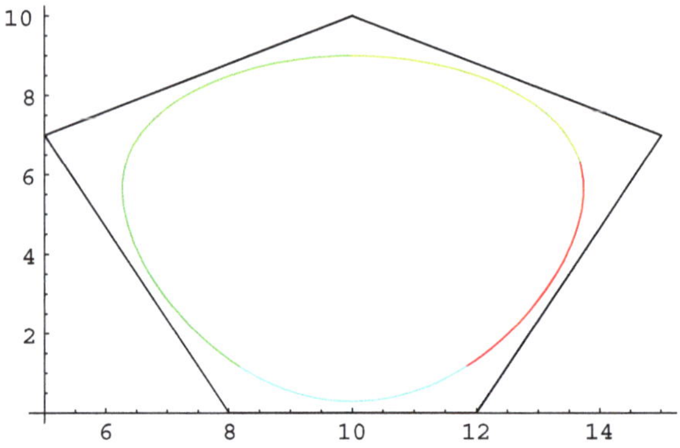

Fig. 16 The curves, defined as piecewise polynomial shapes, are formed as a result of the union of the curve segments that each polynomial shape represent. In the image the segments are shown differentiated

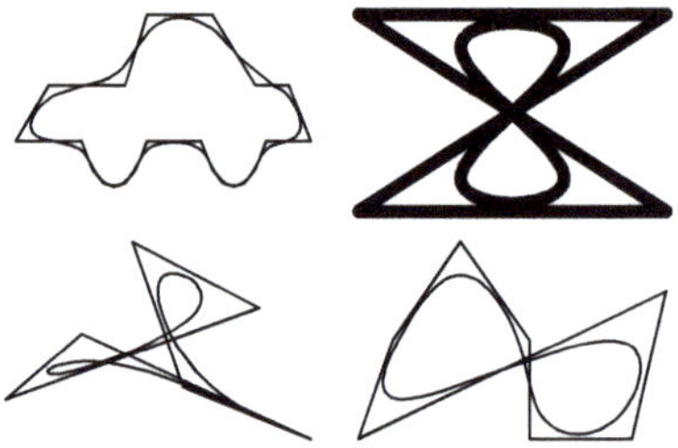

Fig. 17 Different B-Spline curves and their respective control polygons

$$\begin{aligned} N_{i,0}(u) &= \begin{cases} 1 \text{ if } u_i \le u \le u_{i+1}, \\ 0 \text{ otherwise.} \end{cases} \\ N_{i,p}(u) &= \tfrac{u-u_i}{u_{i+p}-u_i} N_{i,p-1}(u) + \tfrac{u_{i+p+1}-u}{u_{i+p+1}-u_{i+1}} N_{i+1,p-1}(u). \end{aligned} \tag{25}$$

Each basis function $N_{i,p}(u)$ is defined in $\mathbb{R}$, although the interest is centered at the interval $[u_0, u_m]$. Usually the knots set is defined such $u_0 = 0$ and $u_m = 1$.

From the B-Spline basis functions, the B-Spline curves and surfaces are defined. In particular, given a knots set

$$U = \{\underbrace{a, \ldots, a}_{p+1}, u_{p+1}, \ldots, u_{m-p-1}, \underbrace{b, \ldots, b}_{p+1}\},$$

and a point set, $\{\mathbf{P}_i \in \mathbb{R}^3, i = 0 \ldots n\}$, the p degree B-Spline curve is defined as:

$$C(u) = \sum_{i=0}^{n} N_{i,p}(u)\mathbf{P}_i, \tag{26}$$

The Bernstein polynomials (23), can be obtained from the B-Spline basis functions (25) if the knots set is defined as:

$$U = \{\underbrace{0, \ldots, 0}_{p+1}, \underbrace{1, \ldots, 1}_{p+1}\}.$$

Therefore, the Bézier curves and surfaces can be considered as a particular case of the B-Spline ones. However, B-Splines have a higher capacity of representation. This is due to some of the characteristics that define them. In fact, B-Spline shapes solve the problems discussed for Bézier forms:

1. The construction of the B-Spline shapes as piecewise polynomials shapes is due precisely to reduce the complexity and the degree of the Bézier shapes.
2. The local support comes from the fact that a certain number of basis function $N_{i,p}(u)$, has not null values only in the interval $[u_i, u_{i+p+1})$. In practice, this makes the influence of each control point to be reduced to only one interval of the shape. Namely, the control point $\mathbf{P}_i$ has influence in the interval $[u_i, u_{i+p+1})$ of the represented shape. Therefore, a change in a control point will only affect to a part, not to the whole shape.
3. Another problem, that of the difficulty of defining complex shapes, is solved since it is possible to define shapes with sharp angles and even without *visual* continuity thanks to the multiplicity of the control points, which can be greater than 1. See Fig. 17.

B-spline surfaces can also be obtained from curves by using tensorial products (24). So a B-Spline surface $S(u, v)$, of degree p in one dimension and q in the other one, with the knot set:

$$U = \{\underbrace{0, \ldots, 0}_{p+1}, u_{p+1}, \ldots, u_{r-p-1}, \underbrace{1, \ldots, 1}_{p+1}\},$$
$$V = \{\underbrace{0, \ldots, 0}_{q+1}, v_{q+1}, \ldots, v_{s-q-1}, \underbrace{1, \ldots, 1}_{q+1}\},$$

and the control points

$$\{\mathbf{P}_{ij} \in \mathbb{R}^3, i = 0 \ldots n, j = 0 \ldots m\},$$

so that $r = n + p + 1$ and $s = m + q + 1$, is defined as:

$$S(u, v) = \sum_{i=0}^{n} \sum_{j=0}^{m} N_{i,p}(u) \cdot N_{j,q}(v) \cdot \mathbf{P}_{ij} \quad 0 \leq u, v \leq 1. \tag{27}$$

Also, it is possible to define solid objects using B-Splines. This can be done from a three-dimensional tensorial products:

$$S(u, v) = \sum_{i=0}^{I} \sum_{j=0}^{J} \sum_{k=0}^{K} N_{k,p}(u) \cdot N_{j,q}(v) \cdot N_{k,r}(w) \cdot \mathbf{P}_{ijk} \quad 0 \leq u, v, w \leq 1. \quad (28)$$

This approach is followed in [107] to represent the human left heart ventricle from medical images. For more details about B-Spline curves and surfaces, see [49, 103].

B-splines are also used in other fields, as for example in registration algorithms, that is classification. In [108] is proposed a diffeomorphic registration method by composing sequences of free-form deformations based on B-splines.

Other interesting articles related to this model are [56] where an analytic solution of an evolution model is proposed in order to deform B-splines parametric surfaces. Also [55] which introduces a dynamic evolution model in order to deform parametric surfaces, finally in [57] a method to deform non-planar parametric surfaces based on B-splines is presented. To develop this method, an energy functional and its variational formulation are introduced.

Although the variety of objects that B-Splines can represent is much broader than using Bézier schemes, still it is not possible to represent the conics and quadrics, shapes commonly used in environments such as simulation, animation, design, and so on. Therefore, it is necessary look for a scheme which, being more general, allow the representation of these shapes.

5.6 NURBS

The NURBS objects appear as a solution to a necessity in the world of computer aided design (*CAD*) and the computer aided manufacturing (*CAM*), that is the need for representation of different geometric elements in an unified way.

Taking into account the formulation established in Sect. 5.5, a NURBS curve, $C(u)$, is defined as:

$$C(u) = \frac{\sum_{i=0}^{n} N_{i,p}(u) \cdot \omega_i \mathbf{P}_i}{\sum_{i=0}^{n} N_{i,p}(u) \cdot \omega_i}, \quad (29)$$

and analogously, a NURBS surface, $S(u, v)$, is defined as:

$$S(u, v) = \frac{\sum_{i=0}^{n} \sum_{j=0}^{m} N_{i,p}(u) \cdot N_{j,q}(v) \cdot \omega_{ij} \cdot \mathbf{P}_{ij}}{\sum_{i=0}^{n} \sum_{j=0}^{m} N_{i,p}(u) \cdot N_{j,q}(v) \cdot \omega_{ij}} \quad 0 \leq u, v \leq 1, \quad (30)$$

being ω_i and ω_{ij} real values, the so-called weights.

The *Non-Uniform Rational B-Spline* (*NURBS*) objects, see Fig. 18, have become a *de facto* standard in the context of objects representation. Some of the characteristics that have promoted this are:

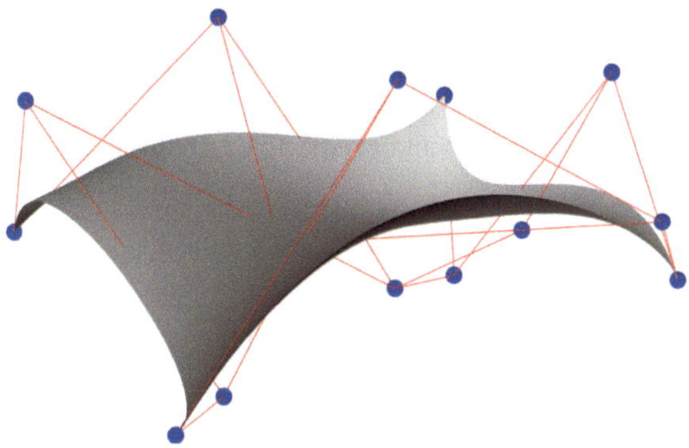

Fig. 18 A NURBS surface, the control points are represented as the *small blue* points connected by the control mesh, the *red lines*

- They allow to represent a wide range of shapes, from analytic shapes to *free-forms*, giving an unifying model. This makes that algorithms whose combining different objects, such as intersection, union, and so on, are general.
- The algorithms are fast and numerically stable [103].
- They can represent discontinuities such as bending, marking, twisting, and so on.
- They are invariant to affine transformations.
- They have a projective semi-invariance.
- The Bézier curves and surfaces, and also the B-Spline ones, are NURBS particular cases.
- They are intuitive, in the sense that the modifications that can be done over the shape to represent, correspond naturally to changes on the elements defining them.

The first references to NURBS are from the mid 1970 [136]. At the beginnings of the 1980 the first NURBS—based modelers appear. The interest aroused in the field of aeronautics, the academic world, and entities related to structures in general, was great.

The incorporation of NURBS in the graphics standards of the time (PHIGS+, IGES), and also for their special features, has allowed a great acceptance in the field of CAD/CAM.

Some of the NURBS characteristics are:

- They are based on a mathematical model allowing to represent both free-forms (*FFD*) and analytics forms (conics, quadrics, ...). They are a B-Splines extension, thus it is also possible to represent parametric forms as Bézier and B-Spline.
- By the manipulation of the control points and the weights, it is possible the representation of a large number of shapes. Furthermore, the manipulation is local, that is: the change of a weight or a control point only affects to a part of the shape, this makes the objects modification become easier.
- In general, the algorithms are numerically stable.

- The geometric interpretation is intuitive, so its use is easier.
- There exists a large amount of operations that can be applied to NURBS. Thus, it is possible to easily adjust and adapt them to multiple forms.
- They are invariant to the affine and perspective transforms, this convert them an ideal design tool.

Working with NURBS has many advantages, but this kind of elements has several drawbacks:

- The parameterization depends on the assigned weights, an incorrect application of the same causes deficiencies in the shape.
- Not all the geometric operation fits correctly to work with NURBS. Determining object intersections, or collision checking, are operations requiring algorithms too expensive computationally.
- Some algorithms, in particular that determines the parameters corresponding to a surface point, are numerically unstable and require algorithms with a high computational cost [103].
- Some elementary representations (circles, squares, ...) require much more data using NURBS than with other kind of representation.

This model has been used widely in many areas shape reconstruction from images [25, 73, 89], Free-Form deformations [35, 67, 72, 73] and in many others [13, 71, 81], also in biomedical images. For example, in [31] is proposed a method of parametric representation and functional measurement of 3D cardiac shapes based on deformable NURBS.

To increase performance, this model can be implemented so that it is possible to display real-time NURBS surfaces by using a Graphics Processing Unit, GPU [142].

5.7 T-Splines

When two NURBS surfaces are joined together, the combination of the two control nets arises represent a problem. To solve this, using NURBS, it is necessary combine both control nets, adding rows and columns where necessary, to generate another control net capable to represent the shape of both surfaces. It is computationally expensive and often requires an additional effort to users. Another problem appears when it is necessary to add a control point. In this case an entire row or column must be added to the control grid. It is due to the necessary regularity of the control net.

T-splines [112] are non-uniform B-spline surfaces with T-junctions. With this add on it is possible to insert control points into the control grid without propagating an entire row or column of control points. It is also possible combine two T-splines without any supplementary effort [110]. This way the regularity of the control net imposed on NURBS can be superseded.

The basis of T-splines are the point-based B-splines, PB-splines. The equation of a PB-spline is

$$P(s,t) = \frac{\sum_{i=1}^{n} P_i \cdot B_i(s,t)}{\sum_{i=1}^{n} B_i(s,t)} \qquad (s,t) \in D, \tag{31}$$

where P_i are the control points, and $B_i(s,t)$ are the basis functions:

$$B_i(s,t) = N_{i,0}^{p}(s) \cdot N_{i,0}^{q}(t),$$

being $N_{i,0}^{p}(s)$ the B-spline basis function of p degree associated with the knot vector $s_i = \{s_{i0}, s_{i1}, \ldots s_{ip}, s_{ip+1}\}$ and $N_{i,0}^{q}(t)$ the B-spline basis function of q degree associated with the knot vector $t_i = \{t_{i0}, t_{i1}, \ldots t_{iq}, t_{iq+1}\}$. Thus, for each control point a pair of knot vectors must be provided.

The domain D in (31) is the union of the domains of each individual control point:

$$D \subseteq D_1 \cup D_2 \cup \ldots D_n = \cup_{i=1}^{n} D_i,$$

D does not has to be rectangular, it only has to be connected and such that:

$$\sum_{i=1}^{n} B_i(s,t) > 0 \quad \forall s,t \in D.$$

T-splines are PB-splines with a control grid, a *T-mesh*, imposing some order to the control points. The T-mesh gives an easier way to handle the control points than the allowed by the independent control points of the PB-splines. It also gives a way to deduce the knot vectors s_i and t_i for each basis function.

The key aspect is that each control point in a NURBS control grid, except those of the edges, has two neighbours in each direction of the grid, horizontal and vertical. But in a T-mesh it is possible to have a control point with only one neighbour vertically o horizontally, this is a T-junction.

A T-mesh is basically a rectangular grid which allows T-junctions. Each edge in a T-mesh is a line segment of constant s, a s-edge or a constant t, a t-edge. A T-junction is a vertex shared by one s-edge and two t-edges or by two s-edges and one t-edge, see Fig. 19.

Each edge in a T-mesh is labeled with a knot interval, constrained by the two rules:

1. The sum of knot intervals on opposing edges of any face must be equal. Then for the face F of the Fig. 19: $r_2 = r_5 + r_6$ and $c_2 + c_4 = c_5$.
2. If a T-junction on one edge of a face can be connected to an opposing edge of the face (thereby splitting the face into two faces) without violating the first rule, that edge must be included in the T-mesh.

For a given control point P_i, to find the knot vectors, s_i and t_i, two lines in the parameter space are used. Let (s_{ij}, t_{ik}) be the knot coordinates of P_i, then

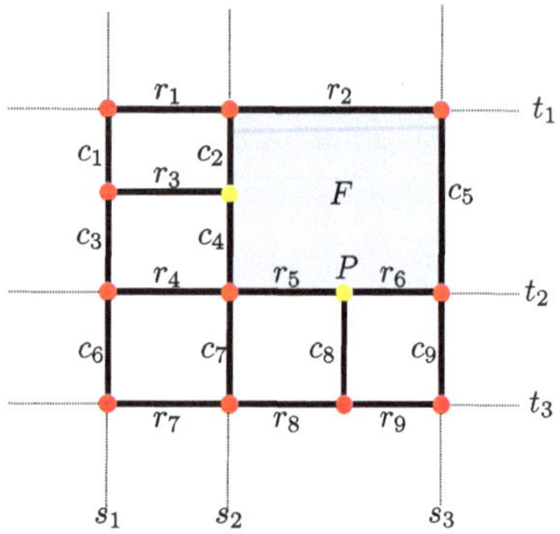

Fig. 19 A T-mesh. The *dotted lines* are the s-edges and the t-edges. The *bullets* are the control points, the *red* ones are *regular* control points and the *yellow* ones are T-junctions

the line $R(\alpha) = (s_{ij} + \alpha, t_{ik})$ will intersect with the s-edges, the intersection will give the s_i knots. The t_i knots can be obtained the same way. For example, in the Fig. 19, for the control point P, if the degree is 2, the knots in the s direction are $s = \{s_1, s_2, s_2 + r_5, s_3\}$.

In conclusion, T-splines have the same advantages as NURBS, but with a simpler way to represent the surfaces. Thus they are a very interesting model to represent objects. In this line, in [96] an alternative to NURBS-based isogeometric analysis that allows for local refinement is presented. The idea is based on polynomial splines and exploits the flexibility of T-meshes for local refinement. Also, [77] investigates higher-order and higher-continuity functions in isogeometric structural analysis under distortion of the control and physical meshes.

An extension of T-splines is proposed in [112], the so-called T-NURCCs. Basically they are an extension of the NURBSS presented in [111]. The Non-Uniform Rational Catmull-Clark (NURCC) surfaces with T-junctions in their control grids, in the same way as been described for the T-splines.

With this approach it is possible to have a model that combine NURBS and subdivision surfaces, as both are particular cases of T-NURCC surfaces, also with a simple formulation.

6 Conclusions

There are many ways in which the objects can be represented. A visual classification of the models described can be seen in Table 1.

The answer to the question of what model has to be used depends on each particular case. Although it is possible to establish a general rule: if the objects can be changed or adjusted to fit some constraint, then it is desirable to have a local control on the deformation. Furthermore it may be taken into account the fact that the objects can

Table 1 A classification of the representation models described

Classification			Properties		
Category	Representation	Model	Shape recovery from data	Synthetic image generation	Local control of the object
		Polygonal meshes	✓	✓	✓
Discrete	Discrete	Particle systems		✓	
		Mass–spring	✓	✓	✓
	Explicit	Generalized cylinders	✓	✓	
		Active contour	✓		
		Continuous objects		✓	
	Implicit	Algebraic surfaces		✓	
		Isosurfaces	✓	✓	✓
		Level sets	✓	✓	✓
Continuous		Sampled object representation	✓		
	Parametric	Superquadrics		✓	
		Modal decomposition	✓	✓	
		Subdivision		✓	✓
		Bézier	✓	✓	
		B-splines	✓	✓	✓
		NURBS	✓	✓	✓
		T-splines	✓	✓	✓

be synthetically generated or, otherwise, can be generated from data gathered from real objects.

Not all the models can be used the same way and each one can be used with advantages in front of the others in some cases. Although, as has been summarized in Table 1, the continuous parametric models, in particular the B-splines, NURBS, and T-splines models, are a good choice and can be considered a general-purpose deformable models.

References

1. Abhau J, Scherzer O (2010) A combinatorial method for topology adaptations in 3D deformable models. Int J Comput Vis 87:304–315. doi:10.1007/s11263-009-0282-5
2. Acton S (2001) Fast algorithms for area morphology. Digit Signal Process 11(3):187–203. doi:10.1006/dspr.2001.0386. http://www.sciencedirect.com/science/article/pii/S105120040 1903860
3. Acton S, Mukherjee D (2000) Area operators for edge detection. Pattern Recognit Lett 21(8):771–777. doi:10.1016/S0167-8655(00)00036-2. http://www.sciencedirect.com/ science/article/pii/S0167865500000362
4. Acton S, Mukherjee D (2000) Scale space classification using area morphology. IEEE Trans Image Process 9(4):623–635. doi:10.1109/83.841939
5. Acton S, Prasad Mukherjee D (2000) Image edges from area morphology. In: Proceedings of IEEE international conference on acoustics, speech, and signal processing, ICASSP '00, vol 6, pp 2239–2242. doi:10.1109/ICASSP.2000.859284
6. Adams B, Wicke M, Ovsjanikov M, Wand M, Seidel HP, Guibas LJ (2010) Meshless shape and motion design for multiple deformable object. Comput Graph Forum 29(1):43–59. doi:10. 1111/j.1467-8659.2009.01536.x
7. Agin GJ (1972) Representation and description of curved objects. PhD thesis, Stanford University of California, Department of Computer Science
8. Agin G, Binford T (1976) Computer description of curved objects. IEEE Trans Comput C-25(4):439–449. doi:10.1109/TC.1976.1674626
9. Bardinet E, Cohen LD, Ayache N (1995) A parametric deformable model to fit unstructured 3D data. Technical report RR-2617, INRIA Institut National de Recherche en Informatique et en Automatique. www.inria.fr/rrrt/rr-2617.html
10. Bardinet E, Cohen LD, Ayache N (1998) A parametric deformable model to fit unstructured 3D data. Comput Vis Image Underst 71(1):39–54
11. Barr A (1981) Superquadrics and angle-preserving transformations. IEEE Comput Graph Appl 1(1):11–23
12. Biegelbauer G, Vincze M, Wohlkinger W (2010) Model-based 3D object detection. Mach Vis Appl 21:497–516. doi:10.1007/s00138-008-0178-3
13. Blanc C, Schlick C (1996) Accurate parametrization of conics by NURBS. IEEE Comput Graph Appl 16(6):64–71
14. Blinn JF (1982) A generalization of algebraic surface drawing. ACM Trans Graph 1(3):235–256
15. Bloomenthal J (1988) Polygonization of implicit surfaces. Compu Aided Geom Des 5:341–355. http://citeseer.ist.psu.edu/bloomenthal88polygonization.html
16. Bloomenthal J, Shoemake K (1991) Convolution surfaces. In: Proceedings of the 18th annual conference on computer graphics and interactive techniques. ACM Press, New York, pp 251–256. doi:10.1145/122718.122757
17. Botsch M, Sorkine O (2008) On linear variational surface deformation methods. IEEE Trans Vis Comput Graph 14(1):213–230. doi:10.1109/TVCG.2007.1054

18. Bouthors A, Neyret F (2004) Modeling clouds shape. In: Alexa M, Galin E (eds) Eurographics '04 (short papers). www.imagis.imag.fr/Publications/2004/BN04
19. Breen D, House D, Getto P (1991) A particle-based computational model for cloth draping behaviour. In: Patrikalakis NM (ed) Scientific visualization of physical phenomena. Spinger, Tokyo, pp 113–134
20. Breen D, House D, Wozny MJ (1994) A particle-based model for simulating the draping behaviour of woven cloth. Text Res J 64(11):663–685. http://citeseer.nj.nec.com/breen94particlebased.html
21. Breen DE, House DH, Wozny MJ (1994) Predicting the drape of woven cloth using interacting particles. In: Computer graphics (Annual conference series), vol 28, pp 365–372. http://citeseer.nj.nec.com/breen94predicting.html
22. Bro-Nielsen M (1994) Active nets and cubes. Technical report 94-13, Intitute of Mathematical Modelling, Technical University of Denmark
23. Bro-Nielsen M (1995) Modelling elasticity in solids using active cubes—application to simulated operations. In: Proceedings of computer vision, virtual reality, and robotics in medicine (CVRMed'95), Lecture notes in computer science, vol 905. Springer, Berlin, pp 535–541
24. Bro-Nielsen M, Cotin S (1996) Real-time volumetric deformable models for surgery simulated using finite elements and condensation. Comput Graph Forum 15(3):57–66. http://citeseer.nj.nec.com/181805.html
25. Brujic D, Ainsworth I, Ristic M, Brujic V (2001) Efficient shape description using NURBS. In: Arcelli C, Cordella LP, Sanniti di Baja G (eds) IWVF-4: proceedings of the 4th international workshop on visual form, vol 2059. Springer, London, pp 643–653
26. Butenuth M, Heipke C (2012) Network snakes: graph-based object delineation with active contour models. Mach Vis Appl 23:91–109. doi:10.1007/s00138-010-0294-8
27. Campbell RJ, Flynn PJ (2001) A survey of free-form object representation and recognition techniques. Comput Vis Image Underst 81:166210
28. Cani MP, Desbrun M (1997) Animation of deformable models using implicit surfaces. IEEE Trans Vis Comput Graph 3(1):39–50. http://www.imagis.imag.fr/Publications/1997/CD97 (Published under the name Marie-Paule Cani-Gascuel)
29. Catmull E, Clark J (1978) Recursively generated B-spline surfaces on arbitrary topological meshes. Comput Aided Des 10(6):350–355. doi:10.1016/0010-4485(78)90110-0. http://www.cs.berkeley.edu/sequin/CS284/PAPERS/CatmullClark_SDSurf.pdf
30. Chaikin G (1974) An algorithm for high speed curve generation. Comput Graph Image Process 3(4):346–349
31. Chen SY, Guan Q (2011) Parametric shape representation by a deformable NURBS model for cardiac functional measurements. IEEE Trans Biomed Eng 58(3):480–487. doi:10.1109/TBME.2010.2087331
32. Chen M, Correa C, Islam S, Jones MW, Shen PY, Silver D, Walton SJ, Willis PJ (2007) Manipulating, deforming and animating sampled object representations. Comput Graph Forum 26(4):824–852 . doi:10.1111/j.1467-8659.2007.01102.x
33. Chikazawa Y, Koshizuka S, Oka Y (2001) A particle method for elastic and visco-plastic structures and fluid-structure interactions. Comput Mech 27:97–106
34. Chuang JH, Ahuja N, Lin CC, Tsai CH, Chen CH (2004) A potential-based generalized cylinder representation. Comput Graph 28(6):907–918. doi:10.1016/j.cag.2004.08.004. http://www.sciencedirect.com/science/article/pii/S0097849304001426
35. Clapés M, González M, Mir A, Palmer PA (2008) Interactive constrained deformations of NURBS surfaces: N-SCODEF. In: Perales FJ, Fisher RB (eds) Articulated motion and deformable objects. Lecture notes in computer science, vol 5098. Springer, Berlin, pp 359–369. http://www.springerlink.com/content/755x0m42567u22k0/
36. Cohen LD, Cohen I (1993) Finite-element methods for active contour models and ballons for 2-D and 3-D images. IEEE Trans Pattern Anal Mach Intell 15(11):1131–1147
37. Cohen I, Cohen LD (1994) A hybrid hyperquadric model for 2-D and 3-D data fitting. Technical report 2188, INRIA Institut Nationale de Recherche en Informatique et en Automatique

38. Delingette H (1994) Simplex meshes: a general representation for 3D shape reconstruction. Technical report 2214, INRIA Institut National de Recherche en Informatique et en Automatique. www.inria.fr/rrrt/rr-2214.html
39. Delingette H (1997) General object reconstruction based on simplex meshes. Technical report 3111, INRIA Institut National de Recherche en Informatique et en Automatique. www.inria.fr/rrrt/rr-3111.html
40. Delingette H (1999) General object reconstruction based on simplex meshes. Int J Comput Vis 32(2):111–146. www.inria.fr/rrrt/rr-3111.html
41. de Vieilleville F, Lachaud JO (2009) Digital deformable model simulating active contours. In: Brlek S, Reutenauer C, Provencal X (eds) Discrete geometry for computer imagery. Lecture notes in computer science, vol 5810. Springer, Berlin, pp 203–216. doi:10.1007/978-3-642-04397-0_18
42. Dias JMS, Galli R, Palmer P, Rebordão JM (1997) Deformable objects with real-time realistic behaviour for virtual scenarios. The Internet in 3D. Academic Press, New York, pp 179–199
43. Diener J, Rodriguez M, Baboud L, Reveret L (2009) Wind projection basis for real-time animation of trees. Comput Graph Forum 28(2):533–540. doi:10.1111/j.1467-8659.2009.01393.x
44. Doo D, Sabin M (1978) Behaviour of recursive division surfaces near extraordinary points. Comput Aided Des 10(6):356–360. doi:10.1016/0010-4485(78)90110-0. http://http://users.cms.caltech.edu/cs175/cs175-02/resources/DS.pdf
45. Dyn N, Levine D, Gregory JA (1990) A butterfly subdivision scheme for surface interpolation with tension control. ACM Trans Graph 9(2):160–169. doi:10.1145/78956.78958
46. Eberhardt B, Weber A, Strasser W (1996) A fast, flexible, particle-system model for cloth draping. IEEE Comput Graph Appl 16(5):52–59
47. Farin G (1983) Algorithms for rational Bézier curves. Comput Aided Des 15(2):73–77
48. Farin G (1992) From conics to NURBS: a tutorial and survey. IEEE Comput Graph Appl 12(5):78–86
49. Farin G (1993) Curves and surfaces for computer aided geometric design, 3rd edn. Academic Press, New York
50. Fatou Ngom N, Monga O, Mohamed MMO, Garnier P (2012) 3D shape extraction segmentation and representation of soil microstructures using generalized cylinders. Comput Geosci 39(0):50–63. doi:10.1016/j.cageo.2011.06.010. http://www.sciencedirect.com/science/article/pii/S0098300411002019
51. Gain J, Bechmann D (2008) A survey of spatial deformation from a user-centered perspective. ACM Trans Graph 27:1–21. doi:10.1145/1409625.1409629
52. Galdames FJ, Jaillet F (2010) From triangulation to simplex mesh: a simple and effcient transformation. Technical report RR-LIRIS-2010-021, LIRIS Laboratoire d'Informatique en Image et Systèmes d'information, UMR 5205 CNRS / INSA de Lyon / Université Claude Bernard Lyon 1 / Université Lumière Lyon 2 / École Centrale de Lyon. http://liris.cnrs.fr/publis/?id=4911
53. Galdames F, Pérez C, Estévez P, Held C, Jaillet F, Lobo G, Donoso G, Coll C (2010) Registration of renal SPECT and 2.5D US images. Technical report RR-LIRIS-2010-022, LIRIS Laboratoire d'Informatiqueen Image et Systèmes d'information, UMR 5205 CNRS / INSA de Lyon / Université Claude Bernard Lyon 1 / Université Lumière Lyon 2 / École Centrale de Lyon. http://liris.cnrs.fr/publis/?id=4912
54. Gibson S, Mirtich B (1997) A survey of deformable modeling in computer graphics. Technical report MERL-TR-97-19, Mitsubishi Electric Research Laboratory, Cambridge, Massachusetts, USA. http://citeseer.nj.nec.com/gibson97survey.html
55. González-Hidalgo M, Mir A, Nicolau G (2006) An evolution model of parametric surface deformation using finite elements based on B-splines. In: Proceedings of CompImage'2006 conference, computational modelling of objects represented in images: fundamentals, methods and applications, Coimbra, Portugal
56. González-Hidalgo M, Jaume Capó A, Mir A, Nicolau-Bestard G (2008) Analytical simulation of B-spline surfaces deformation. In: Perales F, Fisher R (eds) Articulated motion and

deformable objects. Lecture notes in computer science, vol 5098. Springer, Berlin, pp 338–348. doi:10.1007/978-3-540-70517-8_33
57. González-Hidalgo M, Jaume-i Capó A, Mir A, Nicolau-Bestard G (2010) Analytical simulation of non-planar B-spline surfaces deformation. In: Perales F, Fisher R (eds) Articulated motion and deformable objects. Lecture notes in computer science, vol 6169. Springer, Berlin, pp 213–223
58. Greiner G (1994) Surface construction based on variational principles. Laurent P, Le Méhauté A, Schumaker L (eds) Wavelets, images and surface fitting. A. K. Peters, Wellesley, pp 277–286
59. Güdükbay U, Özgüç B, Tokad Y (1997) A spring force formulation for ellastically deformable models. Comput Graph 21(3):335–346
60. Hanson AJ (1988) Hyperquadrics: smoothly deformable shapes with convex polyhedral bounds. Comput Vis Graph Image Process 44(2):191–210
61. Heïgéas L, Luciani A, Thollot J, Castagné N (2010) A physically-based particle model of emergent crowd behaviors. In: CoRR, vol abs/1005.4405
62. Huang QX, Wicke M, Adams B, Guibas L (2009) Shape decomposition using modal analysis. Comput Graph Forum 28(2):407–416. doi:10.1111/j.1467-8659.2009.01380.x
63. Ilic S, Fua P (2003) From explicit to implicit surfaces for visualization, animation and modeling. In: ISPRS workshop on visualization and animation of reality-based 3D Models
64. Jaillet F, Shariat B, Vandorpe D (1998) Deformable object reconstruction with particle systems. Comput Graph 22(2–3):189–194
65. Jaklic A, Leonardis A, Solina F (2000) Superquadrics and their geometric properties, computational imaging and vision, chap. 2, vol 20. Kluwer Academic Publishers, Dordrecht. http://lrv.fri.uni-lj.si/franc/SRSbook/geometry.pdf
66. Jeong IK, Lee I (2004) An oriented particle and generalized spring model for fast prototyping deformable objects. In: Proceedings of the 25th annual conference of the European association for computer graphics, vol 23. Eurographics
67. Juhász I (1999) Weight-based shape modification of NURBS curves. Comput Aided Geom Des 16:377–383
68. Kass M, Witkin A, Terzopoulos D (1988) Snakes: active contour models. Int J Comput Vis 1(4):321–331
69. Kim J, Pollard NS (2011) Fast simulation of skeleton-driven deformable body characters. ACM Trans Graph 30(5):121:1–121:19. doi:10.1145/2019627.2019640
70. Kobbelt L (1996) Interpolatory subdivision on open quadrilateral nets with arbitrary topology. Comput Graph Forum 15(3):409–420
71. Kumar S, Manocha D, Lastra A (1996) Interactive display of large NURBS models. IEEE Trans Vis Comput Graph 2(4):323–336. http://citeseer.nj.nec.com/article/kumar96interactive.html
72. La Gréca R, Raffin R, Gesquière G (2007) Punctual constraint resolution and deformation path on NURBS. In: Graphicon'07. International conference on computer graphics and vision. http://www.graphicon.ru/2007/proceedings/Papers/Paper_38.pdf
73. Lamousin HJ, Waggenspack WN (1994) NURBS-based free-form deformations. IEEE Comput Graph Appl 14(6):59–65
74. Li S, Pu F, Li D (2007) An improved edge detection algorithm based on area morphology and maximum entropy. In: Proceedings of the second international conference on innovative computing, information and control, ICICIC '07. IEEE Computer Society, Washington, DC, USA, pp 536–539. doi:10.1109/ICICIC.2007.148
75. Li H, Leow WK, Chiu IS (2010) Elastic tubes: modeling elastic deformation of hollow tubes. Comput Graph Forum 29(6):1770–1782. doi:10.1111/j.1467-8659.2010.01647.x
76. Liao YL, Lu CF, Sun YN, Wu CT, Lee JD, Lee ST, Wu YT (2011) Three-dimensional reconstruction of cranial defect using active contour model and image registration. Med Biol Eng Comput 49:203–211. doi:10.1007/s11517-010-0720-0
77. Lipton S, Evans J, Bazilevs Y, Elguedj T, Hughes T (2010) Robustness of isogeometric structural discretizations under severe mesh distortion. Comput Methods Appl Mech

Eng 199:357–373. doi:10.1016/j.cma.2009.01.022. http://www.sciencedirect.com/science/article/pii/S0045782509000346 (Computational geometry and analysis)

78. Loop CT (1987) Smooth subdivision surfaces based on triangles. PhD thesis, University of Utah
79. Loop C, Schaefer S, Ni T, Castaño I (2009) Approximating subdivision surfaces with gregory patches for hardware tessellation. In: ACM SIGGRAPH Asia 2009 papers, SIGGRAPH Asia '09. ACM, New York, NY, USA, pp 151:1–151:9. doi:10.1145/1661412.1618497
80. Lorensen WE, Cline HE (1987) Marching cubes: a high resolution 3D surface construction algorithm. In: Proceedings of the 14th annual conference on computer graphics and interactive techniques. ACM Press, New York, pp 163–169
81. Luken WL, Cheng F (1996) Comparison of surface and derivative evaluation methods for the rendering of NURBS surfaces. ACM Trans Graph 15(2):153–178
82. Magnenat-Thalmann N, Bonanni U, Volino P (2009) Physical behavior of deformable hair and clothes: what is common? In: CAD/Graphics. IEEE, pp 12–18 http://dblp.uni-trier.de/db/conf/cadgraphics/cadgraphics2009.html#Magnenat-ThalmannBV09
83. Malcolm JG, Rathi Y, Yezzi A, Tannenbaum AR (2008) Fast approximate surface evolution in arbitrary dimension. In: Reinhardt JM, Pluim JPW (eds) Medical imaging 2008: image processing, vol 6914. SPIE-The International Society for Optical Engineering, Georgia Institute of Technology, Society of Photo-Optical Instrumentation Engineers, Bellingham
84. McDonald J (2001) On flexible body approximations of rigid body dynamics. In: Skala V (ed) WSCG 2001 conference proceedings. http://citeseer.nj.nec.com/446389.html
85. McInerney T, Terzopoulos D (1996) Deformable models in medical image analysis: a survey. Med Image Anal 1(2):91–108
86. Meyer M, Debunne G, Desbrun M, Barr AH (2001) Interactive animation of cloth-like objects in virtual reality. J Vis Comput Animat 12(1):1–12. http://citeseer.nj.nec.com/meyer00interactive.html
87. Mille J, Cohen L (2010) 3D CTA image segmentation with a generalized cylinder-based tree model. In: IEEE international symposium on biomedical imaging: from nano to macro (ISBI), pp 1045–1048. http://liris.cnrs.fr/publis/?id=4666
88. Montagnat J, Delingette H, Scapel N, Ayache N (2000) Representation, shape, topology and evolution of deformable surfaces. Application to 3D medical image segmentation. Technical report 3954, INRIA Institut Nationale de Recherche en Informatique et en Automatique.
89. Moustakides GV, Briassoulis D, Psarakis EZ, Dimas E (2000) 3D image acquisition and NURBS based geometry modelling of natural objects. Adv Eng Softw 31(12):955–969
90. Mueller M, Teschner M, Gross M (2004) Physically-based simulation of objects represented by surface meshes. In: Proceedings of computer graphics international CGI'04, pp 26–33
91. Mukherjee D, Ray N, Acton S (2004) Level set analysis for leukocyte detection and tracking. IEEE Trans Image Process 13(4):562–572. doi:10.1109/TIP.2003.819858
92. Nastar C, Ayache N (1993) Fast segmentation, tracking, and analysis of deformable objects. In: Proceedings of fourth international conference on computer vision, ICCV'93, pp 275–279
93. Natsupakpong S, Cenk Çavuşoglu M (2010) Determination of elasticity parameters in lumped element (mass-spring) models of deformable objects. Graph Models 72(6):61–73. doi:10.1016/j.gmod.2010.10.001
94. Nealen A, Müller M, Keiser R, Boxerman E, Carlson M (2006) Physically based deformable models in computer graphics. Comput Graph Forum 25(4):809–836. doi:10.1111/j.1467-8659.2006.01000.x
95. Nealen A, Igarashi T, Sorkine O, Alexa M (2007) Fibermesh: designing freeform surfaces with 3D curves. ACM Trans Graph 26(3):41. doi:10.1145/1276377.1276429
96. Nguyen-Thanh N, Nguyen-Xuan H, Bordas S, Rabczuk T (2011) Isogeometric analysis using polynomial splines over hierarchical t-meshes for two-dimensional elastic solids. Comput Methods Appl Mech Eng 200:1892–1908. doi:10.1016/j.cma.2011.01.018. http://www.sciencedirect.com/science/article/pii/S0045782511000338
97. Nishimura H, Hirai M, Kawai T, Kawata T, Shirakawa I, Omura K (1985) Object modeling by distribution function and a method of image generation. Trans Inst Electron Commun Eng

Jpn J68-D(4):718–725 (traducido al inglés por T. Fujuwara, Advanced Studies in Computer Aided Art and Design, Middlesex Polytechnic, England, 1989)
98. Osher S, Sethian JA (1988) Fronts propagating with curvature-dependent speed: algorithms based on Hamilton-Jacobi formulations. J Comput Phys 79:12–49. http://citeseer.ist.psu.edu/osher88fronts.html
99. Palmer P, Mir A, González M (2000) Stability and complexity study of animated elastically deformable objects. AMDO'2000. First international workshop on articulated motion and deformable objects, Lecture notes in computer science, vol 1899. Springer, Berlin, pp 58–71
100. Pentland A, Williams J (1989) Good vibrations: model dynamics for graphics and animation. In: Proceedings of the 16th annual conference on computer graphics and interactive techniques. ACM Press, New York, pp 215–222. doi:10.1145/74333.74355
101. Picinbono G, Delingette H, Ayache N (2000) Real-time large displacement elasticity for surgery simulation: non-linear tensor-mass model. In: Third international conference on medical robotics, imaging and computer assisted surgery: MICCAI 2000, pp 643–652. ftp://ftp-sop.inria.fr/epidaure/Publications/Picinbono/miccai2000.ps.gz
102. Piegl L (1991) On NURBS: a survey. IEEE Comput Graph Appl 11(1):55–71. doi:10.1109/38.67702
103. Piegl L, Tiller W (1997) The NURBS book: monographs in visual communications, 2nd edn. Springer, New York
104. Portal R, Sousa L, Dias J, Santos N (2009) Contact detection of convex superquadric using optimization techniques with graphical user interface. In: Proceedings of the 7th EUROMECH solid mechanics conference (ESMC2009)
105. Provot X (1995) Deformation constraints in a mass-spring model to describe rigid cloth behaviour. In: Davis WA, Prusinkiewicz P (eds) Graphics interface '95. Canadian Human–Computer Communications Society, pp 147–154. http://citeseer.nj.nec.com/provot96deformation.html
106. Qin H, Terzopoulos D (1995) Dynamic NURBS swung surfaces for physics-based shape design. Comput Aided Des 27(2):111–127. http://citeseer.nj.nec.com/qin95dynamic.html
107. Radeva P, Amini AA, Huang J (1997) Deformable B-solids and implicit snakes for 3D localizaton and tracking of SPAMM MRI data. Comput Vis Image Underst 66(2):163–178
108. Rueckert D, Aljabar P, Heckemann R, Hajnal J, Hammers A (2006) Diffeomorphic registration using B-splines. In: Larsen R, Nielsen M, Sporring J (eds) Medical image computing and computer-assisted intervention MICCAI 2006, Lecture notes in computer science, vol 4191. Springer, Berlin, pp 702–709. doi:10.1007/11866763_86
109. Schmid J, Iglesias Guitin J, Gobbetti E, Magnenat-Thalmann N (2011) A GPU framework for parallel segmentation of volumetric images using discrete deformable models. Vis Comput 27:85–95. doi:10.1007/s00371-010-0532-0
110. Sederberg MT, Sederber TW (2010) T-splines: a technology for marine design with minimal control points. Technical report
111. Sederberg TW, Zheng J, Sewell D, Sabin M (1998) Non-uniform recursive subdivision surfaces. In: Proceedings of the 25th annual conference on computer graphics and interactive techniques, SIGGRAPH '98. ACM, New York, NY, USA, pp 387–394. doi:10.1145/280814.280942
112. Sederberg TW, Zheng J, Bakenov A, Nasri A (2003) T-splines and T-nurccs. ACM Trans Graph 22(3):477–484. doi:10.1145/882262.882295
113. Selle A, Lentine M, Fedkiw R (2008) A mass spring model for hair simulation. ACM Trans Graph 27(3):64:1–64:11. doi:10.1145/1360612.1360663
114. Sethian JA (1987) Numerical methods for propagating fronts. In: Concus P, Finn R (eds) Variational methods for free surface interfaces. Proceedings of the September 1985 Vallambrosa conference. Springer, New York, pp 155–164
115. Sethian JA (1999) Fast marching methods. SIAM Rev 41(2):199–235
116. Sethian JA, Smereka P (2003) Level set methods for fluid interfaces. Fluid Mech 35:341–372
117. Shetian JA (1999) Level set methods and fast marching methods, 2nd edn. Cambridge University Press, Cambridge (Cambridge monograph on applied and computational mathematics)

118. Shewchuk JR (2002) Delaunay refinement algorithms for triangular mesh generation. Comput Geom 22(1–3):21–74. doi:10.1016/S0925-7721(01)00047-5. http://www.sciencedirect.com/science/article/pii/S0925772101000475 (16th ACM symposium on computational geometry)
119. Shewchuk JR (2005) Theoretically guaranteed delaunay mesh generation—in practice. In: 14th international meshing roundtable, vol. Short Course
120. Silveira M, Nascimento J, Marques J (2007) Automatic segmentation of the lungs using robust level sets. In: Proceedings of the 29th annual international conference of the IEEE engineering in medicine and biology society. EMBS 2007, pp 4414–4417. doi:10.1109/IEMBS.2007.4353317
121. Staib LH, Duncan JS (1996) Model-based deformable surface finding for medical images. IEEE Trans Med Imaging 15(5):720–731
122. Steinemann D, Otaduy MA, Gross M (2009) Splitting meshless deforming objects with explicit surface tracking. Graph Models 71(6):209–220. doi:10.1016/j.gmod.2008.12.004. http://www.sciencedirect.com/science/article/pii/S1524070309000034 (2006 ACM SIGGRAPH/Eurographics Symposium on Computer Animation (SCA 2006))
123. Suri JS, Liu K, Singh S, Laxminarayan S, Zeng X, Reden L (2002) Shape recovery algorithms using level sets in 2-D/3-D medical imagery: a state-of-the-art review. IEEE Trans Inf Technol Biomed 6:8–28
124. Szeliski R, Tonnesen D (1992) Surface modeling with oriented particle systems. Comput Graph 26(2):185–194
125. Terzopoulos D (2011) Deformable and functional models. In: Tavares JMRS, Jorge RMN (eds) Computational vision and medical image processing, computational methods in applied sciences, vol 19. Springer, Netherlands, pp 125–143. doi:10.1007/978-94-007-0011-6_7
126. Terzopoulos D, Fleischer K (1988) Deformable models. Vis Comput 4:306–331
127. Terzopoulos D, Fleischer K (1988) Modelling inelastic deformation: viscoelasticity, plasticity, fracture. Comput Graph 21(4):269–278
128. Terzopoulos D, Metaxas D (1991) Dynamic 3D models with local and global deformations: deformable superquadrics. IEEE Trans Pattern Anal Mach Intell 13(7):703–714
129. Terzopoulos D, Vasilescu M (1991) Sampling and reconstruction with adaptive meshes. In: IEEE computer society conference on computer vision and pattern recognition (CVPR'91), IEEE Computer Society Press, Maui, Hawaii, pp 70–75
130. Terzopoulos D, Witkin A (1988) Physically based models with rigid and deformable components. IEEE Comput Graph Appl 8:41–51
131. Terzopoulos D, Platt J, Barr A, Fleischer K (1987) Ellastically deformable models. Comput Graph (Proceedings (SIGGRAPH) 21(4):205–214
132. Terzopoulos D, Witkin A, Kass M (1988) Constraints on deformable models: recovering 3D shape and nonrigid motion. Artif Intell 36(1):91–123
133. Teschner M, Heidelberger B, Mueller M, Gross M (2004) A versatile and robust model for geometrically complex deformable solids. In: Proceedings of computer graphics international CGI'04, pp 312–319
134. Vasilescu M, Terzopoulos D (1992) Adaptive meshes and shells: irregular triangulation, discontinuities, and hierarchical subdivision. In: IEEE computer society conference on computer vision and pattern recognition (CVPR'92). IEEE Computer Society Press, Champaign, pp 829–832
135. Veltkamp RC, Wesselink W (1996) Variational modeling of triangular Bézier surfaces. http://citeseer.nj.nec.com/387772.html
136. Versprille K (1975) Computer-aided design applications of the rational B-spline approximation form. PhD thesis, Syracuse University
137. Wang Y, Xiong Y, Xu K, Liu D (2012) vKASS: a surgical procedure simulation system for arthroscopic anterior cruciate ligament reconstruction. Comput Anim Virtual Worlds. doi:10.1002/cav.1434
138. Wojtan C, Thürey N, Gross M, Turk G (2009) Deforming meshes that split and merge. In: ACM SIGGRAPH 2009 papers, SIGGRAPH '09. ACM, New York, NY, USA, pp 76:1–76:10. doi:10.1145/1576246.1531382

139. Wyvill G, McPheeters C, Wyvill B (1986) Animating soft objects. Vis Comput 2(4):235–242
140. Wyvill G, McPheeters C, Wyvill B (1986) Data structure for soft objects. Vis Comput 2(4):227–234
141. Xing W, Yuan B (2012) 3D part based structural description extracting and modeling. In: Proceedings of the international multiconference of engineers and computer scientists, IMECS'2012 (2012). http://www.iaeng.org/publication/IMECS2012/IMECS2012_pp209-212.pdf
142. Yau H-T, Lin YK, Yeh CT (2009) A new approach to accelerate NURBS surface rendering on GPU. Comput Aided Des Appl 6(4):529–538. doi:10.3722/cadaps.2009.529-538
143. Yeo SY, Xie X, Sazonov I, Nithiarasu P (2011) Geometrically induced force interaction for three-dimensional deformable models. IEEE Trans Image Process 20(5):1373–1387. doi:10.1109/TIP.2010.2092434
144. Zhang H, Yang L, Foran DJ, Nosher JL, Yim PJ (2009) 3D segmentation of the liver using free-form deformation based on boosting and deformation gradients. In: Proceedings of the sixth IEEE international conference on symposium on biomedical imaging: from nano to macro, ISBI'09. IEEE Press, Piscataway, NJ, USA, pp 494–497. http://dl.acm.org/citation.cfm?id=1699872.1699997
145. Zhu L, Fan B, Tang Y (2009) Active contour method with separate global translation and local deformation. In: Xie M, Xiong Y, Xiong C, Liu H, Hu Z (eds) Intelligent robotics and applications, Lecture notes in computer science, vol 5928. Springer, Berlin, pp 876–884. doi:10.1007/978-3-642-10817-4_86
146. Zorin D, Schröder P (2000) Subdivision for modeling and animation. Technical report, SIGGRAPH 2000—Course Notes

Free Form Deformations or Deformations Non-Constrained by Geometries or Topologies

Romain Raffin

Abstract Free-form deformations are widely used to model 3D objects. In these methods "free-form" designates: "*whatever the object is, whatever its description and topology, we are able to deform it*". They limit the user interaction to pull some points of an embedding rough mesh. From the user point-of-view, it does not matter if the object manipulated is 3-dimensional, of 0-genus or a parametric surface, he or she always uses the same process to model a complex object: load an initial object from a library and deform it via FFD methods to follow his (her) needs. A large number of deformation methods have been published, allowing new deformations, new kinds of controls or enhancing the description of resulting objects. In fact advantage of deformation non-constrained by geometries is also a drawback: as it only manipulates points it could only result in points, so its necessary to use and maintain the neighborhood (the topology) or the surface expression on a second hand, as these methods do not care of object description.

1 Free-Form Deformations from the Beginning

The work of creating an object from an empty scene is quite tedious, first methods of mesh manipulation [28, 31] highlight the future needs of user-oriented, efficient modeling methods. They first deform an object from a single vertex, diffusing the deformation on neighbors according to edges distance weights. The work of Barr [2] initiates free-form deformation techniques. It permits to deform an existing object globally (see Fig. 1), with mathematical functions (taper, bend, twist). Since it was not a "vertex by vertex" displacement method it simplifies the creation process. This method does not act on topology, keeping the neighbors as they were, with the edges

R. Raffin (✉)
Aix-Marseille University, LSIS UMR 7296, Domaine universitaire de Saint Jérôme, Av. Escadrille Normandie-Niemen, Marseille Cedex 13397, France
e-mail: romain.raffin@lsis.org

M. González Hidalgo et al. (eds.), *Deformation Models*, Lecture Notes in Computational Vision and Biomechanics 7, DOI: 10.1007/978-94-007-5446-1_2,

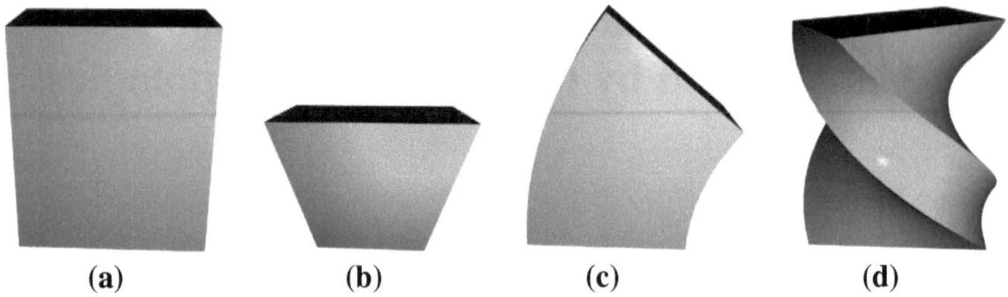

Fig. 1 Barr [2] global deformations. **a** Initial object; **b** taper; **c** bend; **d** twist

and faces (in a B-rep model) providing links between vertices. The deformation functions are described by matrices, as transformations (rotate, scale, translate) and the method allows composition and application of a stack of deformations, compliantly with a CSG modeling environment.

As Barr methods are made to be applied globally, locality of a deformation is obtained by the developer with conditional instructions, as this is not described by the transformation matrices.

2 FFD: Free Form Deformation

The method created by Sederberg and Parry [37] tends to give a virtual sculpture approach. In a global view the method considers the object to be deformed embedded in a parallelepipedical grid (see Fig. 2). Once the local coordinates of all vertices of the object (the teapot in Fig. 2) are expressed in the grid frame, pulling a grid point transmits the deformation to the initial object by a simple re-expression of the coordinates of the object in the world frame.

Authors described (in words) the process of FFD:

> *A good physical analogy for FFD is to consider a parallelpiped of clear, flexible plastic in which is embedded an object, or several objects, which we wish to deform. The object is imagined to also be flexible, so that it deforms along with the plastic that surrounds it.*

More formally expressed, a FFD is a $\mathbb{R}^3 \rightarrow \mathbb{R}^3$ application using a trivariate product of Bernstein polynomials. First step is to define object vertices coordinates in grid space. It needs an origin X_0 and frame vectors: $\overrightarrow{S}$, $\overrightarrow{T}$ and $\overrightarrow{U}$ (in 3D space but 2D or 4D adaptation can be made trivially).

Figure 3a shows the local frame and expression of local object coordinates (s, t and u) of a point X, following:

$$\overrightarrow{X_0X} = s\,\overrightarrow{S} + t\,\overrightarrow{T} + u\,\overrightarrow{U}$$

A way to compute (s, t, u) values is:

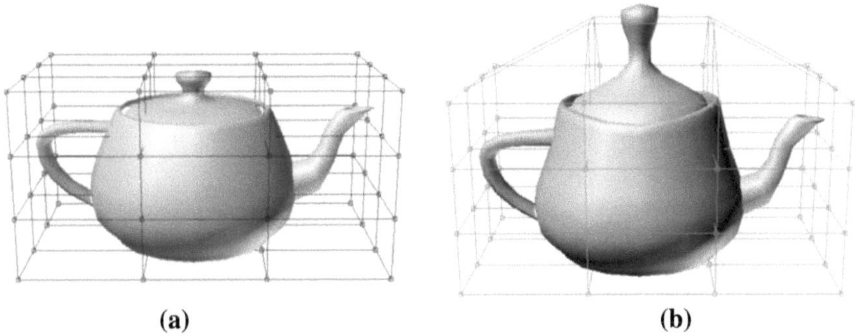

Fig. 2 FFD deformation of a teapot. **a** Initial object and surrounding FFD mesh; **b** changes in mesh configuration involve object deformation

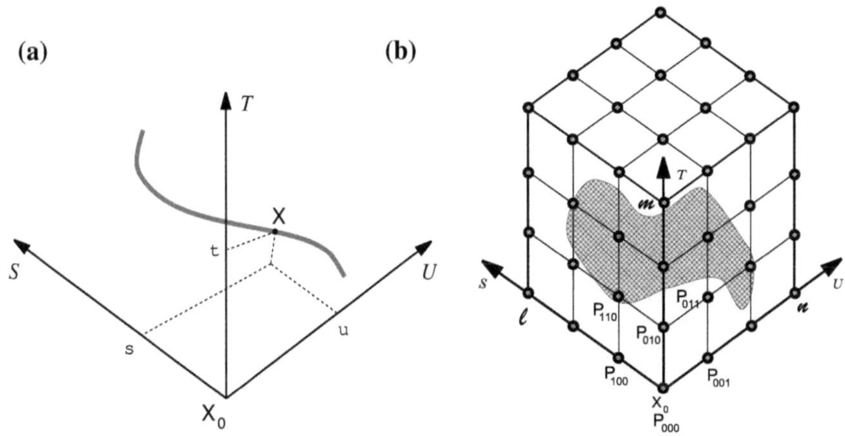

Fig. 3 FFD schema. **a** FFD local associated frame; **b** FFD embedding mesh construction

$$s = \frac{\overrightarrow{T} \wedge \overrightarrow{U} \bullet \overrightarrow{X_0X}}{\overrightarrow{T} \wedge \overrightarrow{U} \bullet \overrightarrow{S}} \quad t = \frac{\overrightarrow{S} \wedge \overrightarrow{U} \bullet \overrightarrow{X_0X}}{\overrightarrow{S} \wedge \overrightarrow{U} \bullet \overrightarrow{T}} \quad u = \frac{\overrightarrow{S} \wedge \overrightarrow{T} \bullet \overrightarrow{X_0X}}{\overrightarrow{S} \wedge \overrightarrow{T} \bullet \overrightarrow{U}}$$

If a point X lies in the grid it verifies: $0 < s < 1$, $0 < t < 1$ and $0 < u < 1$. The grid is cut in l, m, n parts according to $\overrightarrow{S}$, $\overrightarrow{T}$ and $\overrightarrow{U}$ dimensions respectively. In the example of Fig. 3b, we defined a uniform $l = m = n = 3$ grid.

Each vertex P_{ijk} of the embedding grid is associated to a control point of a parametric volume. In the frame $(X_0, \overrightarrow{S}, \overrightarrow{T}, \overrightarrow{U})$, their locations are given by:

$$\overrightarrow{X_0P_{ijk}} = \frac{i}{l}\overrightarrow{S} + \frac{j}{m}\overrightarrow{T} + \frac{k}{n}\overrightarrow{U} \text{ with } i \in [0 \ldots l],\ j \in [0 \ldots m],\ k \in [0 \ldots n]$$

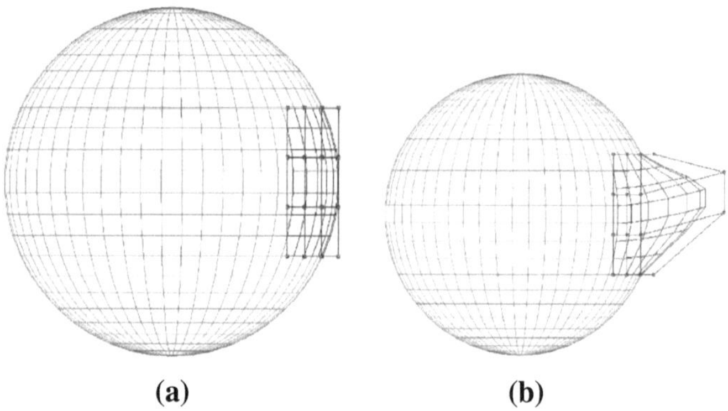

Fig. 4 Local FFD. **a** Initial object and local range grid; **b** deformation by a local grid

We can express the position of X in the grid frame (denoted X_{ffd}) with these control points, and weights given by Bernstein polynomial values:

$$X_{ffd} = \sum_{i=0}^{l} \binom{l}{i} (1-s)^{l-i} s^i \left[\sum_{j=0}^{m} \binom{m}{j} (1-t)^{m-j} t^j \left[\sum_{k=0}^{n} \binom{n}{k} (1-u)^{n-k} u^k P_{ijk} \right] \right] \quad (1)$$

Computing local coordinates of each point of the initial object is made with formulae (1) and attach the object's points to control points of the grid. Local influence of these control points is underlaid in the Bernstein polynomial weights. The more a control point is close from an object vertex, the more it weighs in local coordinates expression, and the more its displacement acts on point deformation. Conversely control points influence all object vertices: even if an extreme control point is displaced, the entire deformed object must be recomputed (an analogy is evident with Bézier curve control points).

Initially the FFD method does not describe local deformation of an object, but authors propose to use a restricted grid on the local part to be deformed (as illustrated at Fig. 4). Continuity between unmoved and deformed parts is not defined (kept as initially), as the deformation continuity exists only in embedding grid. Attention must be paid to not create stretched faces by displacing the whole local grid far from the object, or twisting it until self-intersection.

2.1 FFD *Properties*

FFD method is interesting because:

1. It conserves the object visual aspect. Even though its topology is not managed by the deformation, it is not possible to create or remove holes, even if inverting exterior and interior can be possible.

2. It mainly uses polyhedral B-rep object but one can use parametric object as input to increase potential resulting shapes. If this object obey to control network displacement, deforming such an object consists in deforming the control points net. As it deforms a control net of an approximating parametric surface, the deformation effect is then nothing but user-friendly.
3. An information on the volume of the deformed part of the object can be computed studying the jacobian of the FFD [37] and volume-preserving methods have been published [17, 35].

To conclude, the FFD method is a rapid and user-friendly method to deform an object. Locality can be obtained with a restricted embedding grid. As the method is easy to implement, it is mainly used in a wide range of applications. An important drawback is the parallelepiped shape of the grid with an influence zone of a deformation difficult to place, to localize on the initial object or to manipulate finely. Starting from this, another methods have been implemented, modifying some steps or tools of the process (grids shapes, coordinates functions, objects that can be deformed).

2.2 *Extensions and Classifications of* FFD

The initial method proposed par Sederberg [37] initiate a lot of extensions. In this section, we propose to list some of them sorted by geometry dimension as these methods lie on point "grid", curves or volumes (like the initial FFD).

2.3 *0-Dimensional Methods*

First dimension listed, it involves the deformation tool to be a single point. Vertices of the initial object are expressed regarding this frame by a simple distance value (mostly euclidean). The methods of [3, 4, 18, 36] or [29] can be cited, an other definition of local coordinates can be found in [30]. We will develop some of these methods later in the document. Hsu et al. [18] propose a direct manipulation of the initial mesh, hiding to the user the complexity of the embedding grid. The latter exists but the model interacts with the grid diffusing the initial object point displacement to the grid and computing the resulting deformation on the entire object. B-splines basis functions are used to ensure local action and continuity. Coordinates of a point in space are obtained by:

$$q_{i,j,k}(s,t,u) = \sum_{l,m,n=-3}^{0} P_{i+l,j+m,k+n} B_l(s) B_m(t) B_n(u) \tag{2}$$

As previously seen, local coordinates (s, t, u) are computed for each object point. For one single object point displacement with relative distances it moves the grid

Fig. 5 FFD direct manipulation [18]. **a** Initial object and deformation profile; **b** profile placement; **c** resulting deformation

of control points solving a least-square problem (LSQ). Denoting B the B-spline coefficients matrix and P the control points matrix, this can be written as:

$$q = BP$$

q_{new} new location of point q is given by (ΔP contains control point displacement):

$$\begin{aligned} q_{new} &= B(P + \Delta P) \\ &= BP + B\Delta P \\ &= q + \Delta q \quad \text{with} \quad \Delta q = B\Delta P \end{aligned}$$

Then, finding ΔP is done by:

$$\begin{aligned} \Delta P &= B^{-1}\Delta q \\ &\equiv B^{+}\Delta q \end{aligned}$$

B^{+} is the pseudo-inverse matrix of B, obtained with LSQ method. It may occurs that the resulting control points configuration obtained decrease continuity by repeating control points, authors add a matrix characterizing control points self-dependence that solves this problem.

The preceding method works well if one move two object points that not share the same control point. In the contrary, solving the constraint equations can not be done, and a way to overcome this problem is to refine the initial mesh. It involves more B-splines description but permits to separate their influence zones. Figure 5 shows a deformation obtained by direct manipulation of object points. A tool (profile) is put on the object, once pushed it displaces some object points, updates grid points location and results in the object deformation.

An other method: DFFD [9] stands for *Dirichlet based Free-Form Deformation*. As the preceding method it hides the embedding grid, using Sibson coordinates [13] to control the weights of points. The grid is based on a set of points and influence zones obtained automatically by Voronoï diagram of these points. Once done, the other points of the objects are inserted in this spatial arrangement to compute their locale coordinates. Computing the difference between the two Voronoï configurations before and after point deformation permits to diffuse deformation to regions (sets of object points). This method is more local than the classical FFD as a point acts on its Delaunay-neighbors and support easily multiresolution objects. It has been applied to hand movements in virtual environments [20, 30].

2.4 Dimension 1: Curves

Curves can define deformation tools, local parametrization is obtained by distances computations (so it can raise problems if points are projected on the same site on the curve). Two principal methods are detailed here: AxDF [26] and WIRES [38]. AxDF [26] main idea is to express local coordinates according a curvilinear axis. Apart of this, the process is merely the same:

1. creation of an axis passing through the object,
2. expression of object coordinates in axis ones with projection from $\mathbb{R}^3$ space to $\mathbb{R}$ curve parameter,
3. axis transformation,
4. computation of object's vertices new locations.

End-user interacts with the deformation by defining the axis and its modification. It remains mandatory to keep a well defined sliding frame, whose displacement along the axis is constrained (Frenet based [6, 7]) to avoid axis inversions (see Fig. 6).

To solve this continuity problem, authors use a minimization method [21]. One can retrieve in AxDF deformations the results of classical ones: rotation, torsion (see Fig. 6b, c). An interesting contribution of the method is that it defines an influence area that restricts deformation spatially and preserves continuity, as on Fig. 7c. Three zones are used with two radii R_{min} and R_{max}. In the first one (if distance from the axis is less than R_{min}) the deformation is maximal. From R_{min} to R_{max} influence decreases as the distance increases. Beyond R_{max} the deformation vanishes (Fig. 7a).

The WIRES [38] method is similar to the latter. It deforms a curve in another curve in space, with all the belonging geometries. Two parametric curves are defined by the user: $R(t)$ lies on the object and $W(t)$ figures the destination of $R(t)$. For each point P in space, if we can compute the distance $\|R(p_r) - P\|$ we can assign a deformation value at this point. Denoting p_r the parameter that correspond to the projection of P onto $R(t)$, we have: $\|R(p_r) - P\| = \min(\|R(t) - P\|)$ (see Fig. 9a). The value of p_r is obtained by solving the equation $R'(t) \cdot (R(t) - P) = 0$. The deformation manage the constraint linked to $R(t)$ according to a deformation function f.

An additional parameter s is introduced, allowing to "pinch" the deformation along its axis. It attracts or repulses the deformed points following their distances to $R(t)$ curve.

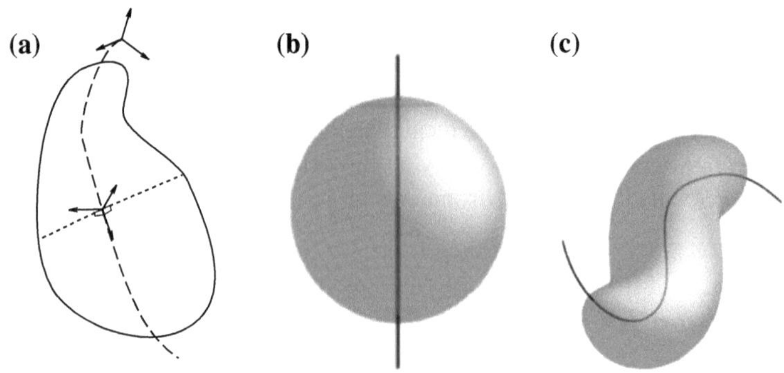

Fig. 6 AxDF construction [26]. **a** Example of construction; **b** object to be deformed and AxDF axis; **c** result of axis transformation

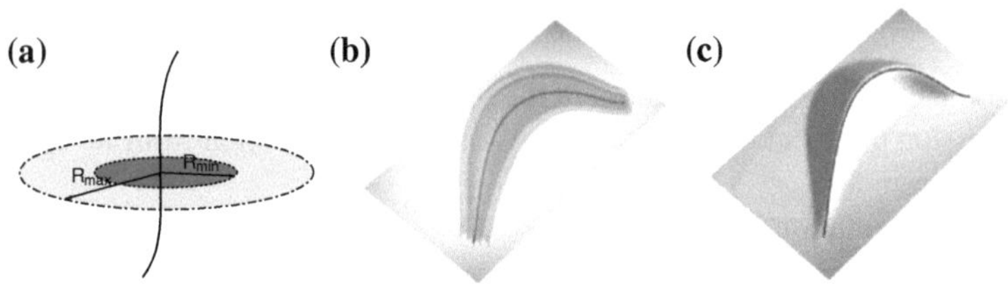

Fig. 7 Influences' radii of AxDF [26]. **a** Radii examples; **b** influence's radii projected on a plane; **c** effects of axis displacement

Curves shown at Fig. 8a–c illustrate the impact of different s values. The location of a deformed point P_s from its initial position P (see Fig. 9b) is now given by:

$$P_s = R(p_r) + (P - R(p_r))\left(1 + (s-1)f\left(\frac{\|R(p_r) - P\|}{r}\right)\right)$$

In the preceding equation r is a radius that creates an isotropic influence around the curve $R(t)$. The path from $R(t)$ curve to $W(t)$ is discretized to a set of positions that preserves tangential continuity and assure the displacement constraint. Tangent $\overrightarrow{R'(p_r)}$ is collinear to $\overrightarrow{W'(p_r)}$ according to the f function. Translation $T_{RW} = (W(p_r) - R(p_r))f\left(\frac{\|R(p_r)-P\|}{r}\right)$ is then applied. The P_{def} image of P by the deformation (see Fig. 9d) is obtained by:

$$P_{def} = P_r + (W(p_r) - R(P_r)).f\left(\frac{\|R(p_r) - P\|}{r}\right)$$

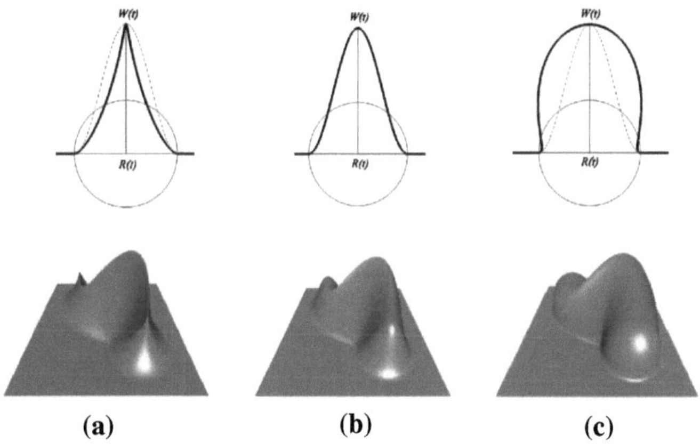

Fig. 8 WIRES deformations obtained by different values of s. **a** $s < 1$; **b** $s = 1$; **c** $s > 1$

P_r is an intermediate image of P_s by the rotations α and θ around the axis $R(p_r)$ and $\overrightarrow{R'(p_r)} \wedge \overrightarrow{W'(p_r)}$ (see Fig. 9c), where θ is the angle $\left(\widehat{\overrightarrow{R'(p_r)}, \overrightarrow{W'(p_r)}}\right)$ and $\alpha = f(P)$.

One can see that if $\overrightarrow{W'(p_r)} = \overrightarrow{R'(p_r)}$ the displacement is reduced to a single translation.

2.5 Dimension 2: Surfacic Deformations

In the third stage of free-form deformations we will study surfacic deformation tools [14, 22, 30]. We can include in this section the DFFD deformations [30] as they use a neighborhood with the edges joining each vertex and describe a discrete mesh. We develop herein the [14] method. If a parametric surface is used as a deformation tool, a parametric surface $S(u, v)$ must be defined and the projection of object's points on these surface gives the local coordinates in the tool space (as in AXDF for curves). The $H(u, v)$ surface ("height" surface) allows the modification of the distance between point P and initial surface $S(u, v)$ (see Fig. 10).

More formally, the deformation is expressed as follow:

- The two parametric surfaces that construct the deformation tool

$$S(u, v) = \sum_{i=0}^{m_s} \sum_{j=0}^{n_s} S_{ij} B_{i,k_{su}}(u) B_{j,k_{sv}}(v) \text{ and } H(u, v) = \sum_{i=0}^{m_h} \sum_{j=0}^{n_h} S_{ij} B_{i,k_{hu}}(u) B_{j,k_{hv}}(v)$$

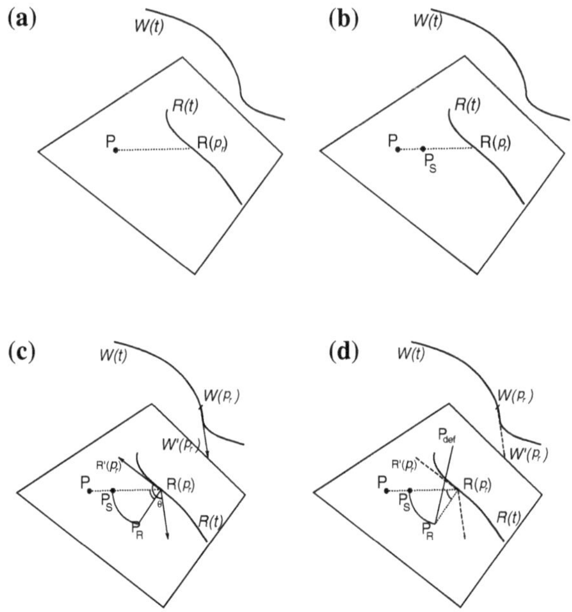

Fig. 9 WIRES deformation schema. **a** Construction of $R(p_r)$; **b** construction of P_s; **c** construction of P_R; **d** deformation

Initially, Feng et al.[14] use B-spline surfaces with node constraints to fix borders curves.

- $P(x, y, z)$ a point of the object to be deformed and $P'(x', y', z')$ the projection of P on $S(u, v)$. With the normal $\overrightarrow{N_S}$ of S at P', P is:

$$P = P' + h_p \overrightarrow{N}_S = S(u_p, v_p) + h_p \overrightarrow{N}_S(u_p, v_p)$$

where h_p is the distance from point P to surface $S(u, v)$. The parameters u_p and v_p are those of P' on S. If we denote P_{new} the new position of P after surface deformations of S and H in S_{new} and H_{new} respectively:

$$P_{new} = S_{new}(u_p, v_p) + H_{new}(u_p, v_p) h_p \overrightarrow{N}_{new}(u_p, v_p)$$

where $\overrightarrow{N}_{new}$ is the normal vector of S_{new}. Figure 10 shows construction and deformations process. This deformation is limited to vertical displacements of the surface H. Its extension to other displacements will increase rapidly the computational costs to minimize displacements.

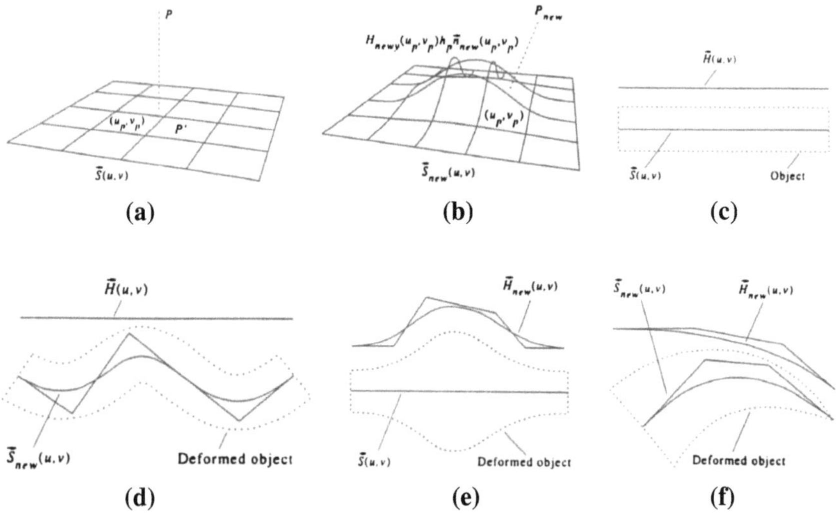

Fig. 10 Surfacic deformation [14]. **a** P' projection of P; **b** deforming S, obtaining P_{new}; **c** configuration of S and H; **d** S displacement; **e** H modification; **f** deformation by S and H

2.6 Dimension 3: Volumes

"Last" tool type (as we can increase easily space dimension of the embedding grid), volumic deformations were the initiators of free-form deformations. Most of works defines new volumes [16, 20, 23, 27], grid construction to define a wide range of tools [8], continuous definition of the deformation [1] or dynamic characterization to animate objects [12]. We will first describe the method of [8] which composes FFD meshes. We will also see [23] for a new embedding grid definition. Last, we will present a method combining implicit objects and deformations [10].

2.6.1 EFFD

Any intuitive it can be, the FFD method is disadvantaged by its simplicity. The construction of a complex object and the definition of a precise deformation are difficult with a parallelepiped embedding grid. As an example, the Fig. 11 shows a deformation applied on a sphere, the rectangular definition of the grid projection on the spherical surface produces a non-isotropic deformation (Fig. 11a, b). Conversely, a cylindrical grid definition would override this problem (see Fig. 11c, d).

EFFD is an extension of FFD (*Extended Free-Form Deformation*), described by Coquillart [8]. The deformation process is kept but it can use non-prismatic grids (see Figs. 11, 12) or a combination of grids. The usable meshes are more interesting for the final user, but their complexity can lead challenges to manipulate them. As the grid manipulate the embedded object, if some details are needed one can increase

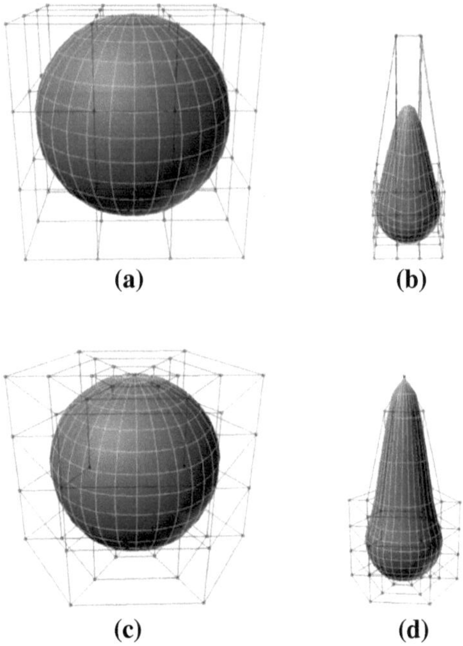

Fig. 11 Isotropy failure in FFD. **a** FFD parallepiped grid; **b** deformation result of this FFD; **c** cylindrical grid; **d** deformation result (EFFD)

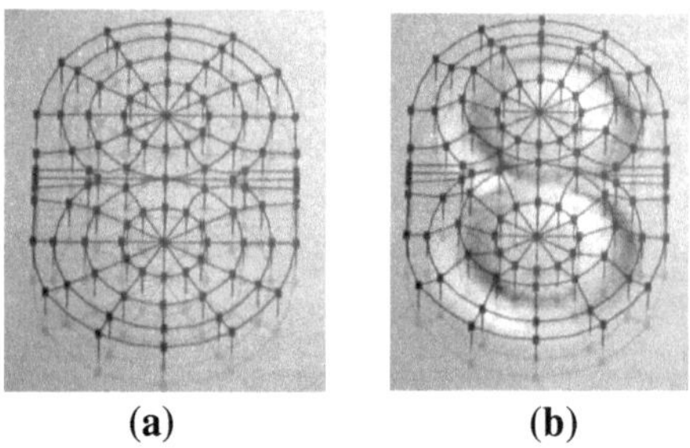

Fig. 12 Non-prismatic grids combination [8]. **a** EFFD mesh building; **b** resulting deformation

locally the grid density, but as joints between grids also manage continuity, a balance must be found (it also increases computational cost).

A hard part of the EFFD process is to express the global coordinates of the embedded object in the grid. It requires to found the chunk an object point lies in and to obtain the local coordinates in the grid volume. Fortunately, the grid is defined by Bézier tensor products and it is possible to use the control networks convex hulls. Once the chunk is found authors use a Newton approximation to compute coordinates in this local grid.

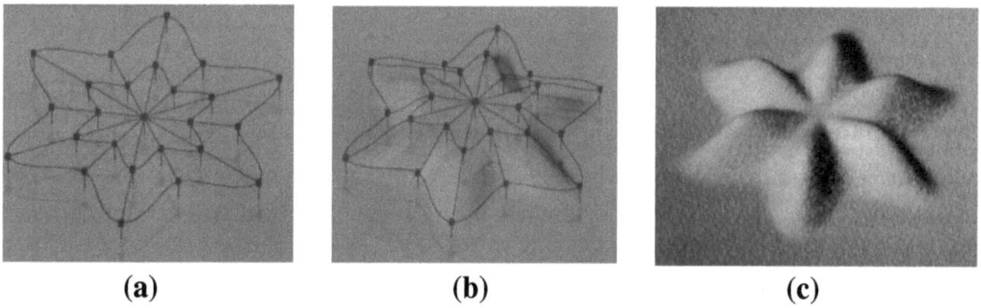

Fig. 13 EFFD deformation example [8]. **a** Initial mesh; **b** grid modification; **c** resulting object

This method is a great improvement of initial FFD. It brings intuitive control, various grid shapes and grids combinations that serve the user wishes. Figure 13 shows the ease of construction of a star obtained from a deformation applied on a plane.

2.6.2 NFFD

To control locality of the deformation Lamousin and Waggenspack [23] propose to use NURBS instead of Bézier volumes (NFFD method). The embedding grid is no longer divided in a uniform way but can be refined in areas where the deformation is important and kept rough where the user want only to conserve the initial object shape. The process is based again on FFD, where the local expression of grid coordinates is the only module that is modified. If the grid mesh $V_{i,j,k} = (x_{i,j,k}, y_{i,j,k}, z_{i,j,k})$ is used, a weight $W_{i,j,k}$ is given to each vertex (initially 1). With u, v and w the axis of the local coordinate system, and a, b and c the divisions of these axis, it gives $(a+1) \times (b+1) \times (c+1)$ grid points that embed the initial object. B-splines basis functions follow the rules:

$$2 \leq p \leq a+1$$
$$2 \leq m \leq b+1$$
$$2 \leq n \leq c+1$$

Nodal vectors are expressed for each coordinates by:

$$U = (u_0, \ldots, u_q) \text{ with } q = a + 2(p-1)$$
$$V = (v_0, \ldots, v_r) \text{ with } r = b + 2(m-1)$$
$$W = (w_0, \ldots, w_s) \text{ with } s = c + 2(n-1)$$

These vectors are non-uniforms and the nodes multiplicity is equal to basis functions order at extremities to ensure borders interpolation. As in EFFD method, a minimization is necessary to obtain, once the cell of the Nurbs volume is determined, the local coordinates of every object vertices.

2.6.3 IFFD

This method is described in [11] and is based on the use of implicit volumes instead of polyhedral ones for the embedding grid (in addition a more global method can be found in [19], manipulating scalar fields). Here again, the main difficulty is raised by the local coordinates computation, as the deformation tool can be defined with $n+1$ objects $P_0, \ldots, P_n$ with for each P_i:

- a local coordinates system and its associated function $\phi_i : \mathbb{R}^3 \to \mathbb{R}^3$ that gives for every point M in global space its coordinates $\widetilde{M}_i$ in P_i by $\widetilde{M}_i = \phi_i(M)$. This function is reversible and bijective to ensure the return from local to global coordinates $M_i = \phi_i^{-1}(\widetilde{M}_i)$,
- a transformation function $\Delta_i : \mathbb{R}^3 \to \mathbb{R}^3$ that permits, in local space, to displace $\widetilde{M}_i$ to $\widetilde{M}'_i$,
- a density function $F_i : \mathbb{R}^3 \to \mathbb{R}^+$ that gives the influence of the implicit primitive. It is a decreasing function with finite support.

The process is the same as in FFD:

$$M \begin{bmatrix} x \\ y \\ z \end{bmatrix} \underset{\text{Freezing}}{\Longrightarrow} \widetilde{M} \begin{pmatrix} \widetilde{M}_0 \\ \vdots \\ \widetilde{M}_n \end{pmatrix} \underset{\text{Deforming}}{\Longrightarrow} \widetilde{M}' \begin{pmatrix} \widetilde{M}'_0 \\ \vdots \\ \widetilde{M}'_n \end{pmatrix} \underset{\text{Combining}}{\Longrightarrow} M' \begin{bmatrix} x' \\ y' \\ z' \end{bmatrix}$$

Influence functions balance displacements according to the following two formulations:

- Globally,

$$M' = M + \frac{\sum_{i=0}^{n} F_i(M)(M'_i - M)}{\sum_{i=0}^{n} F_i(M)}$$

- Locally, a function $\Gamma_i : \mathbb{R} \to [0, 1]$ is added to implement deformation range. The blending function f_i is based on [10]:

$$f_i(t) = \begin{cases} \frac{3-4t}{2} & 0 < t \le \frac{1}{2} \\ 2(1-t)^2 & \frac{1}{2} < t \le 1 \\ 0 & \text{else} \end{cases}$$

with:

$$M' = M + \frac{\sum_{i=0}^{n} \Gamma_i \left(\frac{3}{2} - F(M)\right) F_i(M)(M'_i - M)}{F(M)}$$

where $F(M) = \sum_{i=0}^{n} F_i(M)$

Images of Fig. 14 show some deformations that can be obtained with IFFD.

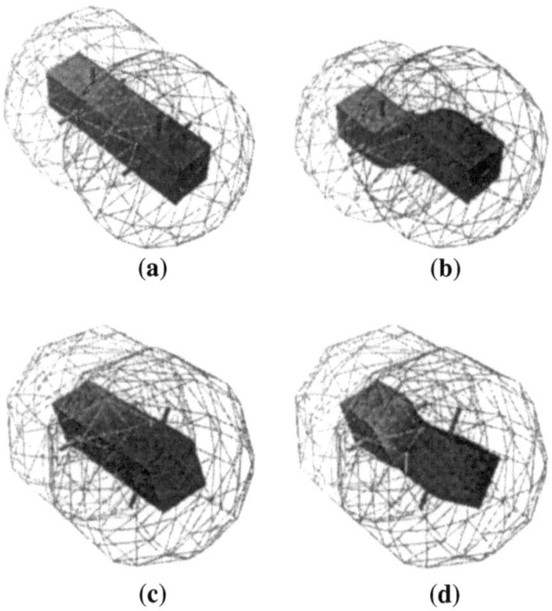

Fig. 14 IFFD: implicit free-form deformation examples [10]. **a** Grid translation; **b** translation result; **c** grid rotation; **d** rotation result

2.7 FFD *Conclusion*

This short survey of FFD methods declination shows that these methods are intuitive, quite easy to implement and easy to use. The complexity of the inner object to be deformed is hidden by the embedding grid. A drawback is due to user needs to go further in the definition of complex grids. Since the grids definitions must ensure that every vertex coordinates of the initial object can be expressed in the grid space, the algorithmic complexity increases as costs of computational process and can raise singularity situation. This has the opposite effect to that intended: the user can not understand this complexity and leaves the method. Another drawback is however the lack of constrained deformations (by positions, continuity, etc). We propose to study in the next section free-form deformations that permits this constraints description.

3 *n*D-Deformation

The previous methods we have compared have in common a description of a deformation but without displacement constraint satisfaction, this prevents the user to easily place objects relatively. Considering the numerous variations based on FFD methods and the hardness to overcome polyhedral grids in 2 or 3 dimensions, Bechmann [3]

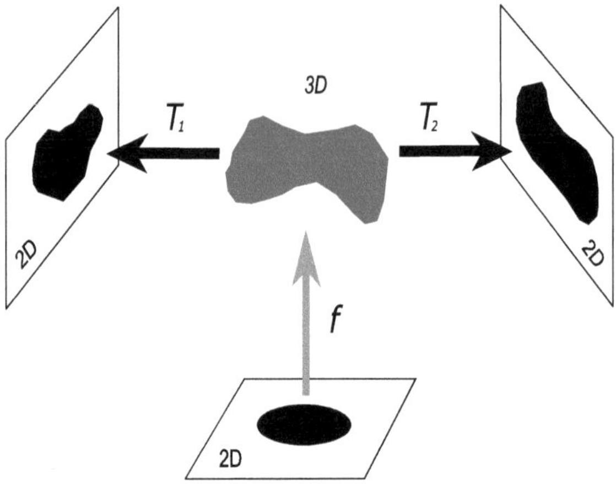

Fig. 15 nD-deformation schema for a 2D deformation

creates a more generic deformation model, named "nD-deformation" that express the deformation whatever the space dimensions. In this system, deformations are considered as constraints satisfactions, initial object is embedded in a space that permits to solve deformation constraints and then reprojected in three or four dimensions (with 4D deformations one can construct animations by deformation of space-time). Figure 15 presents the main principle of nD-deformation. An initial object is embedded in a space where more freedom-degrees exists (an upper dimensional space), in this space constraints solving is made easier and can be processed. Once achieved, the object is projected in a 3D space, eventually by different projection methods that result, for the same constraints solving system, in a set of matching resulting objects.

In the Fig. 15 the deformation is applied on object vertices, and is mathematically described by a polynomial function $f : \mathbb{R}^n \rightarrow \mathbb{R}^m$, with $n = 2$ the initial space dimensions and $m = 3$ the embedding one. There is m degrees of freedom to solve the deformation's constraints and f defines locality controls on the deformation. $T : \mathbb{R}^m \rightarrow \mathbb{R}^n$ is a linear inverse projection, from the upper space $\mathbb{R}^m$ to the initial one. The projection matrix corresponding to T is computed using pseudo-inverse methods to satisfy constraints or least-squares ones to give approximate deformation. If the space $\mathbb{R}^m$ is of sufficient high-order, user can also provides attracting or repulsing constraints.

Formally, if we denote:

- $Q(x_1, x_2, x_3, \ldots, x_n)$ a point in space $\mathbb{R}^n$,
- $\Delta Q(\Delta x_1, \Delta x_2, \Delta x_3, \ldots, \Delta x_n)$ the displacement values of Q, with

$$\Delta Q = d(Q) = (d(x_1), d(x_2), d(x_3), \ldots, d(x_n))^T$$

If M is the projection matrix associated to T, the deformation can be written as:

$$d(Q) = M.f(Q) \tag{3}$$

The deformation process is:

1. User gives r constraints $C_i, i \in [1, r]$ in space $\mathbb{R}^n$ and their new positions. Authors denote $d(C_i) = [d_1(C_i), d_2(C_i), \ldots, d_n(C_i)]^T$ the coordinates variation.
2. Matrix M is obtained by solving a system of $n \times r$ linear equations (each constraint raises n equations):
$$d(C_i) = M.f(C_i) \quad \forall i \in [1, r] \tag{4}$$
3. Once M computed, the displacement $d(Q)$ of a point Q in the initial space is given by:
$$d(Q) = M.f(Q)$$

If we consider each line M_j of M separately (with j a coordinate in $\mathbb{R}^n$), we can solve the lines of M independently:

$$d_j(C_i) = M_j.f(C_i) = f^T(C_i).M_j^T, \quad \forall i \in [1, r] \text{ and } j \in [1, n]$$

For the j th coordinate, the set of constraints gives:

$$D_j = \begin{bmatrix} d_j(C_1) \\ d_j(C_2) \\ \vdots \\ d_j(C_r) \end{bmatrix} = \begin{bmatrix} f^T(C_1) \\ f^T(C_2) \\ \vdots \\ f^T(C_r) \end{bmatrix} . M_j^T = X \cdot M_j^T \tag{5}$$

with:

$$X = \begin{bmatrix} f_1(C_1) \ldots f_m(C_1) \\ \vdots \\ f_1(C_r) \ldots f_m(C_r) \end{bmatrix}$$

We can now aggregate these results to construct the matrix M:

$$M_j^T = X^{-1} \begin{bmatrix} d_j(C_1) \\ d_j(C_2) \\ \vdots \\ d_j(C_r) \end{bmatrix}$$

One difficulty is to find the values for inverse matrix X^{-1}, because of its size $r \times m$. This matrix is generally non-invertible. This led three cases:

- If $m > r$ there is more unknown factors than equations, an infinity of solutions can be computed whatever the rank of matrix X, each of them satisfying the deformation constraints,
- If $m = r$ and rank of X is r, there is only one solution, the set of constraints defines fully the deformation,

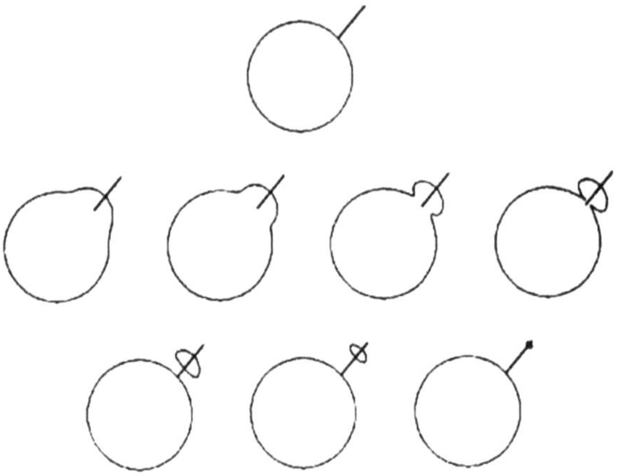

Fig. 16 4D deformation [3] of a *circle*

- If $m < r$ no solution can be found, one can use approximation method to find a "best-satisfying" solution according to the constraints.

The preceding list shows that it is necessary to use a value of m that is greater than (or at least equal to) the constraints number. Authors [3] use a pseudo-inverse matrix and a numerical method [5].

The initial 3D geometrical space can be extended to a 4D one that describe space-time deformations and animations. Figure 16 shows a deformation of a circle defined at initial time (up) and solved at end time (right down). Intermediate times explores the set of transitional shapes from the initial one to the deformed one.

The function f is chosen to preserve deformed (and resulting) object continuity. To obtain independent columns r of X, the f components must also be composed of non-linear combinations of the n coordinates. They can be point-oriented with m sub-functions, or axis-oriented with a tensor product of n vectors, or a mix between these two definitions. Basis formulation of f with m sub-functions is:

$$f(Q) = (f_1(Q), f_2(Q), \dots, f_m(Q))$$

The f_i functions are defined as $f_i : \mathbb{R}^n \rightarrow \mathbb{R}$ for $i \in [1, m]$. If we only use points as constraints, the deformation is defined in $\mathbb{R}^n$. Each component f_i of f is a product of n values:

$$f_i(Q) = f_i^1(x_1) f_i^2(x_2) \dots f_i^n(x_n)$$

With $f_i^j : \mathbb{R} \rightarrow \mathbb{R}$ for $i \in [1, m]$ and $j \in [1, n]$. We can then write f as:

$$f(Q) = f^1(x_1) \dots f^n(x_n) \text{ with } f^j = \left(f_1^j(x_j) \dots f_m^j(x_j) \right)$$

Each function can be managed separately and gives availability to specialize the deformation on a specific axis, for example to favor a direction or a temporal axis. f would be a tensor product of functions $f^k, k \in [1, n]$ defined from $\mathbb{R}^{p_k}$ into u_k if:

$$\forall U \in \mathbb{R}^n, \quad f(U) = \bigotimes_{k=1}^{n} f^k(u_k) \text{ with } f^k : \mathbb{R} \rightarrow \mathbb{R}^{p_k}$$

The dimension of the deformed object is then $m = \prod_{k=1}^{n} p_k$ and we could compute the component of $f(U)$ by:

$$\forall j \in [1, m], f_j(U) = \prod_{k=1}^{n} f^k_{s(j,k)}(u_k) \text{ with } 1 \leq s(j,k) \leq p_k$$

If $m = r$ and B-spline basis functions are used for f_i, we obtain the method SCODEF developed by [4] and presented hereafter.

3.1 SCODEF

Basing their works on the preceding method, Borrel [4] introduces a radius of influence for each constraint: it is the SCODEF model (*Simple Constrained Deformation*). As previously seen, it permits to fix the dimension value of $\mathbb{R}^m$ as the constraints number, facilitating constraint solving and hiding the embedding process (even it is always done). In the SCODEF model defined from $\mathbb{R}^n$ to $\mathbb{R}^n$ we have:

- n the dimension of the space,
- r the number of constraints,
- C_i the initial position of point,
- D_i the point displacement in the deformed space,
- R_i the influence radius associated with the C_i constraint,
- f_i the deformation function linked to C_i.

The deformation is point-oriented, its formulation is the same than Eq. 3, it says that:

$$d(Q) = \sum_{i=1}^{r} M_i . f_i(Q) \tag{6}$$

This can be rewritten with the addition of spherical influences. Each f_i function gives the contribution of the ith constraint to the displacement $d(Q)$ of a point Q. It is a scalar function that depends on each C_i constraint with its radius R_i:

$$f_i(Q) = B_i \left(\frac{\|Q - C_i\|}{R_i} \right)$$

Where B_i is a B-spline basis function, centered at 0 and decreasing to value 0 in 1. If we denote $U_i(Q) = \|Q - C_i\|/R_i$ and $f_i(Q) = B_i(U_i(Q))$ we can consider $U_i(Q)$ as a local parametrization of Q coordinates in C_i constraint influence. The value $f(0) = 1$ ensure constraint satisfaction. In a similar vein, the zero value of the function after passing 1 cancel the deformation outside the range of the influence radius. As in the Eq. 5 we can write for the jth coordinate:

$$d_j(C_i) = D_{ij} = \begin{bmatrix} D_{1j} \\ D_{2j} \\ \vdots \\ D_{rj} \end{bmatrix} = \begin{bmatrix} f^T(C_1) \\ f^T(C_2) \\ \vdots \\ f^T(C_r) \end{bmatrix} . M_j^T = X \cdot M_j^T$$

Following the definition, a constraint C_i influences a point Q if the distance $\|C_i - Q\|$ is lower than the constraint radius R_i: $\|C_i - Q\| < R_i$. Three situations must be evaluated:

- First case, the constraints are disjoints, the influence hull of each constraint does not overlap an other one (see Fig. 17). Influence of each constraint is isotropic. This mean for the constraint C_i:

$$U_j(C_i) = \frac{\|C_i - C_j\|}{R_j} = \begin{cases} 0 & \text{if } i = j \\ 1 & \text{if } i \neq j \end{cases}$$

  As $f_j(C_i) = f(U_j(C_i)) = \delta_{ij}$ this gives $f(C_i) = (0, \ldots, \underbrace{1}_{\text{rank } i}, \ldots, 0)^T$. As M_i is the ith vector of M, $D_i = d(C_i) = (M_1, \ldots, M_r).f(C_i) = M_i$. This column is also the displacement $d(C_i)$ of constraint C_i. Based on Eq. 6 we have:

$$d(Q) = \sum_{i=1}^{r} B_i\left(\frac{\|Q - C_i\|}{R_i}\right) D_i$$

  When the constraints are disjoints the displacement of a point Q is the weighted average of the displacements of the constraints it is under influence. This weight of a constraint C_i is inversely proportional to the distance $\|Q - C_i\|$,
- Second case, the constraints are disjoints but there is an overlapping influence. A point Q will be under control of all of these constraints if it lies inside the overlapping area. It does not compromise the computation of matrix M and can be treated by the first case,
- If two constraints self-influence each other, their radii and centers are overlapping, this situation raises vectors dependance problem for the matrix M (see Fig. 18). If we take an example with two constraints, and identical radius and deformation function, i.e. $R_1 = R_2$ and $f_1(C_2) = f_2(C_1) = \alpha$:

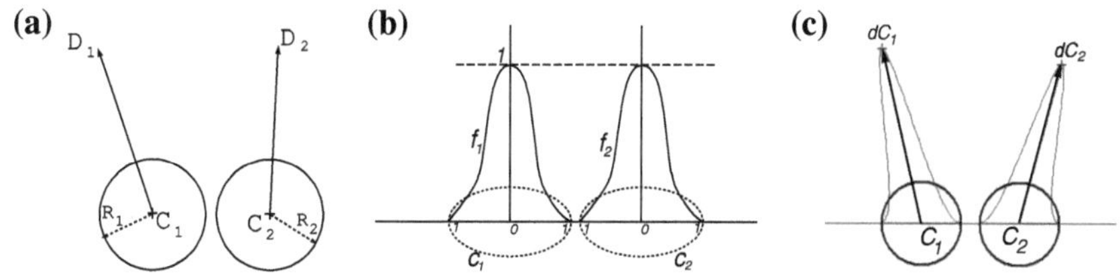

Fig. 17 Disjoints constraints in SCODEF method. **a** Constraints configuration; **b** showing associated functions; **c** deforming a line by these 2 constraints

$$(d(C_1) \;\; d(C_2)) = M \begin{pmatrix} 1 & \alpha \\ \alpha & 1 \end{pmatrix}$$

It implies:

$$M_1 = \frac{1}{1-\alpha^2} d(C_1) - \frac{\alpha}{1-\alpha^2} d(C_2) \text{ and } M_2 = \frac{-\alpha}{1-\alpha^2} d(C_1) + \frac{1}{1-\alpha^2} d(C_2)$$

In the worst case $d(C_1) = -d(C_2)$, that gives:

$$M_1 = -M_2 = \left(\frac{1}{1-\alpha}\right)$$

And finally the displacement of a point Q is given by:

$$d(Q) = \frac{d(C_1)}{1-\alpha} (f_1(Q) - f_2(Q))$$

As much as the distance between the constraints decreases, i.e. $\alpha \to 1$, displacement vanishes, i.e. $d(Q) \to \infty$. The deformation raises a singularity, named as "Space-tearing" in [4].

To solve the "Space-tearing" problem, authors propose to add another influence radius for each constraint, and it looks like a constraint duplication (see Fig. 18d). The larger radius permits to define the constraint influence on space, the smaller one manage the co-influences between constraints. In the preceding example, if C_{1bis} is the new constraint, duplicated from C_1, we have $f_1(C_{1bis}) = 1$, $f_2(C_{1bis}) = \alpha$, $f_{1bis}(C_1) = 1$ and $f_{1bis}(C_2) = 0$, that gives for calculations of f:

$$f = \begin{pmatrix} 1 & \alpha & 1 \\ \alpha & 1 & 0 \\ 1 & \alpha & 1 \end{pmatrix}$$

This matrix can be inverted by the same method we see in [3], a large number of solutions can be found varying the smaller radius of influence.

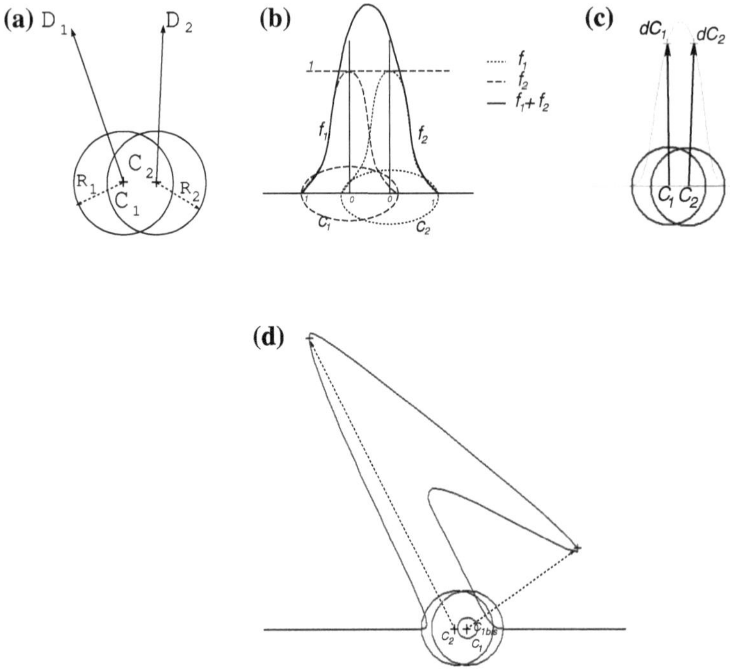

Fig. 18 Non-disjoints constraints. **a** Constraints; **b** Functions; **c** *line* deformation; **d** constraint duplication avoids SCODEF singularity

3.1.1 Extensions of the SCODEF Model

Extensions of the SCODEF model with geometric tools to handle the deformation have been made by [24, 32, 34] to form "Extended-SCODEF". Authors propose a generic expression of SCODEF model. They add new functions definitions, to vary the pinch of the object part deformed. As another extension of the deformation function, authors propose to consider the displacement to be curvilinear. The example of Fig. 19a presents a Bézier curve to construct the deformation path from the constraint point C to the displacement constraint $d(C)$.

The deformation is then expressed by:

$$T\left(d(Q)\right) = T\left(\sum_{i=1}^{r} M_i f_i(Q)\right)$$

Where T is a transformation of the deformation according to curve displacement. To prevent twisted curve authors use a sliding Frenet-frame along the displacement curve (see Fig. 19). This approach is quite near of sweeping techniques (see [39]).

They use this formal expression to complete influence hull's definition and to permit non-isotropic influences, Raffin et al. [32] use star-shaped shapes as a proof of concept. As the center of this shape is always defined and interior, the computations

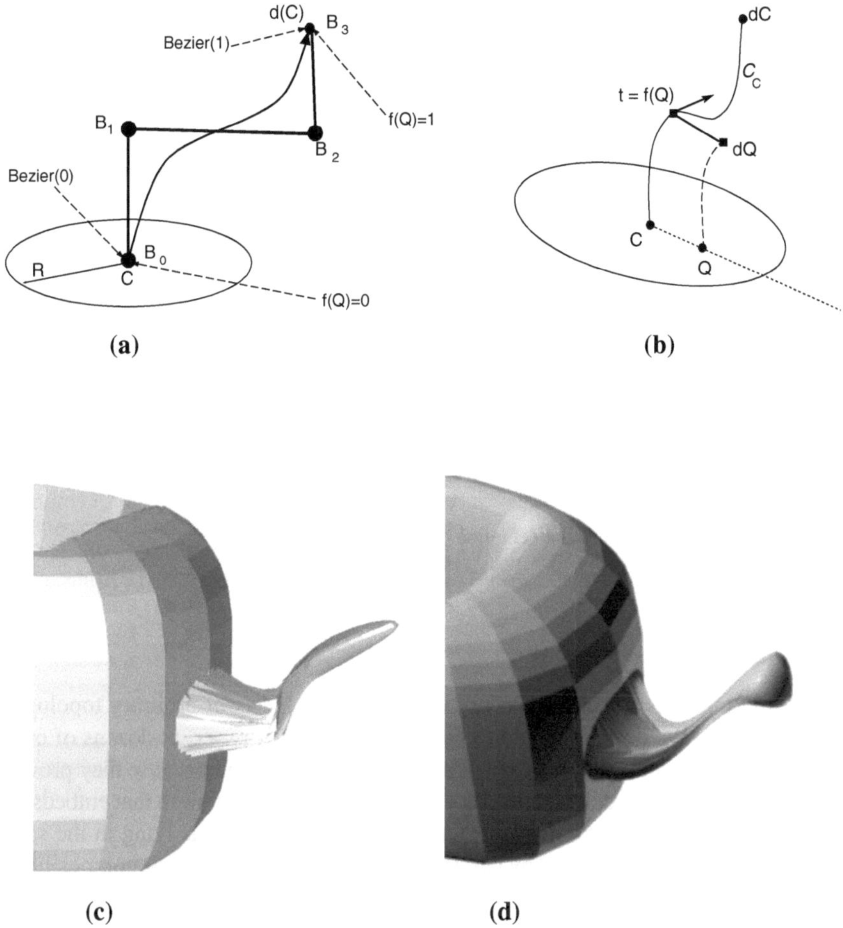

Fig. 19 Avoiding stretching in a curvilinear displacement. **a** Curvilinear displacement using a Bézier curve. $B_0 \ldots B_3$ are the control points of the parametric curve; **b** frenet sliding frame; **c** initial method; **d** frenet based curvilinear construction

of relatives distances is possible and extended-scodef can be applied. They also work with polyhedral or implicit hulls.

Another important add is the use of curvilinear constraints (see Fig. 21). As in WIRES's method, user can describe a curve to specify the trace of the deformation on the object's surface, taking into account of some neighboring points with influence hull, and associated displacement (or curvilinear path [25, 33]). This method is more user-friendly, as one can easily see the relation between deformation functions and deformed object pinch, influence hull with object's part deformed, etc.

22. Kobayashi K, Ootusubo K (2003) t-FFD: free form deformation by using triangular mesh. In: Proceedings of the ACM symposium on solid modeling and application, pp 226–234
23. Lamousin H, Waggenspack W (1994) Nurbs-based free-form deformations. IEEE Comput Graphics Appl 14(6):59–65
24. Lanquetin S, Raffin R, Neveu M (2006) Generalized scodef deformations on subdivision surfaces. In: Proceedings of 4th international conference articulated motion and deformable objects AMDO, LNCS 4069, pp 132–142
25. Lanquetin S, Raffin R, Neveu M (2010) Curvilinear constraints for free form deformations on subdivision surfaces. Math Comput Model 51(3–4):189–199
26. Lazarus F, Coquillart S, Jancène P (1994) Interactive axial deformations: an intuitive deformation technique. Comput Aided Des 26(8):607–613
27. MacCraken R, Joy K (1996) Free-form deformations with lattices of arbitrary topology. In: Proceedings of Siggraph'96 conference, Comput Graph, pp 181–188
28. Mäntylä M (1988) An introduction to solid modeling. Principles of computer science series. Computer Science Press, Rockville
29. McDonnell KT, Qin H (2007) PB-FFD: a point-based technique for free-form deformation. J Graph Tools 12(3):25–41
30. Moccozet L, Magnenat-Thalmann N (1997) Dirichlet free-form deformations and their application to hand simulation. In: Proceedings computer animation 97, IEEE Computer Society, pp 93–102
31. Parent R (1977) A system for sculpting 3D data. Comput Graph 11(2):138–147
32. Raffin R, Neveu M, Derdouri B (1998) Constrained deformation for geometric modeling and object reconstruction. In: 6th international conference in central europe on computer graphics and visualization (WSCG), vol 2, pp 299–306
33. Raffin R, Neveu M, Jaar F (1999) Curvilinear displacement of free-from based deformation. The Visual Computer, vol 16. Springer, pp 38–46
34. Raffin R, Neveu M, Jaar F (1999) Extended constrained deformations: a new sculpturing tool. In: Proceedings of international conference on shape modeling and applications, pp 219–224
35. Rappoport A, Sheffer A, Bercovier M (1996) Volume-preserving free-form solids. IEEE Trans Vis Comput Graph 2(1):19–27
36. Hu S, Zhang HCT, Sun J (2000) Direct manipulation of ffd. The Vis Comput 17:370–379
37. Sederberg T, Parry S (1986) Free-form deformation of solid geometrics models. ACM Trans Graph 20(4):151–160
38. Singh K, Fiume E (1998) Wires: a geometric deformation technique. In: Proceedings of Siggraph'98 conference, Comput Graph, pp 405–414
39. Yoon SH, Kim MS (2006) Sweep-based freeform deformations. Comput Graph Forum (Eurographics) 25(3):487–496

Cage Based Deformations: A Survey

Jesús R. Nieto and Antonio Susín

Abstract Cage based deformation techniques aims to be an easy to use tool for graphics modeling, texturing and animation. In this paper we describe the most important methods, their foundations, and the desirable properties that they should satisfy. We also present a comparative to show the strong and weak points of each one, taking into account their distinctive utilities. Finally, we discuss some applications that exploit cage capabilities in order to create a more complex deformation system or to simplify other deformation techniques.

1 Introduction

Mesh deformation is a common process in geometry modeling and computer animation. Modeling can be very accurate, like in engineering design, or be flexible to allow the artist to freely express his creative ideas. Similarly, in computer animation we may want a realistic behavior, simulating physics, or rather a stylized and artistic animation far from what is really possible. But, regardless of the preferred approach, we need flexible tools for mesh deformation to achieve the desired results easily. In the past years there have been many efforts in this direction, from different points of view: Free-form deformation (FFD), generalized barycentric coordinates, Radial Basis Functions (RBF), curve based deformation, skeletons and physics simulation [1–3, 20, 26, 27, 29, 32]

Character articulation, also called rigging, has a significant place in the field of mesh deformations; It is an important component in high-end applications used in film and audiovisual content. Professional softwares, specially Softimage XSI,

J. R. Nieto · A. Susín (✉)
Universitat Politecnica de Catalunya, Barcelona, Spain
e-mail: toni.susin@upc.edu

J. R. Nieto
e-mail: jrodrigueznieto@gmail.com

M. González Hidalgo et al. (eds.), *Deformation Models*, Lecture Notes in Computational Vision and Biomechanics 7, DOI: 10.1007/978-94-007-5446-1_3,

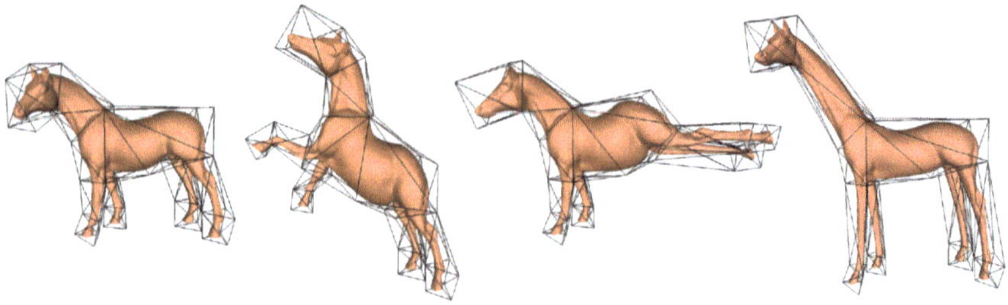

Fig. 1 An object wrapped by a cage. In the *left*, the cage and the object are in rest pose for coordinates computing. Then we can transform cage vertexes for producing a deformation in the volume and consequently in the object. From [18]

Maya and 3DStudio, provide a wide range of character articulation methods. Some classic examples are Enveloping [20] and blend shapes [16], but there are also many chainnable deformers that achieve amazing results.

Regarding character animation, there are a number of constraints that some of the previous methods do not fulfill. Firstly we need a deformation method able to work in real-time, for interactive applications, which limits the computation time. Also, it is desirable to have a convenient and easy to use system which makes it simple for a broad array of users to get quickly familiar with it. Cage based methods are good candidates for this purpose. Even if nowadays these methods are not the most used in professional animation applications, they have undergone significant developments so that could push forward the usual rigging techniques (Fig. 1).

Research in cage based methods were born at the beginning of feature film animation. The pioneers were Sederberg and Parry [27], in 1986, with the Free Form Deformation. This development become popular for several reasons. First, because it offers a smooth and intuitive control over the character using limited parameters: free form lattice control points. Besides, the model to be deformed does not need to satisfy other geometric constraint apart from being inside the control lattice. On the other hand, this kind of deformer has a well number of drawbacks: if the deformations are complex, such as character articulated with several limbs, it becomes difficult or even impossible to implement. A lattice will never fit perfectly the character shape, since the topologic rigidity of lattice is not flexible enough for that, and combining several cages would be a hard task [17].

Next, we introduce the methods that will be the main line of discussion throughout the paper. We need a cage which better fits the character shape. The first complete environment that allow to do this is Mean Value Coordinates (a solution proposed and developed simultaneously by to papers: Floater et al. [9] and Ju et al. [18], in 2005). This method lead mesh deformation to the world of generalized barycentric coordinates. The original concepts were introduced by Möbius in 1827, and have been developed by many mathematicians since then [7, 10, 12, 24, 25, 33]. The main point to solve is the relationship between a cage and its interior. If there is an object inside a cage, we can deform it using the cage no matter the complexity of the object. For this purpose, Floater [8] proposed to use the mean value theorem, but his formulation has some problems; the main one, being that the coordinates could be negative, which will produce anti-intuitive and annoying results.

Even if this problem only applies to certain cases, the relevance of this issue advises to discard this method for character animation, as Joshi [17] claims. He proposed an alternative solution adopting a slightly different approach which does not produce negative coordinates: Harmonic coordinates. This approach produces a more local deformation, which is a very positive feature, but this also has drawbacks, i.e. computation time and discretization accuracy.

Similarly, Lipman [21] proposes an improvement to Mean Value Coordinates that leads to positive coordinates. He uses GPU visibility render techniques for analyzing the volume inside the cage in order to cut off negative coordinates. In spite of the improvements of this method, also has some smoothness problems with concave shapes (as the original Mean Value Coordinates does).

Then, Lipman [22] realized that the surface details were not preserved, especially if the deformation is large, thereby suggesting that more data should be used for being able to relate the cage to the object. While Mean Value Coordinates and Harmonic Coordinates only use cage vertex positions, Lipman's Green Coordinates also uses face normals.

These are strictly cage based deformation methods, but there are other deformation methods that use cages in other ways. This is the case for Biharmonic weights [15] and subspace gradient domain deformations [14]. There have also been improvements and new developments for using these concepts, such as volume-preserving deformation [4], Cage-based deformation transfer [5] and skinning templates [19].

In the following section we are going to describe some basic concepts for deeper understanding cage based methods, which we analyze afterwards in Sect. 3. Then, in Sect. 4, the main conclusions are summarized and, finally, in the appendix we collect some pseudocode algorithms that can be very useful for a better understanding of these methods.

2 Barycentric Coordinates and Cage Based Deformation

Let us denote as a cage C any triangle mesh, or more generally a polyhedric mesh, convex or not, which is closed, and that envelop a model to be deformed. For perfect model fitting the cage must be topologically flexible, and may be manually or automatically generated [35]. As far as there is a well defined relationship between the cage surface and its inside volume, the deformation applied to the cage will also affect the volume, and therefore any object it contains. This procedure endows us with an easy to use control handles (cage vertexes) to deform whatever complex model inside the cage.

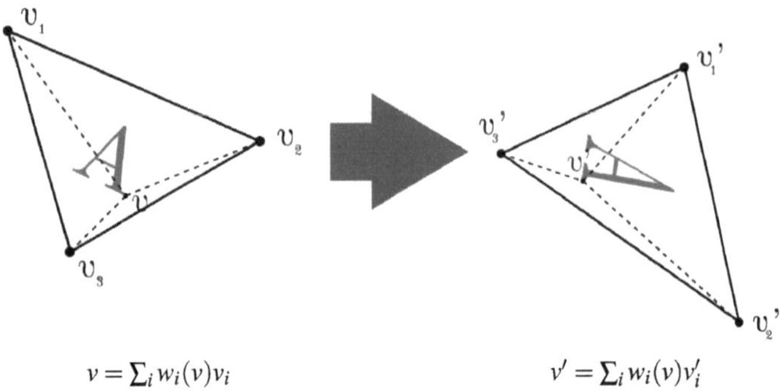

Fig. 2 The coordinates w_1, w_2, w_3 define the relationship between v_1, v_2, v_3 and the interior point v. After having distorted the triangle vertexes, the inside points are relocated constrained with the same relationship

2.1 Barycentric Coordinates Definition

The first mathematical approach defining a cage-to-model relationship was introduced by Möbius in 1827 with barycentric coordinates for triangles [25]. The issue was set out in the following terms: Which weights w_1, w_2, w_3 must be given to the vertexes A, B, C of a triangle to obtain P as the center of gravity of these weights? Thus, point P is the barycenter, while the value of the vertex weights are the barycentric coordinates of P for each vertex. Generalizing the problem, v can be considered as the barycenter of points $v_1, \ldots, v_n$ if and only if

$$v = \frac{w_1(v)v_1 + \cdots + w_n(v)v_n}{w_1(v) + \cdots + w_n(v)}. \tag{1}$$

More precisely, these coordinates are homogeneous and need to be normalized by

$$\lambda_i(v) = \frac{w_i(v)}{\sum_i w_i(v)}, \quad \sum_{i=0}^{n} \lambda_i(v) = 1. \tag{2}$$

Where all the values λ_i are between 0 and 1. For the case of triangles, there is also a relation between the coordinate w_i (from vertex v_i) of the interior point v and the area of the triangle $[v, v_{i+1}, v_{i+2}]$ (Fig. 2). Therefore, they are also called *area coordinates*. The barycentric coordinates are a linear transformation of Cartesian Coordinates over a triangle. Then, they vary linearly over the boundary, inside and outside the triangle. This means that we can use it as an interpolation function ϕ inside the polygon using the values defined at the vertexes $\phi(v_i)$ (Eq. 3)

$$\phi(v) = \sum_{i=0}^{n} \lambda_i(v)\phi(v_i). \tag{3}$$

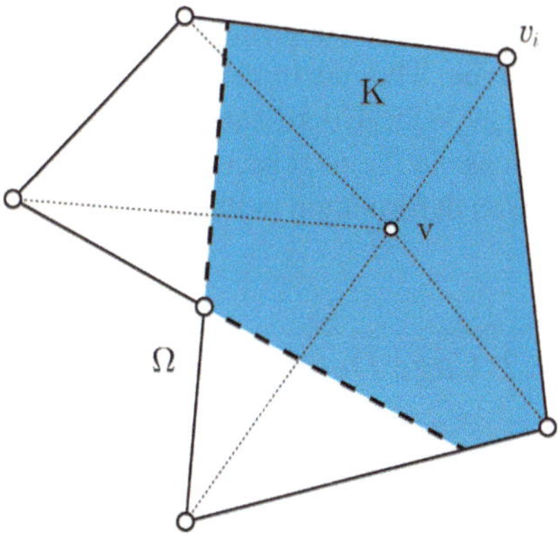

Fig. 3 The kernel of this polygon is the region K

An example of this interpolation is Phong and Gouraud shading, a rendering algorithms widely used in computer graphics [12]. These interpolations can also be employed for geometry deformation. The coordinates define the relationship between the triangle vertexes and the triangle inside. After having distorted the triangle vertexes, we can relocate the inside points constrained with the same relationship (Fig. 2). In fact, by using identity as interpolation function $\phi(v_i) = v_i$ in Eq. 3, we have an interpolation of the vertex space positions.

2.1.1 Barycentric Coordinates Generalization

The above formulation works well with simplexes, but a number of difficulties arise when the issue is generalized to more complex polygons. Let us see now some of the many approaches that have been tried to implement such a generalization. The first attempt was proposed by Wachspress [33], who tried to deal with finite element methods. The goal of this kind of methods is to solve equations, approximating continuous values as a set of discrete points, usually distributed into a grid. The Wachspress's definition is only suited to the case of convex polygons. Other studies have developed these methods for quasi-convex polygons [23] and arbitrary polygons [31].

We distinguish quasi-convex polygons from convex ones to denote this kind of polygons which are not convex but have a convex kernel, and therefore are convex in some sense. The kernel of a polygon is the region K of a polygon Ω such that, taking any point $v \in K$, for all vertex v_i of the cage, the segment $[v, v_i]$ only intersects with the polygon boundary at v_i (Fig. 3).

Another approach to barycentric coordinates generalization is point cloud interpolation. The objective of this method is to define the interior volume of the point's convex hull [6, 11, 28].

Finally, we consider that the most interesting approach as far as this paper is concerned, is piecewise surface interpolation [7, 8]. It is particularly relevant because it enables volume deformation by cage control. In the same way that a triangle is able to define a deformation for its support plane, a mesh can define one for its inside volume. The most interesting works in this area have been done in turn, ending in Mean Value Coordinates, Harmonic Coordinates and Green Coordinates, that we discuss in the next section, but first we will summarize the main properties of cage based deformation methods.

2.2 Cage Deformation Desired Properties

The final goal is to develop a procedure for producing natural and intuitive deformations by using some control handle vertexes that form a cage. For this purpose, some properties may be defined.

The first concept to be analyzed is *deformation domain*, which is the space region influenced by the cage. To start with, deformation needs to be well defined inside the mesh volume, without singular points. To this aim, we can constraint the object to be totally inside the cage. However if only a part of the object is aimed to deform, we need the cage to cover just that part, intersecting the object, which is a problem when the domain is not well defined also at the boundary and outside the cage. There are different approaches for outside coordinate's extension from cage, as we will see in the next sections.

Coordinates can be managed as a function, so that, if coordinates are well defined (ensuring continuity all over the domain) we also expect *smoothness*, which means continuity at first derivatives.

Interpolation, that is, coincidence values for the vertexes, is the main application. As long as interpolation is ensured, the fact that $\phi(v) = 1$ implies $\sum_{i=1}^{n} \lambda_i(v) = 1$, for all points v in the domain. This fact is known as *affine invariance* (Eq. 3).

Besides, *local deformation* will be needed to restrict vertexes influence to their neighborhood only at local state. The implication follows that a control point (cage vertex) must cut off the influence of other control points over a part of the cage volume, if it is placed between them.

In deformation methods, *conformality* is an important property that relates transmission of local rotation transformations and, accordingly, surface detail preservation. While conformal mapping maps infinitesimal spheres to other infinitesimal spheres, quasi-conformal mapping is able to map infinitesimal spheres into infinitesimal ellipsoids (with a certain ratio between axis). This property works towards a natural deformation without losing shape semantic (Fig. 4).

Another important property for character articulation is *positiveness*. This feature ensures that the coordinate values are positive in the entire domain, or at least in some controlled parts. In as much as this is achieved, deformation will be successfully intuitive.

Moreover, if we want to develop a real-time interactive deformation tool, *computation time* needs to be reduced as much as possible. All the implementations segment deformation process in two parts to break off the heaviest computations into a preprocess (getting coordinates), so that the actual application of deformation is isolated

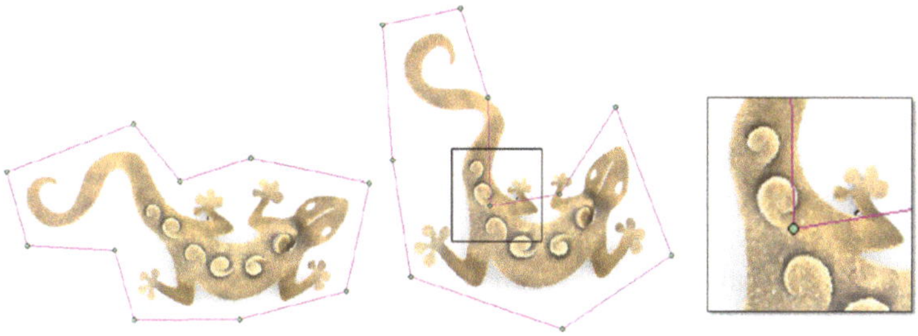

Fig. 4 Rotation transformation is naturally applied keeping the object shape semantic, extracted from [22]

in a simple computation to achieve real-time. Using GPU parallel techniques speed up the process, since transformations for each vertex can be done simultaneously. In order to develop such a low time-consuming computations we need to allocate in memory coordinates for every point of the object. Depending on the method, a fast access to memory for many little sets of data (sets of vertex coordinates) will be needed.

3 Mean Value Coordinates (MVC)

The first approach to MVC comes from Floater et al. [9] and Ju et al. [18]. Both studies aimed at looking for an interpolation method for surfaces. Initially, Floater [8] presented a 2D mean value coordinates over quasi-convex polygons, where the coordinates were well defined inside and outside the polygon. Even if there was a problem of discontinuity at boundaries, it could be solved easily [12] by extending coordinates from the interior domain. In this case, the coordinates were well defined, but it was not ensured that the values were necessarily positive for the entire domain. In fact, positive coordinates can only be ensured inside the kernel of the polygon (a convex or star-shaped one) [9]. Afterwards, Floater developed a 3D version of the algorithm, but the negativity problem remained unsolved. Then, Ju et al. developed another solution by using the Floater's 2D approach but addressing the issue from a slightly different point of view. Their attempt to solve the matter as a 3D interpolator provided us with a solution which is more robust than the 3D one presented by Floater (the pseudocode of Ju's MVC is included in the appendix). In Summary, these coordinates satisfy most of the properties enumerated in the previous section [18] (mainly the fact of being a good interpolation method [12]) but the lack of positiveness and locality are serious drawbacks for mesh deformation.

MVC uses mean value theorem to relate the points of a cage with its interior, which is oriented by harmonic coordinates theory. Next, we present some concepts that may help understanding the process, and also be helpful for discussing further methods.

3.1 Harmonic Functions

The main problem to solve the desired cage-object relationship starts with a more theoretical problem, the approximation of harmonic functions by piecewise linear functions over triangulations, in such way that the injective property is preserved. An harmonic function u is a function over the reals for which the second derivative is continuous and satisfy Laplace's equation

$$\Delta u = 0 \Longleftrightarrow \frac{\partial^2 u}{\partial x^2} + \frac{\partial^2 u}{\partial y^2} + \frac{\partial^2 u}{\partial z^2} = 0. \tag{4}$$

Dirichlet raised the problem of finding a relation between a continuous function f over a boundary $v|_{\partial\Omega}$ of a region Ω and its interior, by using a harmonic function u assuming as known the values of f at $v|_{\partial\Omega}$. Basically, he explores whether such a function u can exist and, if it does, if it is unique. The conclusion of such an attempt is to establish the existence of a function with these characteristics and named his solution as Dirichlet's principle.

Following Dirichlet development, Floater approximates a solution of u that satisfies the Dirichlet's boundary conditions, $u|_{\partial\Omega} = f_i$, by means of a linear piecewise function $u_{\mathcal{T}}$ over the triangulation. This leads us to a linear system where the values of $u_{\mathcal{T}}$ are at the inside vertexes of the triangulation T, (observe the similitude with Eq. (3))

$$u_{\mathcal{T}}(v) = \sum_{i=1}^{k} \lambda_i u_{\mathcal{T}}(v_i) \tag{5}$$

3.2 Mean Value Theorem

Floater carries out this approximation by using the mean value theorem, which states that for a circumference $B = B(v, r) \subset \Omega$ (where r is the radius of a circumference centered at point v, completely inside the region Ω) and its circumference perimeter Γ, the Eq. (6) approximates the function $u_{\mathcal{T}}$ fitting the Dirichlet's conditions, as is expressed in Eq. (5)

$$u(v) = \frac{1}{2\pi r} \int_{\Gamma} u(v|_{\partial\Omega}) dS. \tag{6}$$

It is also relevant to analyze the geometric interpretation. Recall *area coordinates*, we compute the i coordinate of interior point v as a ratio between the area of the opposite sub-triangle (v, v_{i+1}, v_{i+2}) from vertex v_i and the total triangle area. Similarly, there is also a ratio to compute Mean Value Coordinates, this time between partial perimeter of a circumference centered at point v, obtained from the projection of the cage segment $[v_i, v_{i+1}]$ over the circumference, and the total circumference

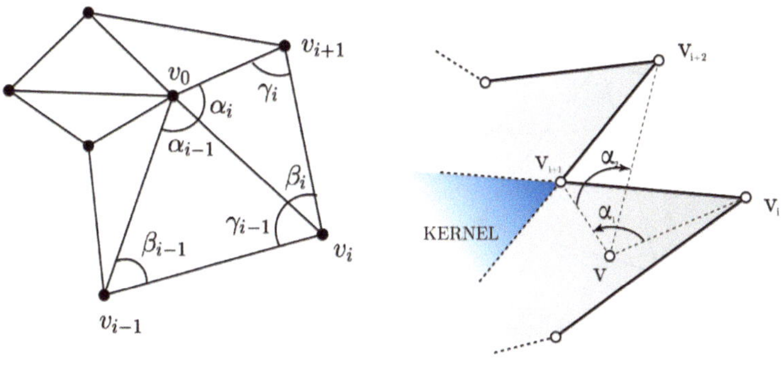

Fig. 5 **a** Parameter interpretation to obtain mean value coordinates. It is easy to see that if we consider angles with sign there may be negative values for some coordinates, this is the case of concave polygons as in **b**. The angle α_1 corresponding to segments $[v, v_i]$ and $[v_i, v_{i+1}]$ is positive while the angle α_2 defined from segments $[v, v_{i+1}]$ and $[v_{i+2}]$ is negative

perimeter. Note that Floater, as an extension of his analysis, uses the angles between segments $[v, v_i]$ and $[v_i, v_{i+1}]$ to obtain the same ratio (Fig. 5a). It is easy to see that, by considering angles (or perimeters) with sign there may be negative values for the coordinates. This is the case of concave polygons (Fig. 5b). The following expressions are derived by Floater to obtain Mean Value Coordinates w_i and their normal form λ_i :

$$\lambda_i = \frac{w_i}{\sum_{j=1}^{k} w_j}, \quad w_i = \frac{tan(\alpha_{i-1}/2) + tan(\alpha_i/2)}{\|v_i - v_0\|} \tag{7}$$

In a similar way, mean value coordinates in R^3 can be derived. In this case, the coordinates are obtained as the area ratio over a sphere. With a sphere S centered at point v, the coordinates of point v are equal to the ratio of the projected area of every cage simplex over the sphere and the total sphere surface area. To get a closed form expression, Floater uses other tools. He claims that, in fact, 2D coordinates can be computed as a proportion of normals integration of the projected cage piece over the circumference. Thus, we can integrate in 3D the normals over the partial sphere. The integration of sphere normals over the entire sphere is equal to 0, from which we can derive:

$$0 = \int_S n(p) = \int_S (p - v) = \sum_{T \in \mathcal{T}} \int_{T'} (p - v) \tag{8}$$

and define the Spherical Barycentric Coordinates of every partial sphere that comes from the cage triangles projection over the sphere. These coordinates can be used as a $u_{\mathcal{T}}$ approximation, which is what we need. The derivation of these coordinates in a closed form expression is given by:

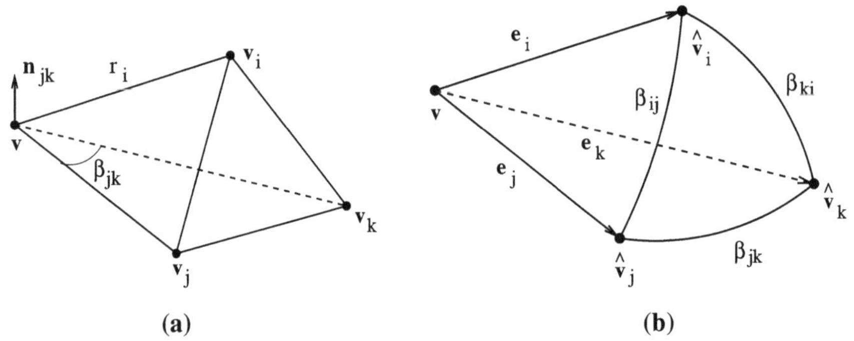

Fig. 6 **a** Tetrahedron **b** "tetrahedron" defined by a *spheric triangle*. From [9]

$$0=\sum_{i=1}^{n}\sum_{v_i\in T}\mu_{i,T}e_i=\sum_{i=1}^{n}w_i(v_i-v) \tag{9}$$

$$w_i=\frac{1}{r_i}\sum_{v_i\in T}\mu_{i,T}>0,\quad \mu_{i,T}=\frac{\beta_{jk}+\beta_{ij}n_{ij}\cdot n_{jk}+\beta_{ki}n_{ki}\cdot n_{jk}}{2e_i\cdot n_jk} \tag{10}$$

Following the scheme of Fig. 6, where $\beta_{rs}\in(0,\pi)$ is the angle between two segments $[v, v_r]$ and $[v, v_s]$, and n_{rs} as the unitary vector of the face $[v, v_r, v_s]$ pointing into the tetrahedron. (r and s take the values of i, j, k)

Floater analyzes this case carefully and concludes that the Eq. (10) is not well defined on the boundary. Besides, he proves that the coordinates can be extended to the boundary and beyond without losing of continuity and smoothness. The closed form expressions proposed by Floater to obtain coordinates is the most appropriate context for computing with parallel GPU techniques.

4 Harmonic Coordinates (HC)

Joshi et al. [17] developed another method to set the coordinates successfully avoiding some of the MVC drawbacks. Their concern was focussed on articulation in feature film animation, so that the mentioned issues were annoying for that purpose. The main problem is the fact that MVC are based in euclidean distances, which neglects the visibility between cage points and object points. Then, the deformation results are non-intuitive and require additional work to solve or to avoid the problem. Consider the case of Fig. 7: the cage's left limb acts over the object's left and right limbs, resulting a non-expected displacement.

When analyzing further the matter, two important issues are found. Firstly, the *negative coordinates*. The influence of the vertexes of the cage's left limb deliv-

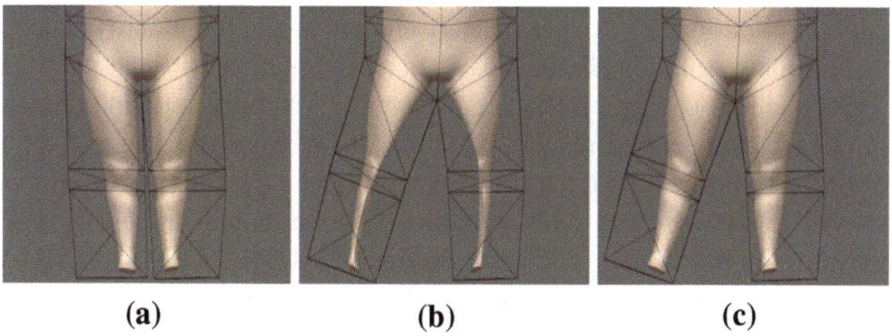

Fig. 7 Comparative of MVC and HC in a conflictive deformation case of character articulation. **a** Bind pose. **b** Mean value coordinates. **c** Harmonic coordinates. From [17]

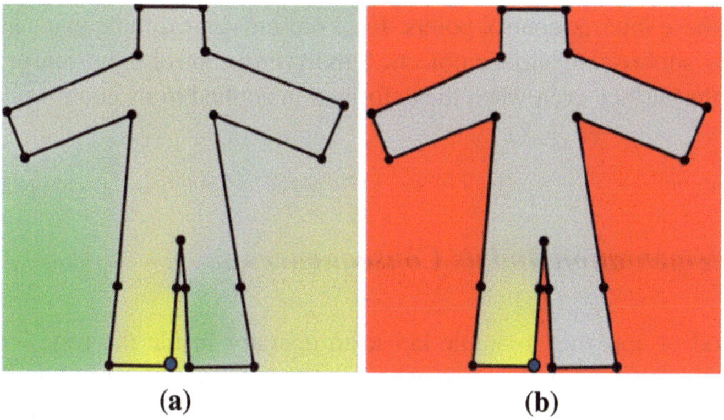

Fig. 8 Coordinate values for the *blue* marked vertex, all over the domain. Note that Mean value coordinates (**a**) takes two colors for coordinate representation: *yellow* for positive values and *green* for negative ones. In Harmonic coordinates (**b**), the domain only complains cage interior, but values are always positive. From [17]

ers positive values for points inside the cage's left limb, but negative ones for the cage's right limb. Figure 8 show coordinate's value distribution for a marked vertex.

The second issue is related to *global influence*: every cage vertex affects all object vertexes producing a general deformation, even if control vertex and object vertex are in opposite parts of the cage. Note though that in this case the influence is small.

Joshi exposes a local method, like heat diffusion, to decay control vertexes influence as it flows through the interior of the cage. The coordinates values inside the cage have a top bound value that corresponds to control vertexes values, usually 1, and decay to 0 as they are placed farther from the cage vertexes. Hence, all the coordinates will be within the range $[0 \dots 1]$, but are constrained to be inside the cage, since the boundary breaks off the influence diffusion. This approach ensures positive coordinate values all over the domain: the cage interior, as shown in Fig. 8b.

Joshi's approximation for piecewise linear function $u_{\mathcal{T}}$ is the identity function, the simplest harmonic function. It satisfies Laplace's equation and can propagate the values from boundary to cage interior as we want. By simply applying the laplacian operator, we get the desired results. Note that if the cage is a triangle the resulting coordinates are the simple barycentric coordinates. Unfortunately, this process encloses the domain to cage interior, allowing no possible influence outside, so that continuity at boundaries will only be preserved by Dirichlet boundary conditions. Therefore, to get a smooth result, the cage needs to wrap completely the object without intersections, or to trick the process for achieve some smoothness at boundaries. An interesting feature of this method is that we can use interior control points that behave like cage pieces. These will propagate the influence to the cage volume in the same way as the boundary cage does. But in accordance with that, there will be a lack of smoothness in deformation at points where the object intersects these interior control points. Joshi presents an example that sorts out this problem by subdivision surface objects. Subdivision control points ensure smoothness over the surface even when the deformation applied to its control points is not smooth.

4.1 Implementation and its Consequences

The method of applying a simple laplacian operator inside the cage works theoretically (with infinitesimal equations) but computationally we need to discretize a volume big enough to cover the cage, and to compute a coordinate value for every volume division cell. Joshi implemented some empirical tests to fit the best volume subdivision and then propose dividing it into 2^s cells, being $s = 6$ in the 2D version, and $s = 7$ in the 3D case. To compute the coordinates he proposed the next process: first, to mark every cell as it belongs to cage boundary (BOUNDARY), inside (INTERIOR) or outside (EXTERIOR), and then assign cage vertex values to its corresponding cells. Next, through an iterative process these values will be propagated through the volume if cells are "INTERIOR" marked, with a 4 cell-connected laplacian operator in two dimensions, and 6 cell-connected one for 3D. Joshi proposes some optimizations for reducing computation time, since convergence of laplacian operators is too slow and requires many iteration-steps before it reaches the end process condition. This condition consist of imposing a threshold that sets a top bound for cell values variation between two laplacian operator successive steps. He proposes to use the threshold $t = 10^{-5}$.

One of the main advantages of this approach is the fact that the inside object can be changed whenever is needed, given that the coordinates are computed over the grid cells and not over the object. The coordinates can be reassigned to the new object with no more computation than reading values from the grid. Nevertheless, saving the coordinates every cell is a waste of memory, unless the number of cells is smaller than the number of object vertexes. On the other hand, discretization causes some precision errors that need to be managed to deliver right deformation outcomes.

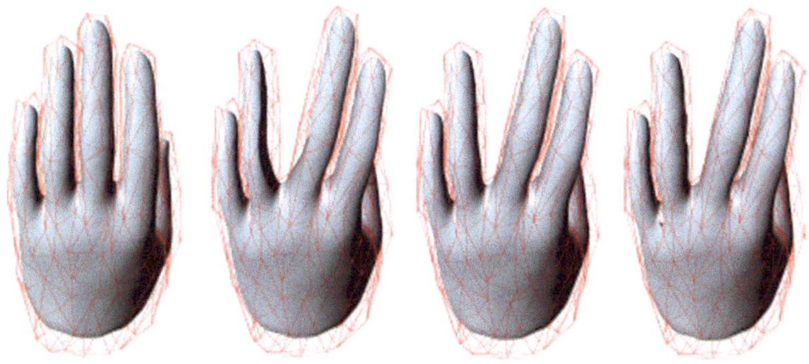

Fig. 9 This example, created by Lipman, shows clearly how negative value coordinates are presented on every kind of coordinates. From [21]

In some sense, the problem may be reduced if the implementation is done by applying an adaptive resolution grid.

5 Positive MVC (PMVC)

The Harmonic Coordinates procedure solves the negativity problem of Mean Value Coordinates, but the fact that it lacks closed form expressions makes it very consuming time, even with optimizations. While MVC and Green Coordinates have a closed form formulation that enables parallel computation, in the case of HC is too hard to apply parallelism. To prevent negative values Lipman proposes applying a local Mean Value Coordinates which is accomplished with visibility testing adopting GPU rendering techniques. The normals integration over the sphere (or circumference) centered at point v is computed if the corresponding cage's simplex for the coordinate w_i is viewed by point v. This process allows us to localize the influence of each cage vertex. The results are very good (as Fig. 9 illustrates), but there are some singularities, specially in concave polygons. Lipman trivializes the problem claiming that such a distortion is not significant enough to be taken into account, as his experiments prove. Yet, the problem arises as a consequence of visibility computation, since the visibility region from point v (similar to a kernel centered at point v) has a really sharp boundary. This fact introduces a little lack of smoothness which as long as the vertex forms a concavity becomes a problem. Thus, as a way to overcome this difficulty, Lipman proposes to split out the cage at the points where this problem is presented.

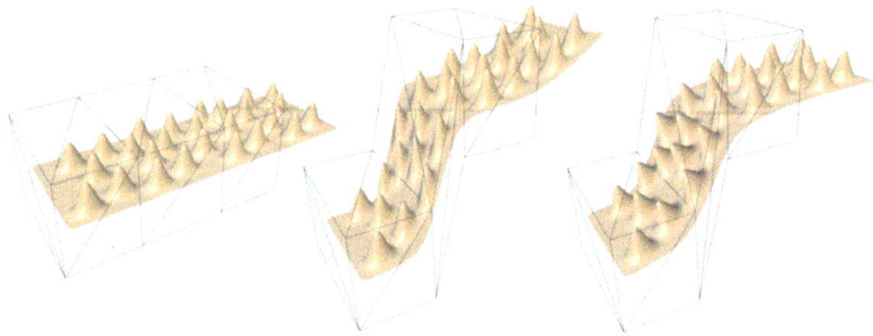

Fig. 10 Detail preservation is exhibited using Green Coordinates (on the *right*), where the details adhere to the surface deformation and rotate accordingly. In the *middle*, the MVC result is depicted where the details maintain their original orientation and therefore shear. From [22]

6 Green Coordinates (GC)

The main advantage of deformation methods based on cages are their simplicity, flexibility and fastness. Every object vertex is deformed independently under these techniques, which are transparent to object surface representation and are "discretization-error-free" in most of the cases. Unfortunately, the methods discussed previously do not preserve surface details, which is a drawback compared to other deformation techniques (such as those based in surface differentials [30]), especially if deformation is too big. Green Coordinates, developed by Lipman [22], try to apply generalized barycentric coordinates taking into account this fact.

MVC and HC uses only cage vertex positions which, as Lipman noticed, induce an axis independent deformation (i.e., any translation over x axis of a cage vertex does not translate into the object in the y and z axis, resulting an unnatural deformation (Fig. 10)).

To achieve a natural deformation with shape preserving, Lipman adds cage faces data to the deformation operator.

$$v = F(v; C) = \sum_{i \in I_V} \omega_i(v) v_i + \sum_{j \in I_T} \psi_j(v) n(t_j) \tag{11}$$

Being the cage $C = (V, T)$, where V are the cage vertexes and T the cage simplexes (edges or faces). ω_i values are the coordinates based in cage's vertexes and ψ_i are the ones based in the normal of the simplexes, which are the face data added. And $n(t_j)$ is the normal outward unitary vector of the simplex t_j. The deformation over the object is obtained in this way:

$$v \mapsto F(v; C') = \sum_{i \in I_V} \omega_i(v) v'_i + \sum_{j \in I_T} \psi_j(v) s_j n(t'_j) \tag{12}$$

where v_i' and t_j' are the modified vertex and faces of C' respectively. The expression introduces a new term $\{s_j\}_{j \in I_T}$ to ensure some properties such as scale invariance. The result is very good as it preserves better and more natural than other methods the object shape and details (Figs. 4, 10, 12) producing a conformal mapping in 2D and quasi-conformal in 3D.

6.1 Coordinates Derivation

To obtain closed form expressions Lipman realizes that the harmonic functions also follows the Green's theory, thereby permitting to apply the third Green identity to solve the problem.

$$u(v) = \int_{\partial D} \left(u(\varepsilon) \frac{\partial G(\varepsilon, v)}{\partial n(\varepsilon)} - G(\varepsilon, v) \frac{\partial u(\varepsilon)}{\partial n(\varepsilon)} \right) d\sigma_\varepsilon \tag{13}$$

where ε is a point over the surface and G is the fundamental solution of the Laplace equation, which can be solved by

$$G(\varepsilon, v) = \frac{1}{(2-d)A_d} \|\varepsilon - v\|^{2-d} \text{ dimension } d \geq 3 \tag{14}$$

$$G(\varepsilon, v) = \frac{1}{2\pi} \log \|\varepsilon - v\| \text{ dimension } d = 2 \tag{15}$$

where A_d is the area of a unit sphere in R^d, which in the 3D case is equal to 4π. By deriving these equations, we obtain the closed form expression of the two coordinate sets for each object vertex, $\omega_i(v)$ and $\psi_j(v)$, which are used in the Eq. (12) to obtain the deformation.

$$\omega_i(v) = \int_{\varepsilon \in N\{v_i\}} \Gamma_k(\varepsilon) \frac{\partial G}{\partial n} d\sigma_\varepsilon, \quad i \in I_V \tag{16}$$

$$\psi_i(v) = -\int_{\varepsilon \in t_j} G(\varepsilon, v) d\sigma_\varepsilon, \quad j \in I_T \tag{17}$$

where $N\{v_i\}$ is the neighborhood 1-connected vertexes of vertex v, which are points used for computing $\Gamma_k(\varepsilon)$, a linear combination to get ε depending on the vertexes of $N\{v_i\}$, in such way that it satisfies $\varepsilon = \sum_{k=1}^{d} \Gamma_k(\varepsilon) v_k$.

The scale factor s_j depends on the growing or decreasing of every cage simplex, and is computed in real-time when the deformation is applied:

$$s_j = \|t_j'\| / \|t_j\| \quad \text{in 2D case.} \tag{18}$$

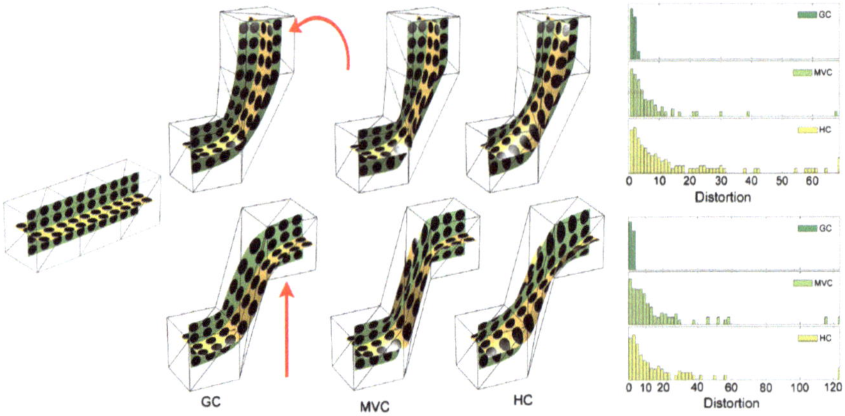

Fig. 11 There are two successive deformations applied to the same object using all methods. The graphics show the distortion error accumulation at every deformation. Only GC presents an upper bounded error. From [22]

$$s_j = \frac{\sqrt{\|u_1'\|^2\|u_2\|^2 - 2(u_1' \cdot u_2')(u_1 \cdot u_2) + \|u_1\|^2\|u_2'\|^2}}{\sqrt{8}\,\text{area}(t_j)} \quad \text{in 3D case,} \tag{19}$$

where vectors u_1 and u_2 define the edges of the triangle t_j, and where u_1' and u_2' are the corresponding edges of the modified cage triangle t_j'. This parameter ensures a natural deformation, which can though be modified to distort slightly the uniform scaling (see the pseudocode in the appendix).

Lipman has studied the existing distortion in MVC, HC and GC, with the ratio $\frac{\sigma_{max}(v)}{\sigma_{min}(v)}$, where $\sigma_{max}(v)$ and $\sigma_{min}(v)$ are the maximum and minimum singular value of the differential F (jacobian matrix) at point v. Any map with a top bounded distortion is named quasi-conformal, but only Green Coordinates present such upper bound (except in some degenerated cases). MVC and HC distortions are proportional to the deformation applied (Fig. 11).

The coordinates expressed in Eqs. 11, 12, 16 and 17 are smooth and well defined all over its limited domain: inside the cage. Besides, $\omega_i(v)$ as a function (Eq. 16), present derivate discontinuities over the simplexes at vertexes intersections and, therefore is not smooth at the boundaries. In fact, this function cannot be used as a surface interpolation method. In spite of that, it seems to be a really good system for character articulation. Lipman does not mention anything about the negativity of his coordinates, but attending to the corresponding closed expressions, we expect positive values at the entire domain.

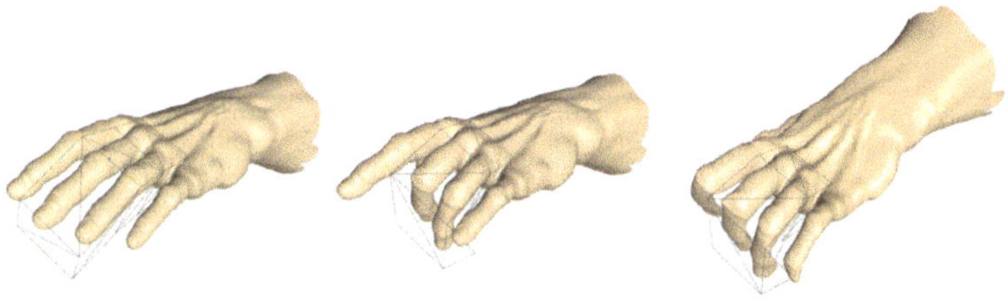

Fig. 12 GC cage deformation affects only *center* fingers of the model, leaving deformation-free the rest of the model. MVC affect all the model with a natural extension of the coordinates, but producing unwanted results. **a** Bind pose. **b** Green coordinates. **c** Mean value coordinates. From [22]

6.2 Outwards Cage Extension

The lack of smoothness introduced by the mentioned discontinuities at the boundaries entails limitations for the expressed formulation to be inside the cage. That is why Lipman adds some terms to extend the inside coordinates to reach beyond the cage. He considers the outside region as a set of subregions linked to some simplexes. Whenever an object crosses through a simplex, the whole part of the object that is outside the cage will be affected by the transformations of this simplex in the way of affine transformations. Then, a set of simplexes that works with the same similarity transformation could be used as a unique simplex that propagates a unique affine transformation outside the cage. Therefore, we will compute as many affine transformations as separated parts of the object are placed outwards the cage (Fig. 12).

7 Other Developments of Cage Based Methods

There are two early approaches of barycentric coordinates. The first one, presented by Hormann and Sukumar [13], is called Maximum entropy coordinates (MEC). These coordinates are non-negative for arbitrary polytopes, are smooth inside the domain, and can be computed locally at any point v inside the cage like MVC and GC, but they are only defined inside the convex hull of the polytope, and not everywhere in R^d. MEC do not present a significant improvement over the already discussed methods. Weber [34] presents another kind of coordinates named Complex Barycentric Coordinates. This interesting approach reformulates the barycentric coordinates definition with expressions based in complex numbers. Unfortunately, it only works for 2D, and seems to be very difficult to be extended to a higher dimension.

There are some studies that make use of cage based methods and reach further. We have already discussed direct cage based deformation, but enriching a more complex behavior to this deformation will be interesting. Since the cage provides a simplified version of an object for simpler deformation, it enables computing complex constraints over the object with a simpler computation over the cage. Actually, this constraint assignment is further complex, we should apply constraints to the object rather than to the cage. Therefore, in a previous step, we group object constraints on cage vertexes for collapse computations. Generalized barycentric coordinates can do this for us. An interesting contribution in this regard is the one made by Cervero et al. [4], who develop a method for object volume preservation over any kind of generalized barycentric coordinates. Volume preservation is a highly esteemed skill for creating believable deformations in character animation. Another paper that follows such kind of scheme is Huang et al. [14], in which object deformation is obtained by subspace gradient domain techniques constrained by some conditions. They uses this constraint grouping method for improving the computations which would not otherwise be made in real or at least close-to-real time.

There is still another interesting work that has a close relationship with the essence of cage based methods: biharmonic coordinates [15]. The goal of all the mentioned methods consist of defining a volume from its cage surface, in the case of biharmonic coordinates the cage is the object itself. As long as we have a good knowledge of the volume (the relationship between points within) we will be able to process deformations by modifying some points inside or over the surface while preserving these relations as a constraint. Control points could be vertexes, segments or faces. Besides, they can be inside, on the surface or outside the object. These coordinates, properly created, generate smooth deformations everywhere and can be used with other character articulation tools, such as point deformers, skeletons and cages.

Animation transfer using cages is another approach that also uses generalized barycentric coordinates [5]. Whenever for the cases in which two characters are using exactly the same cage (or different cages but with the same cage topology), every deformation applied to one cage can be translated into the other one easily. A different approach pursuing a similar goal is the one proposed by Ju et al. [19]. They use templates, as cage predefined setups, for applying similar animations to several characters easily.

8 Conclusions

In this paper we have described cage based deformation methods and some of the applications that exploit their utility. Each of these methods has strong and weak points, which are more or less relevant depending on the purpose. In the following section, we discuss the most significant points in a schematic way, as a summary of the concepts previously exposed. In the appendix we collect some pseudocode implementations, extracted from their corresponding papers, that could be useful for further clarifying these concepts.

8.1 Discussions

The three methods described have been developed by gradually approximating the issue. Floater [9] first proposed a generalization for barycentric coordinates, by introducing harmonic functions theory, which proves mean value theorem. The main purpose of Mean Value Coordinates was to interpolate values from some points over a surface into its volume (i.e. vertex color). The same formulation is used to interpolate the space position of v based on cage vertex positions. Then, a cage could be created wrapping a model, and deform the model maintaining the same interpolation values (coordinates) before and after the cage transformation. Note that until this point there have been used only point data to compute the interpolation, regardless of the cage topology. This is important since the lack of data will cause problems like negative coordinates and no-local deformations. These drawbacks are specially annoying for character animation. The Harmonic Coordinates approach solves these problems by adding more data such as intra-cage-space point relations, but to satisfy the neighboring relationships at the time of computing the coordinates, is too time and memory consuming. On the other hand, grid discretization process proposed for HC computation is very flexible for character articulation because we can reassign coordinates to different models, composed with a unique mesh or in several parts, without recomputing anything else, provided they use the same cage. Green Coordinates was born as a solution of MVC and HC drawbacks, taking into account also surface details preservation. To achieve it, is added more data into the computation (cage simplexes normal) for describing better the behavior of deformation at surface, and translate it accordingly to the interior, producing a conformal or quasi-conformal mapping. On Table 1 we summarize the features that characterize each method.

MVC are the best solution for value interpolation, given that are well defined inside the volume and over the surface, and their simple closed form expressions encourages to use they in the fastest real-time applications, specially if they are computed by GPU techniques. The fact of having negative values and global deformation makes this method less interesting to character articulation.

HC is a good solution for character articulation, specially to add a specific and controlled deformation to a part of an object, due to its flexibility to control influence propagation. Although we must take into account the smooth discontinuities at boundaries to achieve a good overcome. In a rigging process is really valuable that the model could change after the cage creation and coordinates computation, so that HC become a better solution in that cases.

Table 1 Cage based methods properties comparative

Property	Mean value coords	Harmonic coords	Green coords
Cage topology	Triangles (vertex)	No matter (vertex)	Triangles (vertex + normal)
Conformality	No	No	conformal(2D) quasi-conformal(3D)
Interpolation	Boundary + inside	Inside	Inside
Closed formulation	Yes	No	Yes
Control point influence	Global (euclidean distances)	local	Global
Positiveness	No	Yes	Yes
Outside cage extension	Natural formulation extension	No	Segmented affine transformation

Table 2 A comparative of the main cage based methods exposed (MVC, HC and GC) working in different scenarios (Column, Bust and Horse)

Model name	Model vertexes	Model faces	Cage vertexes	Cage faces	MVC	HC	GC
Column	2202	4400	24	44	0.051	31.905	0.241
Bust	255358	510712	32	60	8.232	229.534	38.348
Horse	19851	39698	90	176	1.614	513.549	8.692

Finally, GC is the best approach for general purpose in character articulation, since surface detail preservation with few control points makes it effective and easy to use. Its formulation is more complex than the other two methods, so that the deformation application will be slower. Besides, its global deformation is a small drawback that must be solved in further research (Table 2).

We present a table for analyzing the preprocessor computation time. It has been prepared by a non-optimized implementation of the MVC, HC, and GC papers discussed. The timings were measured on a 2.8 Ghz Quad-Core Intel Xeon with 4 GB of RAM and expose clearly that MVC is the fastest algorithm, followed by GC with a near to 5 times factor, and far behind, the HC with several magnitude orders over. Note that while MVC and GC increases time computation proportionally with model and cage complexity, HC depends on the grid size, specifically the INSIDE marked cells count.

Acknowledgments We are grateful to Pedro García, John Grieco and Guillermo Posadas for helping us to enrich the text with their corrections. This work was partially supported by TIN2010-20590-C02-01.

Appendix

First, we expose Green coordinates pseudocode, extracted from Lipman's paper [22]. The 2D and 3D version for deforming object vertexes $\Lambda \subset C^{in}$. We have changed $\phi_i(v)$ from the original paper by $\omega_i(v)$ to keep coherence all over the document. We note that for exterior or boundary points one should add to these coordinates the $\{\alpha_k\}$ and β as is introduced in Sect. 6.2, and explained in depth in Sect. 4 of Lipman's paper. Note that α_k and β also posses a simple closed-form formula employing the regular barycentric coordinates in triangles (3D) or edges (2D).

Algorithm 1: 2D version of Lipman's Green Coordinates.

Input: cage C = (V,T), set of points $\Lambda = \{\eta\}$
Output: 2D GC $\omega_i(\eta)$, $\psi_j(\eta)$, $i \in I_V$, $j \in I_T$, $\eta \in \Lambda$
/* Initialization
set all $\omega_i = 0$ and $\psi_i = 0$
/* Coordinate computation
for each point $\eta \in \Lambda$ **do**
  for each face $j \in I_T$ with vertices v_{j1}, v_{j2} **do**
    $a := v_{j2} - v_{j1}$; $b := v_{j1} - \eta$
    $Q := a \cdot q$; $S := b \cdot b$; $R := 2a \cdot b$
    $BA := b \cdot \|a\| \, n(t_j)$; $SRT := \sqrt{4SQ - R^2}$
    $L0 := log(S)$; $L1 := log(S + Q + R)$
    $A0 := \frac{tan^{-1}(R/SRT)}{SRT}$
    $A1 := \frac{tan^{-1}((2Q+R)/SRT)}{SRT}$
    $A10 := A1 - A0$: $L10 := L1 - L0$
    $\psi_j(\eta) := -\|a\| / (4\pi)[(4S - \frac{R^2}{Q})A10 + \frac{R}{2Q}L10 + L1 - 2]$
    $\omega_{j2}(\eta) := \omega_{j2}(\eta) - \frac{BA}{2\pi}[\frac{L10}{2Q} - A10\frac{R}{Q}]$
    $\omega_{j1}(\eta) := \omega_{j1}(\eta) - \frac{BA}{2\pi}[\frac{L10}{2Q} - A10(2 + \frac{R}{Q})]$
  end for
end for

Harmonic Coordinates implementation is exposed in the corresponding Sect. 4.1, there is no pseudocode in the paper. Finally, Mean Value Coordinates pseudocode from Ju's paper [18] is presented. It is written for value interpolation, but with some modifications could be adapted for mesh deformation.

Algorithm 2: 3D version of Lipman's Green Coordinates.

Input: cage C = (V,T), set of points $\Lambda = \{\eta\}$
Output: 3D GC $\omega_i(\eta)$, $\psi_j(\eta)$, $i \in I_V$, $j \in I_T$, $\eta \in \Lambda$
/* Initialization
set all $\omega_i = 0$ and $\psi_i = 0$
/* Coordinate computation
for each point $\eta \in \Lambda$ **do**
  for each face $j \in I_T$ with vertices v_{j1}, v_{j2}, v_{j3} **do**
    for each l = 1,2,3 **do**
      $v_{jl} := v_{jl} - \eta$
    end for
    $p := (v_{j1} \cdot n(t_j) n(t_j))$
    for each l = 1,2,3 **do**
      $s_l := sign(((v_{jl} - p) \times (v_{jl+1} - p))n(t_j))$
      $I_l := GCTriInt(p, v_{jl}, v_{jl+1}, 0)$
      $II_l := GCTriInt(0, v_{jl+1}, v_{jl}, 0)$
      $q_l := v_{jl+1} \times v_{jl}$
      $N_l := q_l / \|q_l\|$
    end for
    $I := -|\sum_{k=1}^{3} s_k I_k|$
    $\psi_j(\eta) := -I$
    $w := n(t_j)I + \sum_{k=1}^{3} N_k II_k$
    if $\|w\| > \varepsilon$ **then**
      for each l = 1, 2, 3 **do**
        $\omega_{jl}(\eta) := \omega_{jl}(\eta) + \frac{N_{l+1} \cdot w}{N_{l+1} \cdot v_{jl}}$
      end for
    end if
  end for
end for

proc **GCTriInt(p, v₁, v₂, η)**
$\alpha := cos^{-1}\left(\frac{(v_2 - v_1)(p - v_1)}{\|v_2 - v_1\| \|p - v_1\|}\right)$
$\beta := cos^{-1}\left(\frac{(v_1 - p)(v_2 - p)}{\|v_1 - p\| \|v_2 - p\|}\right)$
$\lambda := \|p - v_1\|^2 sin(\alpha)^2$
$c := \|p - \eta\|^2$
for each $\theta = \pi - \alpha, \pi - \alpha - \beta$ **do**
  $S := sin(\theta)$; $C := cos(\theta)$
  $I_\theta := \frac{-sign(S)}{2}\left[2\sqrt{c}\,tan^{-1}\left(\frac{\sqrt{c}C}{\sqrt{\lambda + S^2 c}}\right) + \sqrt{\lambda}\left(\frac{2\sqrt{\lambda}S^2}{(1-C)^2}\left(1 - \frac{2cC}{c(1+C) + \lambda\sqrt{\lambda^2 + \lambda c S^2}}\right)\right)\right]$
end for
return $\frac{-1}{4\pi}|I_{\pi-\alpha} - I_{\pi-\alpha-\beta} - \sqrt{c}\beta|$
end proc

Algorithm 3: Ju's version of Mean Value Coordinates in 3D

```
for each vertex p_j with values f_j do
  d_j ← ‖p_j − x‖
  if d_j < ε return f_i
  u_j ← (p_j − x)/d_j
end for
totalF ← 0
totalW ← 0
for each triangle with vertices p_1, p_2, p_3 and values f_1, f_2, f_3 do
  l_i ← ‖u_{i+1} − u_{i−1}‖ // for i = 1, 2, 3
  θ ← 2arcsin[l_i/2]
  h ← (∑θ_i)/2
  if (π - h < ε) then
    //x lies on t, use 2D barycentric coordinates
    w_i ← sin[θ_i]d_{i−1}d_{i+1}
    return (∑ w_i f_i)/(∑ w_i)
  end if
  c_i ← (2sin[h]sin[h − θ_i])/(sin[θ_{i+1}]sin[θ_{i−1}]) − 1
  s_i ← sign[det[u_1, u_2, u_3]] √(1 − c_i^2)
  if ∃i, |s_i| ≤ ε then
    // x lies outside t on the same plane, ignore t
    continue
  end if
  w_i ← (θ_i − c_{i+1}θ_{i−1} − c_{i−1}θ_{i+1})/(d_i sin[θ_{i+1}]s_{i−1})
  totalF+ = ∑ w_i f_i
  totalW+ = ∑ w_i
end for
f_x ← totalF/totalW
```

References

1. Barr AH (1984) Global and local deformations of solid primitives. SIGGRAPH Comput Graph 18:21–30
2. Botsch M, Kobbelt L (2004) An intuitive framework for real-time freeform modeling. In: ACM SIGGRAPH 2004 papers, SIGGRAPH '04. ACM, New York, NY, USA, pp 630–634
3. Botsch M, Kobbelt L (2005) Real-time shape editing using radial basis functions. Comput Graph Forum 24(3):611–621
4. Cerveró MÀ, Vinacua Á, Brunet P (2010) Volume-preserving deformation using generalized barycentric coordinates. In: Procc XX Congreso Español de Informática Gráfica (CEIG-2010) pp 86–92
5. Chen L, Huang J, Sun H, Bao H (2010) Cage-based deformation transfer. Comput Graph 34(2):107–118
6. Farin G (1990) Surfaces over dirichlet tessellations. Comput Aided Geom Des 7(1–4):281–292
7. Floater MS (1997) Parametrization and smooth approximation of surface triangulations. Comput Aided Geom Des 14(3):231–250
8. Floater MS (2003) Mean value coordinates. Comput Aided Geom Des 20(1):19–27
9. Floater MS, Kós G, Reimers M (2005) Mean value coordinates in 3d. Comput Aided Geom Des 22(7):623–631

10. Floater M, Hormann K, Kós G (2006) A general construction of barycentric coordinates over convex polygons. Adv Comput Math 24:311–331
11. Hiyoshi H, Sugihara K (2000) Voronoi-based interpolation with higher continuity. In: Proceedings of the sixteenth annual symposium on computational geometry, SCG '00. ACM, New York, NY, USA, pp 242–250
12. Hormann K, Floater MS (2006) Mean value coordinates for arbitrary planar polygons. ACM Trans Graph 25(4):1424–1441
13. Hormann K, Sukumar N (2008) Maximum entropy coordinates for arbitrary polytopes. Comput Graph Forum 27(5):1513–1520. doi:10.1111/j.1467-8659.2008.01292.x. http://dx.doi.org/10.1111/j.1467-8659.2008.01292.x
14. Huang J, Shi X, Liu X, Zhou K, Wei LY, Teng SH, Bao H, Guo B, Shum HY (2006) Subspace gradient domain mesh deformation. In: ACM SIGGRAPH 2006 papers, SIGGRAPH '06. ACM, New York, NY, USA, pp 1126–1134
15. Jacobson A, Baran I, Popović J, Sorkine O (2011) Bounded biharmonic weights for real-time deformation. ACM Trans Graph 30:78:1–78:8
16. Joshi P, Tien WC, Desbrun M, Pighin F (2005) Learning controls for blend shape based realistic facial animation. In: ACM SIGGRAPH 2005 courses, SIGGRAPH '05. ACM, New York, NY, USA
17. Joshi P, Meyer M, DeRose T, Green B, Sanocki T (2007) Harmonic coordinates for character articulation. ACM Trans Graph 26:71
18. Ju T, Schaefer S, Warren J, (2005) Mean value coordinates for closed triangular meshes. In: ACM SIGGRAPH 2005 papers, SIGGRAPH '05. ACM, New York, NY, USA, pp 561–566
19. Ju T, Zhou QY, van de Panne M, Cohen-Or D, Neumann U (2008) Reusable skinning templates using cage-based deformations. ACM Trans Graph 27:122:1–122:10
20. Lewis JP, Cordner M, Fong N (2000) Pose space deformation: a unified approach to shape interpolation and skeleton-driven deformation. In: Proceedings of the 27th annual conference on computer graphics and interactive techniques, SIGGRAPH '00. ACM Press/Addison-Wesley Publishing Co., New York, NY, USA, pp 165–172
21. Lipman Y, Kopf J, Cohen-Or D, Levin D (2007) Gpu-assisted positive mean value coordinates for mesh deformations. In: Proceedings of the fifth Eurographics symposium on geometry processing. Eurographics Association, Aire-la-Ville, Switzerland, pp 117–123
22. Lipman Y, Levin D, Cohen-Or D (2008) Green coordinates. ACM Trans Graph 27:78:1–78:10
23. Malsch EA, Dasgupta G (2004) Interpolations for temperature distributions: a method for all non-concave polygons. Int J Solids Struct 41(8):2165–2188. doi:10.1016/j.ijsolstr.2003.11.037
24. Meyer M, Lee H, Barr A, Desbrun M (2005) Generalized barycentric coordinates on irregular polygons. Computer 7(1):0–4
25. Möbius AF (1827) Der barycentrische Calcül. Georg Olms Verlag
26. Peng Q, Jin X, Feng J (1997) Arc-length-based axial deformation and length preserved animation. Comput Animat'97. Conference Publications, pp 86–92
27. Sederberg TW, Parry SR (1986) Free-form deformation of solid geometric models. SIGGRAPH Comput Graph 20:151–160
28. Sibson R (1981) A brief description of natural neighbour interpolation. In: Barnett V (ed) Interpreting multivariate data. Wiley, Chichester, pp 21–36
29. Singh K, Eugene F (1998) Wires: a geometric deformation technique. In: Proceedings of the 25th annual conference on computer graphics and interactive techniques, SIGGRAPH '98. ACM, New York, NY, USA, pp 405–414
30. Sorkine O (2006) Differential representations for mesh processing. Comput Graph Forum 25(4):789–807
31. Sukumar N, Malsch E (2006) Recent advances in the construction of polygonal finite element interpolants. Arch Comput Methods Eng 13:129–163
32. Terzopoulos D, Platt J, Barr A, Fleischer K (1987) Elastically deformable models. In: Proceedings of the 14th annual conference on computer graphics and interactive techniques, SIGGRAPH '87. ACM, New York, NY, USA, pp 205–214

33. Wachspress E (1975) A rational finite element basis. Academic Press, New York
34. Weber O, Ben-Chen M, Gotsman C (2009) Complex barycentric coordinates with applications to planar shape deformation. Comput Graph Forum 28(2):587–597
35. Xian C, Lin H, Gao S (2011) Automatic cage generation by improved obbs for mesh deformation. Vis Comput 28(1):21–33

Image Gradient Based Level Set Methods in 2D and 3D

Xianghua Xie, Si Yong Yeo, Majid Mirmehdi, Igor Sazonov and Perumal Nithiarasu

Abstract This chapter presents an image gradient based approach to perform 2D and 3D deformable model segmentation using level set. The 2D method uses an external force field that is based on magnetostatics and hypothesized magnetic interactions between the active contour and object boundaries. The major contribution of the method is that the interaction of its forces can greatly improve the active contour in capturing complex geometries and dealing with difficult initializations, weak edges and broken boundaries. This method is then generalized to 3D by reformulating its external force based on geometrical interactions between the relative geometries of the deformable model and the object boundary characterized by image gradient. The evolution of the deformable model is solved using the level set method so that topological changes are handled automatically. The relative geometrical configurations between the deformable model and the object boundaries contribute to a dynamic vector force field that changes accordingly as the deformable model evolves. The geometrically induced dynamic interaction force has been shown to greatly improve the deformable model performance in acquiring complex geometries and highly concave boundaries, and it gives the deformable model a high invariancy in initialization configurations. The voxel interactions across the whole image domain provide a global view of the object boundary representation,

X. Xie (✉)
Department of Computer Science, Swansea University, Faraday Tower, Singleton Park, Swansea SA2 8PP, UK
e-mail: x.xie@swansea.ac.uk

S. Y. Yeo · I. Sazonov · P. Nithiarasu
College of Engineering, Swansea University, Talbot Building, Singleton Park, Swansea SA2 8PP, UK

M. Mirmehdi
Department of Computer Science, Bristol University, Merchant Ventures Building, Bristol BS8 1UB, UK
e-mail: majid@cs.bris.ac.uk

M. González Hidalgo et al. (eds.), *Deformation Models*, Lecture Notes in Computational Vision and Biomechanics 7, DOI: 10.1007/978-94-007-5446-1_4,

giving the external force a long attraction range. The bidirectionality of the external force field allows the new deformable model to deal with arbitrary cross-boundary initializations, and facilitates the handling of weak edges and broken boundaries.

1 Introduction

Depending on the assumption of how object boundary is described, active contours can be classified into edge based [3, 14, 17, 25], region based [5, 7, 16], and hybrid approaches [4, 9, 23]. For edge based methods, it is assumed that object boundaries collocate with image intensity discontinuities which is widely adopted, for example, in depth estimation from stereo [2]. Region based techniques, on the other hand, assume that object boundaries collocate with discontinuities in regional characteristics, such as color and texture. In other words, each object has its own distinctive and continuous regional features.

Region based techniques have some obvious advantages over edge based methods in that object boundary description based on image gradient can often be compromised by noise and weak edges. They are also less sensitive to initialization, while edge based active contours are prone to local minima. Thus, it is often desirable for edge based techniques to carefully place the initial contour. This assumes that the prior knowledge of the object location is available, which is not always true in reality. Existing techniques can only reduce this initialization dependency to a very limited extent. The balloon force [3] can only expand or shrink the contours. The bidirectionality of GVF can sometimes cause the contours to collapse on approach to the same boundary. Moreover, it has convergence issues caused by critical points. [8, 17, 24]. It is evidently clear that initialization invariance is particularly difficult to achieve for edge based methods. More recent attempts, such as [8, 12, 14, 17], showed promising but limited success.

In this chapter, we present an image gradient based approach to perform 2D and 3D deformable model segmentation using level set. Section 2 presents the 2D method which uses an external force field that is based on magnetostatics and hypothesized magnetic interactions between the active contour and object boundaries. The major contribution of the method is that the interaction of its forces can greatly improve the active contour in capturing complex geometries and dealing with difficult initializations, weak edges and broken boundaries. This method is then generalized to 3D in Sect. 3 by reformulating its external force based on geometrical interactions between the relative geometries of the deformable model and the object boundary characterized by image gradient. The relative geometrical configurations between the deformable model and the object boundaries contribute to a dynamic vector force field that changes accordingly as the deformable model evolves. Experimental results are shown in Sect. 4. The proposed dynamic interaction force has been shown to greatly improve the deformable model performance in acquiring complex geometries and highly concave boundaries, and it gives the deformable model a high invariancy in initialization configurations. The voxel interactions across the whole image domain provide a global view of the object boundary representation, giving the external force a long attraction range. The bidirectionality of the external force field allows the new deformable model to deal with arbitrary cross-boundary initializations, and facilitates the handling of weak edges and broken boundaries.

2 MAC Model: A 2D Approach

Fittings based on local intensity discontinuity can often lead to undesired local minima. The CPM [12] assigns opposite charges to edges and free particles so that the particles are pulled towards edges while repelling each other. This global interaction provides much freedom of initialization. However, particles on weak edge can be gradually pulled towards neighboring strong edges, resulting in broken boundaries. Particle addition and deletion and contour reconstruction can also be difficult in practice.

Instead of assigning fixed charges, we allow the charges flow through the edges. These flows of charges will then generate a magnetic field. The active contour, carrying similar flow of charges, will be attracted towards the edges under this magnetic influence. Without losing generality, let us consider the image plane as a 2D plane in a 3D space whose origin coincides with the origin of the image coordinates. Additionally, the third dimension of this 3D space is considered perpendicular to the image plane.

The direction of the currents, flows of charges, running through object boundary can be estimated based on edge orientation, which can be conveniently obtained by a 90° rotation in the image plane of the normalized gradient vectors $(\hat{\boldsymbol{I}}_x, \hat{\boldsymbol{I}}_y)$, where I denotes an image. Let $\mathbf{x}$ denote a point in the image domain. Thus, the object boundary current direction, $\mathbf{O}(\mathbf{x})$, can be estimated as: $\mathbf{O}(\mathbf{x}) = (-1)^{\lambda}(-\hat{\boldsymbol{I}}_y(\mathbf{x}), \hat{\boldsymbol{I}}_x(\mathbf{x}), 0)$, where $\lambda = 1$ gives an anti-clockwise rotation in the image coordinates, and $\lambda = 2$ provides a clockwise rotation. However, we show later by using the proposed level set updating scheme different λ values lead to the same result. Since the active contour is embedded in a signed distance function, the direction of current for the contour, denoted as υ, can be similarly obtained by rotating the gradient vector $\nabla\Phi$ of the level set function. Similar to $\mathbf{O}$, υ is also three dimensional and lies in the image domain, i.e. $\upsilon(\mathbf{x}) = (-\hat{\Phi}_y(\mathbf{x}), \hat{\Phi}_x(\mathbf{x}), 0)$.

Let $f(\mathbf{x})$ be the magnitude of edge pixel and the magnitude of boundary current be proportional to edge strength, that is, the electric current on object boundary is defined as $f(\mathbf{x})\mathbf{O}(\mathbf{x})$. The magnetic flux $\mathbf{B}(\mathbf{x})$ generated by gradient vectors at each pixel position $\mathbf{x}$ can then be computed as:

$$\mathbf{B}(\mathbf{x}) \propto \sum_{\mathbf{s}\in\mathbf{S},\mathbf{s}\neq\mathbf{x}} f(\mathbf{s})\mathbf{O}(\mathbf{s}) \times \frac{\hat{\mathbf{R}}_{\mathbf{xs}}}{R_{\mathbf{xs}}^2}, \tag{1}$$

where $\mathbf{s}$ denotes an edge pixel position, $\mathbf{S}$ is the set containing all the edge pixel positions across the image, $\hat{\mathbf{R}}_{\mathbf{xs}}$ denotes a 3D unit vector from $\mathbf{x}$ to $\mathbf{s}$ in the image plane, and $R_{\mathbf{xs}}$ is the distance between them. Thresholding can be applied to remove some erroneous edge pixels with very small gradient magnitude [12, 24]. The active contour is assigned with unit magnitude of electric current. The force imposed on it can be derived as:

$$\mathbf{F}(\mathbf{x}) \propto \upsilon(\mathbf{x}) \times \mathbf{B}(\mathbf{x}). \tag{2}$$

From (1) and (2), we can see that $\mathbf{B}$ intersects the image plane perpendicularly and $\mathbf{F}$ is always perpendicular to both υ and $\mathbf{B}$. Thus, $\mathbf{F}$ also lies in the image domain and its third element equals to zero. For simplicity, from now on, we shall ignore its third dimensional component and denote $\mathbf{F}(\mathbf{x})$ as a 2D vector field in the image domain. The basic model can then be formulated as:

$$\frac{\mathrm{d}C}{\mathrm{d}t} = \alpha g(\mathbf{x})\kappa\hat{\mathbf{n}} + (1-\alpha)(\mathbf{F}(\mathbf{x}) \cdot \hat{\mathbf{n}})\hat{\mathbf{n}}, \tag{3}$$

where $g(\mathbf{x}) = 1/(1 + f(\mathbf{x}))$, κ denotes the curvature, and $\hat{\mathbf{n}}$ is inward unit normal. Its level set representation then takes this form:

$$\frac{\partial \Phi}{\partial t} = \alpha g(\mathbf{x})\nabla \cdot \left(\frac{\nabla\Phi}{|\nabla\Phi|}\right)|\nabla\Phi| - (1-\alpha)\mathbf{F}(\mathbf{x}) \cdot \nabla\Phi. \tag{4}$$

We can see from (1) and (2) that the image force is derived from global interactions among rotated gradient vectors, i.e. $f(\mathbf{x})\mathbf{O}(\mathbf{x})$. Thus, it is more robust than fittings based on local gradient towards weak edges (where $f(\mathbf{x})$ is small) and noise (where $\mathbf{O}(\mathbf{x})$ is locally inconsistent). It is worth noting, however, that general contrast consistency along the object boundaries is important to the model. Large contrast variation can disrupt the force field, e.g. half of the object appears brighter than background and the other half appears to be darker. However, this does not mean that the entire object has to be brighter or darker than background. Those regions away from object boundary can be continuously varying in intensity. The model also can tolerate a fair amount of local contrast inconsistency, in the same way as to image noise and weak/broken edges.

As aforementioned, due to cross product computation the external force, $\mathbf{F}$, is always perpendicular to υ which is tangent to the contour, i.e. the external force is imposed along the normal direction. Note the internal force due to curvature flow is enforced in the inward normal direction. Thus, the total force is always perpendicular to active contour. In other words, it dynamically updates itself according to contour evolution to push and pull the contours along the normal direction until they reach object boundaries where forces from both sides are in balance. As a result, the propagating contour will not suffer from those convergence issues related to static force fields, such as GVF, in which evolving contours may become tangent to underlying force vectors resulting in false convergence. This force field is also significantly different from others used in edge based methods. For example, in CPM, the force

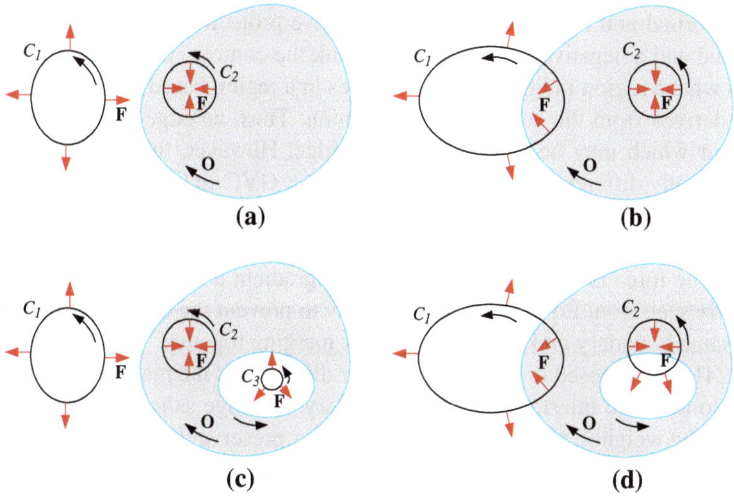

Fig. 1 Preventing contour collapsing. **a** Two contours, C_1 and C_2, are placed on each side of an object boundary with current directions indicated by arrows. Contour C_1 is attracted by the object boundary and expands itself in the outward normal direction. It eventually will wrap around and capture the object boundary. Contour C_2 however is repelled and forced to shrink in the inward normal direction. Thus, two contours will not collapse to each other. **b** Similar to (a), however contour C_1 is placed across the object boundary. Those contour segments of C_1 that are inside object boundary will be pulled towards object boundary and the rest of contour C_1 will expand and wrap around the object boundary. The segments inside object boundary and outside will not collapse to each other. **c** The object in this case contains an internal boundary. The behavior of C_1 and C_2 is similar to that in (a). Contour C_3 will expand itself to capture the internal boundary. Three contours will not collapse to each other, while capturing both boundaries. **d** Contours C_1 and C_2 are now initialized across external and internal boundaries, respectively. The behavior of C_1 is similar to that in (b). The contour segments of C_2 that are inside the object (*gray area*) will be attracted to the object internal boundary that is initially inside contour C_2. The other contour segments of C_2 will expand to capture the rest internal boundaries. No contour collapsing will occur, either. GVF contours, as an example, will collapse to each other in all above scenarios

between an edge pixel **s** and an infinitesimal contour segment **c** lies in a straight line between these two, regardless the orientation of the contour segment. In our model, the orientation of the edge pixel and the contour segment also have influence on the resulting force interaction. This ability to adapt is very important since it ensures the active contour, once initialized, overcome deep concavities and narrow regions to reach object boundaries (Fig. 1).

By incorporating (2), Eq. (3) can be re-written as:

$$\begin{aligned}\frac{\mathrm{d}C}{\mathrm{d}t} &= \alpha g(\mathbf{x})\kappa\hat{\mathbf{n}} + (1-\alpha)\big(\upsilon(\mathbf{x}) \times \mathbf{B}(\mathbf{x}) \cdot (\hat{\mathbf{n}}, 0)\big)\,\hat{\mathbf{n}} \\ &= \alpha g(\mathbf{x})\kappa\hat{\mathbf{n}} + (1-\alpha)\big(\mathbf{B}(\mathbf{x}) \cdot ((\hat{\mathbf{n}}, 0) \times \upsilon(\mathbf{x}))\big)\,\hat{\mathbf{n}}.\end{aligned} \quad (5)$$

The external force in the second term is in fact a projection of the magnetic flux onto a binormal unit vector which is computed from a cross product of the contour inward normal and its tangent vector. A positive projection will force the contour to expand and a negative projection will shrink the contour, which acts in a similar way as what a region indication function does in a region based approach, however, this is derived from the edge based assumption. Thus, an edge can attract or push a contour which may lie either side of the edge. However, this bidirectionality is fundamentally different from that in, for example, GVF. In GVF, the force imposed on the contour is independent of the contour itself, which can cause the contours to collapse to each other when reaching to the same object boundary. For the proposed method, the force is related to both the image gradient and the contour (which can be clearly seen from Eq. (5)). It has the ability to prevent the contour from reaching to the same boundary and disappearing after merging together.

In [24], we proposed to perform nonlinear diffusion of the magnetic field in order to overcome noise interference when necessary. An edge saliency measure can be added to the weighting function in order to better preserve the edges [22]. Let $B(\mathbf{x})$ denote the signed magnitude of $\mathbf{B}(\mathbf{x})$. The diffused field $\hat{\mathscr{B}}(\mathbf{x})$ is obtained by solving:

$$\frac{\mathrm{d}\mathscr{B}}{dt}(\mathbf{x}) = p(B(\mathbf{x}))\nabla^2\mathscr{B}(\mathbf{x}) - q(B(\mathbf{x}))(\mathscr{B}(\mathbf{x}) - B(\mathbf{x})), \tag{6}$$

where $p(B(\mathbf{x})) = e^{-\frac{|B(\mathbf{x})|\mathscr{S}(\mathbf{x})}{K}}$, $q(.) = 1 - p(.)$, and $\mathscr{S}(.)$ is a n edge saliency measure which is measured based on edge strength and orientation coherency, i.e. $\mathscr{S}(\mathbf{x}) = f(\mathbf{x})v(\mathbf{x})$ where $v(.)$ is the variance of orientation in a local neighborhood, e.g. 9×9 as used here. More sophisticated saliency measures, e.g. [11], can be used. Weighting the flux magnitude with $\mathscr{S}(.)$ further ensures as little diffusion as possible at object boundaries, while areas lack of consistent support from edges result in substantial diffusion.

3 Extension to 3D

Shape segmentation from volumetric data has an important role in applications such as medical image analysis. Volumetric image segmentation remains an intricate process, due to the complexity and variability of image data and shapes (i.e. anatomical structures). There have been applications of simple techniques such as thresholding and region growing in the extraction of 3D objects from volumetric images [20, 21]. However, these techniques are very sensitive to noise and intensity inhomogeneities which exist in real images, and often produce leakages and regions which are not contiguous. Statistical approaches [10, 19] are also used to identify different tissue structures from medical images. It usually involves manual interaction to segment images in order to obtain a sufficiently large set of training samples. Such strategies are often restricted to problems where there is sufficient prior knowledge about the shape or appearance variations of the relevant structures. Also, the use of the same training set for a large number of image scans may lead to biased results that do not take sufficient consideration of the variability within

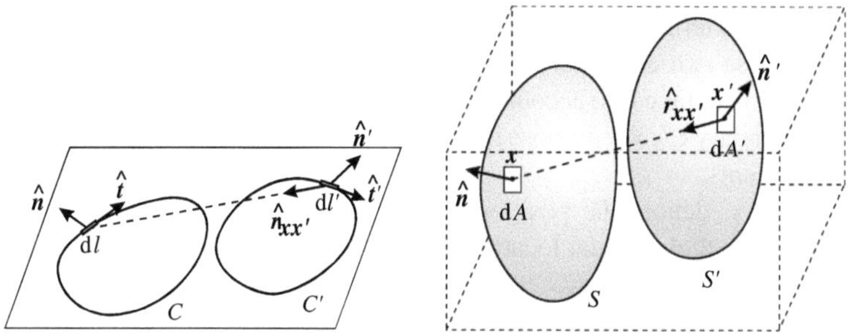

Fig. 2 Relative position and orientation between geometries in 2D and 3D

individuals. Atlas based approaches perform segmentation based on image registration techniques [15], whereby an image can be segmented by finding a transformation that maps a template image to the target image. It is however generally difficult for atlas based techniques to accurately extract complex geometries such as those from volumetric medical images due to the variability of anatomical structures.

The external force field presented previously is based on the hypothesized magnetic force between the active contour and object boundaries. This formulation can be applied directly in the magnetostatic active contour to compute the magnetic field and force required to draw the active contour towards object boundaries in 2D images. This image gradient based method showed significant improvements on convergence issues, e.g. reaching deep concavities, and in handling weak edges and broken boundaries. When applying the analogy directly to deformable modeling, it requires estimation of tangent vectors for the deformable contours, which is convenient in 2D case, however, not possible in 3D. Our approach is to define a novel external force field that is based on hypothesized geometrically induced interactions between the relative geometries of the deformable model and the object boundaries (characterized by image gradients). In other words, the magnitude and direction of the interaction forces are based on the relative position and orientation between the geometries of the deformable model and image object boundaries, and hence, it is called the *geometric potential force (GPF)* field [27]. The bidirectionality of the new external force field can facilitate arbitrary cross-boundary initialization, which is a very useful feature to have, especially in the segmentation of complex geometrics in 3D. It also improves the performance of the deformable model in handling weak edges. In addition, the proposed external force field is dynamic in nature as it changes according to the relative position and orientation between the evolving deformable model and object boundary. This GPF force however is in fact a 3D extension of the 2D MAC model.

3.1 Geometric Potential Force

First, consider a deformable contour C and an ideal object boundary C' in the image plane (see Fig. 2). Let $\mathrm{d}l$ and $\mathrm{d}l'$ denote the infinitesimal elements of contour C and object boundary C', respectively. In the existing force field based models such as [13, 26], the interaction between dl and dl' is inversely proportional to the distance separating these two elements and the derived force lies in a straight line between them. They do not take into account the local geometry of the deformable contour C or object boundary C'. We propose to incorporate the mutual location and orientation of these elements.

Let $\mathbf{x}$ and $\mathbf{x}'$ denote the positions of elements dl and dl', respectively. Thus, $\mathbf{r}_{\mathbf{xx}'} = \mathbf{x} - \mathbf{x}'$ is their mutual location of those two elements, $r_{\mathbf{xx}'} = |\mathbf{x} - \mathbf{x}'|$ is the distance between them, and $\hat{\mathbf{r}}_{\mathbf{xx}'} = (\mathbf{x} - \mathbf{x}')/r_{\mathbf{xx}'}$ is the unit vector pointing dl from dl'. The directions of these elements can be represented by their unit tangent vectors $\hat{\mathbf{t}}$ and $\hat{\mathbf{t}}'$. However, a unique tangent vector is no longer available for infinitesimal surface elements in 3D. Thus, we use unit outward normal vectors $\hat{\mathbf{n}}$ and $\hat{\mathbf{n}}'$ to characterize the orientations of these elements instead (see Fig. 2). In 2D, they are simply 90° rotated tangent vectors.

We are now ready to introduce the hypothesized interaction force d$\mathbf{F}$ dl which acts on element dl by virtue of the hypothesized force field induced by element dl'. It is desirable to combine the element orientation vectors and distance vector in deriving the force. We propose a simple but effective combination of these three vectors as $\hat{\mathbf{n}}\,(\hat{\mathbf{r}}_{\mathbf{xx}'} \cdot \hat{\mathbf{n}}')$, unlike CPM [12] as an example where only the distance vector $\hat{\mathbf{r}}_{\mathbf{xx}'}$ is used. The multiplication of contour normal $\hat{\mathbf{n}}$ ensures that the force is always imposed in the normal direction so that the deformable model does not suffer from convergence issues (i.e. stationary points, saddle points and extreme boundary concavities), which are often associated with other vector force field based methods such as GVF [25]. The dot product of the object boundary element normal with the distance vector allows the force on the contour in the normal direction to diminish as the contour reaches the object boundary. Similar to other physics-inspired force field, it is also desirable to decay the force interaction with the increase of distance between the elements, i.e. the force is designed proportional to $\hat{\mathbf{n}}\,(\hat{\mathbf{r}}_{\mathbf{xx}'} \cdot \hat{\mathbf{n}}')/r^{\lambda}_{\mathbf{xx}'}$ where $\lambda > 0$. Thus, the contribution of element dl' of object boundary C' to the total force acting on dl in accordance with their distance and mutual orientation can be formulated as

$$\mathrm{d}\mathbf{F}\,\mathrm{d}l = \hat{\mathbf{n}}\,\mathrm{d}G\,\mathrm{d}l, \qquad \mathrm{d}G = \left(\frac{\hat{\mathbf{r}}_{\mathbf{xx}'}}{r^{\lambda}_{\mathbf{xx}'}} \cdot \hat{\mathbf{n}}'\right) \mathrm{d}l' \tag{7}$$

where $\mathbf{F}$ is defined as force per unit length, dG is the contribution of element dl' of object boundary C' into the scalar field $G(\mathbf{x})$, which can be considered as an intermediate potential field, and λ is a positive constant that affects the magnitude of the interaction force based on the distance between the elements. In our study, we obtained the best results when λ coincides with the dimension of the image data, i.e. $\lambda = 2$ in the 2D case. Furthermore, we show later that when λ coincides with data dimension in 2D, the proposed force interaction has an explicit link to the magnetostatics theory and thus the spatial decay of the magnitude of the interaction force is analogous to that of the magnetic field.

As shown in (7), the computation of the new force field only requires unit normal vectors and relative position of the two elements, which is convenient to acquire. Thus, this new force field can be easily extended to higher dimensions, e.g. 3D. Let dA

belong to the deformable surface S whereas $\mathrm{d}A'$ belongs to the object boundary S' (see Fig. 2). The generalized 3D version of force $\mathrm{d}\mathbf{F}\,\mathrm{d}A$ acting between these two area elements can be readily given as

$$\mathrm{d}\mathbf{F}\,\mathrm{d}A = \hat{\mathbf{n}}\,\mathrm{d}G\,\mathrm{d}A, \qquad \mathrm{d}G = \left(\frac{\hat{\mathbf{r}}_{\mathbf{xx}'}}{r^{\lambda}_{\mathbf{xx}'}} \cdot \hat{\mathbf{n}}'\right)\mathrm{d}A' \tag{8}$$

where $\mathbf{F}$ is defined as force per unit area, G is the corresponding 3D potential field, $\hat{\mathbf{n}}$ and $\mathbf{n}'$ are unit surface normals of the deformable model and object boundary, respectively, and $\lambda = 3$. Again, the magnitude and direction of the induced force $\mathbf{F}$ is handled intrinsically by the relative position and orientation between the geometries of the deformable model determined by the evolving surface S and object boundary determined by S'. Since the force is derived geometrically and its interaction is a function of inverse distance, we name it geometric potential force (GPF).

3.2 GPF Deformable Model

The GPF force in (8) is derived using geometrical information from ideal object boundaries. Next, we extend this to deal with real image data and formulate it in 3D deformable modelling. In this work, we adopt an edge based approach, that is using image intensity discontinuity to estimate the presence and strength of object boundaries.

Let $I(\mathbf{x})$ denote the 3D image, where $\mathbf{x}$ is a voxel location in the image domain. Temporarily, we consider $\mathbf{x}$ as a continuously varying point. One may treat this as an interpolation between voxel grid points to obtain a continuous image $I(\mathbf{x})$. To compute the force acting on $\mathrm{d}A$, we first compute the total potential field for an arbitrary point $\mathbf{x}$:

$$G(\mathbf{x}) = P.V. \oiint_{S'} \mathscr{W}(\mathbf{x}')\left(\frac{\hat{\mathbf{r}}_{\mathbf{xx}'}}{r^{\lambda}_{\mathbf{xx}'}} \cdot \hat{\mathbf{n}}'(\mathbf{x}')\right)\mathrm{d}A'. \tag{9}$$

where $\mathscr{W}(\cdot)$ is a weighting function that is defined later, and $P.V.$ means *'Principal Value'*: the contribution of infinitesimal circular vicinity of singular point $\mathbf{x}' = \mathbf{x}$ into the integral is disregarded, which occurs when surfaces S and S' intersect.

First, we consider the case, in which S' can be defined rigorously on an ideal object O, i.e. $S' = \partial O$. The object O can be specified by a binary image:

$$\mathscr{I}(\mathbf{x}) = \begin{cases} \mathscr{I}_0 & \mathbf{x} \in O \\ 0 & \mathbf{x} \notin O, \end{cases} \tag{10}$$

where $\mathscr{I}_0$ is a nonzero constant. For such an image, $\nabla\mathscr{I}$ is infinite on S' and can be represented through the 3D Dirac's delta as

$$\nabla \mathscr{I}(\mathbf{x}) = \Delta \mathscr{I}\, \delta(\mathbf{x} - \mathbf{x}')\, \hat{\mathbf{n}}'(\mathbf{x}') \tag{11}$$

where $\Delta \mathscr{I}$ is the jump in function $\mathscr{I}(\mathbf{x})$ at the boundary of O; $\mathbf{x}' \in S'$ and $\hat{\mathbf{n}}'(\mathbf{x}')$ is the unit normal vector to the surface S'. Setting $\mathscr{W}$ equal to the jump of $\mathscr{I}$ at the boundary, i.e. $\mathscr{W} = \Delta \mathscr{I}$, we can re-write (9) as a volume integral

$$G(\mathbf{x}) = P.V. \iiint_{\Omega} \mathscr{W}(\mathbf{x}') \left(\frac{\hat{\mathbf{r}}_{\mathbf{x}\mathbf{x}''}}{r^{\lambda}_{\mathbf{x}\mathbf{x}''}} \cdot \hat{\mathbf{n}}'(\mathbf{x}') \right) \delta(\mathbf{x}'' - \mathbf{x}')\, \mathrm{d}V'' \tag{12}$$

Here, $\mathbf{x}''$ is the integration variable and dV'' denotes a volume element. The Dirac's delta is used to obtain the area element from the volume element, i.e. $dA' \rightarrow \delta(\mathbf{x}'' - \mathbf{x}')\, dV''$.

Taking into account (11) and $\mathscr{W} = \Delta \mathscr{I}$, we can replace the product $\mathscr{W}(\mathbf{x}')\, \hat{\mathbf{n}}'(\mathbf{x}')$ $\delta(\mathbf{x}'' - \mathbf{x}')$ in the integral of (12) by $\nabla \mathscr{I}(\mathbf{x}'')$. Thus, (12) can be re-formulated as

$$G(\mathbf{x}) = P.V. \iiint_{\Omega} \left(\frac{\hat{\mathbf{r}}_{\mathbf{x}\mathbf{x}''}}{r^{\lambda}_{\mathbf{x}\mathbf{x}''}} \cdot \nabla \mathscr{I}(\mathbf{x}'') \right) \mathrm{d}V''. \tag{13}$$

It is now readily generalizable to real 3D data.

In real images, ∇I is a smooth function reaching maximum magnitude in the vicinity of the object boundary. The natural generalization of (13) is to substitute Dirac's delta by this smoothed function analog into (13), i.e. $\mathscr{W}(\mathbf{x}')\, \hat{\mathbf{n}}'(\mathbf{x}')\, \delta(\mathbf{x}'' - \mathbf{x}')$ $\rightarrow \nabla I(\mathbf{x}'')$, where I denotes a real image. The geometric potential field in a continuous form can then be formulated as

$$G(\mathbf{x}) = P.V. \iiint_{\Omega} \left(\frac{\hat{\mathbf{r}}_{\mathbf{x}\mathbf{x}'}}{r^{\lambda}_{\mathbf{x}\mathbf{x}'}} \cdot \nabla I(\mathbf{x}') \right) \mathrm{d}V'. \tag{14}$$

Note, due to the substitution of $\mathscr{W}(\mathbf{x}')\, \hat{\mathbf{n}}'(\mathbf{x}')\, \delta(\mathbf{x}'' - \mathbf{x}')$ by $\nabla I(\mathbf{x}'')$, the $\mathbf{x}'$ defined on the ideal surface S' is no longer needed. Hence, the notation is simplified by replacing the integral variable $\mathbf{x}''$ with $\mathbf{x}'$. Finally, its discrete form can be written as

$$G(\mathbf{x}) = \sum_{\mathbf{x}' \in \Omega, \mathbf{x}' \neq \mathbf{x}} \left(\frac{\hat{\mathbf{r}}_{\mathbf{x}\mathbf{x}'}}{r^{\lambda}_{\mathbf{x}\mathbf{x}'}} \cdot \nabla I(\mathbf{x}') \right). \tag{15}$$

This can be considered as a convolution of the image gradient with the vector kernel $\mathbf{K}_{\lambda}(\mathbf{x})$

$$\begin{cases} \mathbf{K}_{\lambda}(\mathbf{x}) = P.V. \frac{\hat{\mathbf{x}}}{|\mathbf{x}|^{\lambda}} = P.V. \frac{\mathbf{x}}{|\mathbf{x}|^{\lambda+1}} \\ G = \mathbf{K}_{\lambda} * \nabla I = \iiint_{\Omega} \left(\mathbf{K}_{\lambda}(\mathbf{x} - \mathbf{x}') \cdot \nabla I(\mathbf{x}') \right) dV' \end{cases} \tag{16}$$

which can be computed efficiently using the fast Fourier transform (FFT). Note that the potential field G is computed as a convolution of two vector functions.

The total force acting on the unit area element of the deformable surface S is thus given as $\mathbf{F} = \hat{\mathbf{n}}\, G(\mathbf{x})$. where $\hat{\mathbf{n}}$ is the outward unit normal of level set surface. Note, an inward normal can also be used, i.e. $\mathbf{F} = -\hat{\mathbf{n}}\, G(\mathbf{x})$, which will result in opposite deformable model propagation since the force field is exactly in the opposite direction. Hence, the force can be re-written in a generalized form:

$$\mathbf{F} = \mathscr{I}\, \hat{\mathbf{n}}\, G(\mathbf{x}). \tag{17}$$

where $\mathscr{I}$ is a constant taking values of ± 1. Note this is different from the constant force in the geodesic model, where the force is monotonically expanding or shrinking. The sign convention $\pm$ is merely used to determine whether outward and inward normals of the deformable surface are considered.

The general contrast consistency along the object boundaries however is important to the model. Large contrast variation can disrupt the force field, e.g. half of the object appears brighter than background and the other half appears to be darker. However, this does not mean that the entire object has to be brighter or darker than background. Those regions away from object boundary can be continuously varying in intensity.

Once the force field $\mathbf{F}(\mathbf{x})$ is derived from the hypothesized interactions based on the relative geometries of the deformable model and object boundary is determined, the evolution of the deformable model $S(\mathbf{x}, t)$ under this GPF field can be given as

$$\frac{\mathrm{d}S}{\mathrm{d}t} = (\mathbf{F} \cdot \hat{\mathbf{n}})\hat{\mathbf{n}}. \tag{18}$$

Since surface smoothing is usually desirable, the mean curvature flow can be incorporated and the complete GPF deformable model evolution can be formulated as

$$\frac{\mathrm{d}S}{\mathrm{d}t} = \alpha\, g\, \kappa\, \hat{\mathbf{n}} + (1 - \alpha)(\mathbf{F} \cdot \hat{\mathbf{n}})\hat{\mathbf{n}} \tag{19}$$

where $g(\mathbf{x}) = 1/(1 + |\nabla I|)$ is the edge stopping function. Note that in our case, the flow of $\mathbf{F}$ is directed by definition normal to surface S, therefore $(\mathbf{F} \cdot \hat{\mathbf{n}})\hat{\mathbf{n}} = \mathbf{F}$. Notation $(\mathbf{F} \cdot \hat{\mathbf{n}})\hat{\mathbf{n}}$ is inherited from the traditional methods, e.g. GGVF. The level set representation of the proposed deformable model based on GPF can then be written as

$$\frac{\partial \Phi}{\partial t} = \alpha\, g\, \kappa\, |\nabla \Phi| - (1 - \alpha)(\mathbf{F} \cdot \nabla \Phi) \tag{20}$$

where $\Phi(t, \mathbf{x})$ is the level set function, such that the deformable surface S is defined as $\Phi(t, \mathbf{x}) = 0$. Note, the GPF force field is defined on the deformable surface, which is implicitly embedded in the level set function, i.e. the force field computed at the propagating front needs to be extended across the computational domain so that the full level set function can be continuously evolved. Although direct force extension

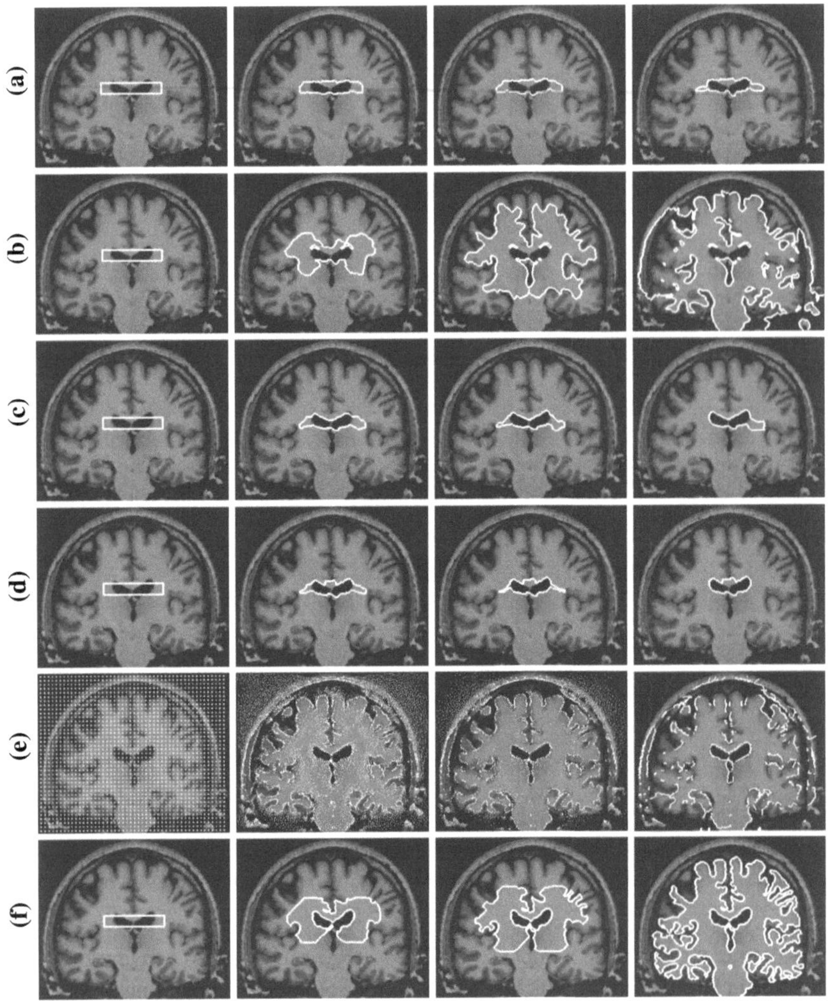

Fig. 4 Comparative study—results by row: **a** DVF, **b** geodesic, **c** GGVF, **d** GeoGGVF, **e** CPM, **f** MAC

their stationary vector force fields exhibit stationary and saddle points, e.g. the saddle point at the entrance of the concave shape which prevents the snake converging to the object boundaries. Again, given an arbitrary cross-boundary initialization, the geodesic snake suffers severe problems and the constriction on the left side of the concave shape causes difficulties for the CVF active contour. MAC was the only active contour model that could successfully recover the shape in both initializations. When dealing with complex geometries, such as the swirl shape and the text "PAMI" shown in the last two rows in Fig. 3, MAC was the only model that managed to fully recover the shapes. The latter example further illustrates MAC's ability in dealing with multiple objects with complex topology.

Fig. 5 GPF: *first row* from *left* to *right*—input image and initial deformable model, corresponding edge map and computed geometric potential field, *second row*—initial and evolving deformable models, and the *third row*—associated GPF vector field

Figure 4 shows a brain MRI image and its comparative segmentation results. For the active contour models, the snake was initialized across the left and right hemispheres, while for the particle model a grid of charges was used. The static vector force based methods (DVF, GGVF, and GeoGGVF) failed to evolve through the tortuous structures and collapsed to nearby edges as shown in rows (a), (c), and (d). The geodesic snake, in row (b), stepped across the weak edges but also failed to localize the boundaries. The free charges of CPM initially reached most of the object boundaries, but later failed to stabilize at weaker edges resulting in incomplete boundary description (row (e)). The MAC contours succeeded in evolving through the narrow and twisted structures as shown in row (f). Multiple regions were captured simultaneously.

Next, we demonstrate the results of the 3D model. The first row of Fig. 5 shows a substantially blurred image with linearly varying intensity, and the corresponding edge map and computed geometric potential field. In addition, as the deformable model evolves, the unit vector $\hat{\mathbf{r}}_{\mathbf{x}\mathbf{x}'}$ changes accordingly based on the relative geometries. This contributes to a vector force field that changes dynamically as the deformable model evolves, as depicted in the second row of Fig. 5. Therefore, the proposed model has much better invariance to its initial position and can deal with complex geometries and extreme boundary concavities.

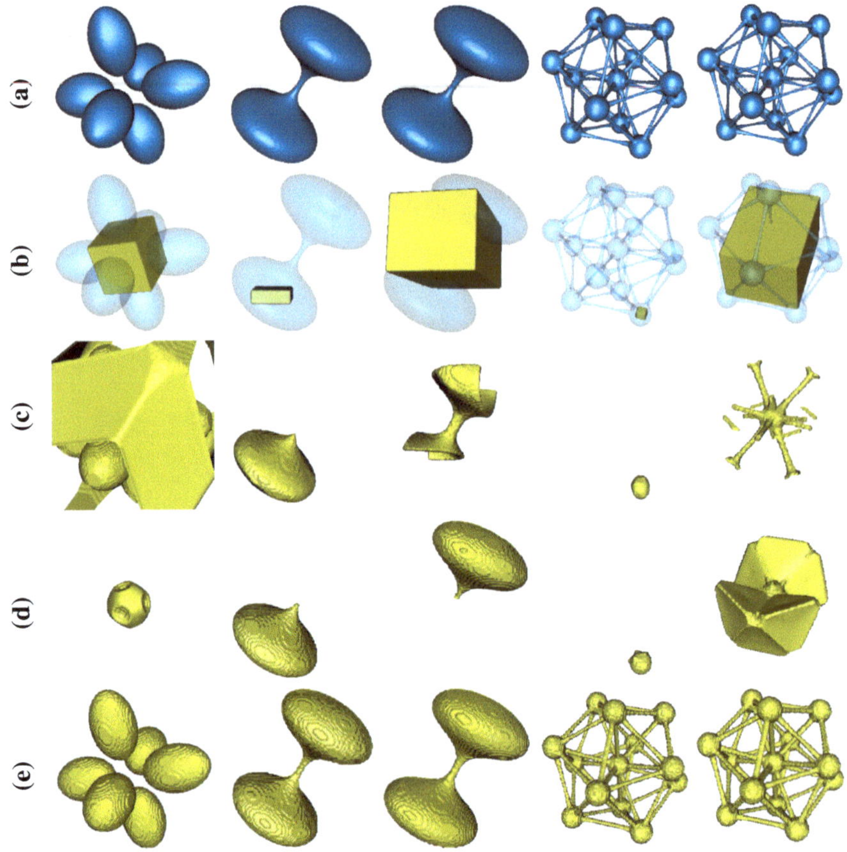

Fig. 6 Shape recovery from synthetic images: **a** isosurfaces of various shapes to be recovered from synthetic images (128×128×128), **b** initial deformable models (*yellow*) with input shapes (*blue*, semi-transparent), **c** recovered shape using geodesic, **d** GGVF, **e** proposed GPF

Figure 6 shows comparative results of extracting 3D shapes. The first column shows the shape extraction results for the six-ellipsoids problem. Given an arbitrary initialization across all the ellipsoids, only GPF could accurately recover the shapes. The geodesic model, given the same initialization configuration, simply expanded outwards and reached the image borders. This is due to the fact that the geodesic model cannot handle cross-boundary initialization as the constant pressure term can only monotonically shrink or expand the contour. Although the bidirectionality of the GGVF model enables it to handle cross-boundary initialization, the saddle and stationary points in this example prevented GGVF from extracting the ellipsoids. The second and third columns show the geometrical object to be recovered consists of two flattened ellipsoids connected by a narrowing tube with a constriction in the middle. With the deformable models initialized inside one of the ellipsoid, only GPF could propagate through the narrowing tube to accurately extract the shape. Also, with a more arbitrary cross-boundary initialization, GPF was the only successful model to

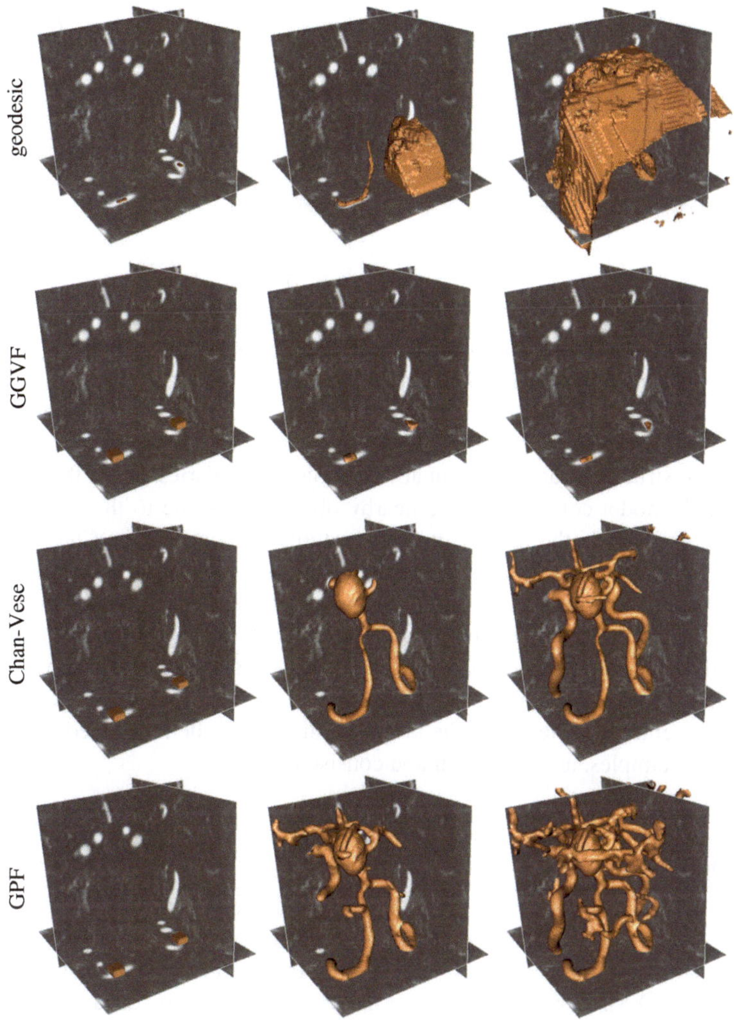

Fig. 7 Segmentation of cerebral arterial structure using different deformable models—*first row* geodesic, *second row* GGVF, *third row* Chan-Vese, *fourth row* proposed GPF

extract the exact shape. The fourth and fifth columns in Fig. 6 compares the shape extraction results on a complex geometry with different initialization configurations. When the initial surface is placed inside one of the sphere of the molecular structure, GPF is the only model that managed to extract the geometry successfully. These examples demonstrate the superior performance of the GPF deformable model in resolving deep concavities and handling complex geometries and topologies. This is mainly due to the dynamic nature of the vector force field. In addition, we show that the bidirectionality of the new force field gives GPF the flexibility to deal with arbitrary cross-initializations.

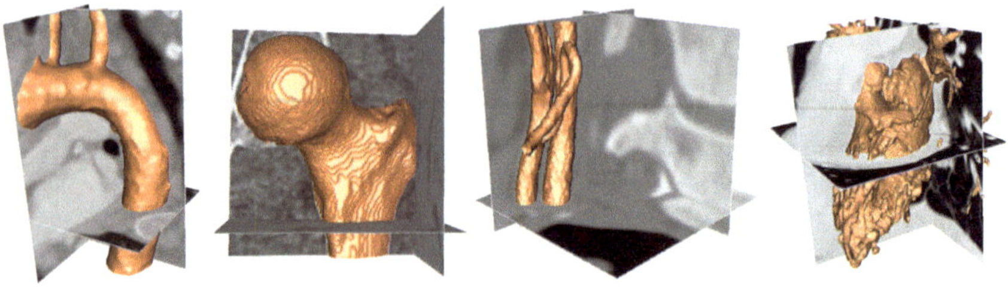

Fig. 8 More examples of the proposed method on real 3D medical data

Figure 7 shows comparative results on the segmentation of cerebral arterial structure from magnetic resonance (MR) imaging. Two initial surfaces are placed inside the object of interest for the geodesic model, and across the object boundaries for GGVF, Chan-Vese and GPF. The geodesic model cannot propagate through the narrow tubular structures, and leaks out at weak object boundaries during the evolution. The GGVF model collapsed to the nearby object edges due to the saddle or stationary points inside the narrow image structures. In contrast, the Chan-Vese and GPF models are able to propagate through the long tubular structures to extract the cerebral arterial geometry. Further examples of the 3D method on real data are given in Fig. 8. The examples above have shown that the GPF deformable model can efficiently segment thin and complex structures, and can handle inhomogeneity in image intensities, noises and weak edges, which are often present in real images. The improvements achieved by the proposed method, as demonstrated extensively in various examples, are significant and consistent.

5 Conclusions

We have presented two image gradient based deformable models that are both based on global image gradient vector integrations, instead of conventional local edge fitting. The 2D MAC model can attract the contour into deep concave regions and does not suffer from saddle point and stationary point problems. Our comparative study showed significant improvement in initialization invariancy and convergence capability on existing state-of-the-art techniques. Its extension to 3D, known as the geometric potential force (GPF) model, utilizes pixel or voxel interactions across the whole image. The derived geometric potential field is thus more informative and exhibits spatial and structural characteristics of image objects which are more coherent than image cues that are based solely on local edge or regional information. This makes the model more robust towards image noise and weak object edges. The relativity between geometries gives the proposed deformable model its distinctive bidirectionality, which facilitates the handling of arbitrary cross-boundary initializations. The straightforward generalization of the proposed model to higher dimensions allows the framework to be applied on N-dimensional images, and opens up to a wide range of potential applications.

References

1. Adalsteinsson D, Sethian J (1998) The fast construction of extension velocities in level set methods. J Comp Phy 148:2–22
2. Birchfield S, Tomasi C (1999) Depth discontinuities by pixel-to-pixel stereo. IJCV 35(3):269–293
3. Caselles V, Kimmel R, Sapiro G (1997) Geodesic active contour. IJCV 22(1):61–79
4. Chakraborty A, Staib H, Duncan J (1996) Deformable boundary finding in medical images by integrating gradient and region information. IEEE T-MI 15(6):859–870
5. Chan T, Vese L (2001) Active contours without edges. IEEE T-IP 10(2):266–277
6. Cohen L, Cohen I (1993) Finite-element methods for active contour models and balloons for 2-D and 3-D images. IEEE T-PAMI 15(11):1131–1147
7. Cremers D, Rousson M, Deriche R (2007) A review of statistical approaches to level set segmentation: integrating color, texture, motion and shape. IJCV 72(2):195–215
8. Gil D, Radeva P (2003) Curvature vector flow to assure convergent deformable models for shape modelling. In: EMMCVPR, pp 357–372
9. Haddon J, Boyce J (1990) Image segmentation by unifying region and boundary information. IEEE T-PAMI 12:929–948
10. Hao J, Li M (2007) A supervised bayesian method for cerebrovascular segmentation. WSEAS Trans Signal Process 3(12):487–495
11. Heidemann G (2005) The long-range saliency of edge- and corner-based saliency points. IEEE T-IP 14:1701–1706
12. Jalba A, Wilkinson M, Roerdink J (2004) CPM: a deformable model for shape recovery and segmentation based on charged particles. IEEE T-PAMI 26(10):1320–1335
13. Li C, Kao C, Gore J, Ding Z (2007) Implicit active contours driven by local binary fitting energy. In: CVPR, pp 1–7
14. Li C, Liu J, Fox M (2005) Segmentation of edge preserving gradient vector flow: an approach toward automatically initializing and splitting of snakes. In: CVPR, pp 162–167
15. Maintz JB, Viergever MA (1998) A survey of medical image registration. Med Image Anal 2(1):1–36
16. Paragios N, Deriche R (2002) Geodesic active regions and level set methods for supervised texture segmentation. IJCV 46(3):223–247
17. Paragios N, Mellina-Gottardo O, Ramesh V (2004) Gradient vector flow geometric active contours. IEEE T-PAMI 26(3):402–407
18. Park HK, Chung MJ (2002) External force of snake: virtual electric field. Electron Lett 38(24):1500–1502
19. Ruan S, Jaggi C, Xue J, Fadili J, Bloyet D (2000) Brain tissue classification of magnetic resonance images using partial volume modeling. IEEE T-MI 19(12):1179–1187
20. Smith CM, Smith J, Williams SK, Rodriguez JJ, Hoying JB (2007) Automatic thresholding of three-dimensional microvascular structures from confocal microscopy images. J Microsc 225(3):244–257
21. Wu J, Ye F, Ma J, Sun X, Xu J, Cui Z (2008) The segmentation and visualization of human organs based on adaptive region growing method. In: International conference on computer and information technology, pp 439–443
22. Xie X (2010) Active contouring based on gradient vector interaction and constrained level set diffusion. IEEE T-IP 19(1):154–164
23. Xie X, Mirmehdi M (2004) RAGS: region-aided geometric snake. IEEE T-IP 13(5):640–652
24. Xie X, Mirmehdi M (2008) MAC: magnetostatic active contour. IEEE T-PAMI 30(4):632–646
25. Xu C, Prince J (1998) Snakes, shapes, & gradient vector flow. IEEE T-IP 7(3):359–369
26. Yang R, Mirmehdi M, Xie X (2006) A charged active contour based on electrostatics. In: ACIVS, pp 173–184
27. Yeo S, Xie X, Sazonov I, Nithiarasu P (2011) Geometrically induced force interaction for three-dimensional deformable models. IEEE T-IP 20(5):1373–1387
28. Zhu G, Zhang S, Zeng Q, Wang C (2008) Anisotropic virtual electric field for active contours. Pattern Recognit Lett 29(11):1659–1666

A Fast Geometric Deformation Method to Adapt a Foot to a Platform

J. M. Buades, M. González-Hidalgo, Francisco J. Perales, S. Ramis-Guarinos, A. Oliver and E. Montiel

Abstract The main goal of this research work is to develop a new system that designs shoes that adapts exactly to the foot shape. This research is based on a biomechanical anatomical structure of the foot and of the deformable shape. The system automatically selects significant foot points. We consider several anthropometrical parts of the foot in order to apply a global deformation with different axis. Also an interpolation process is designed to combine the several parts of the foot in an efficient and accurate manner. We consider different criteria in the deformation process because the top is rigid and the sole is assumed non-rigid. The system is implemented in an optimized software version in order to control the computational cost of the deformation process. A prototype of oriented commercial Application Programming Interface (API) is developed for non specialized users of the system. The results presented evaluate

J. M. Buades (✉) · M. González-Hidalgo · F. J. Perales · S. Ramis-Guarinos · A. Oliver
Computer Graphics, Vision and Artificial Intelligence Group, Department of Mathematics and Computer Science, University of Balearic Island, Anselm Turmeda Building, Crta. Valldemossa Km 7.5, 07122 Palma de Mallorca, Spain
e-mail: josemaria.buades@uib.es

M. González-Hidalgo
e-mail: manuel.gonzalez@uib.es

F. J. Perales
e-mail: paco.perales@uib.es

S. Ramis-Guarinos
e-mail: silvi.ragu@gmail.com

A. Oliver
e-mail: antoni.oliver.tomas@gmail.com

E. Montiel
INESCOP, Footwear Technological Institute, Polig. Ind. Campo Alto, C/. Alemania 102, Aptdo. Correos 253, 03600 Elda-Alicante, Spain
e-mail: emontiel@inescop.es

M. González Hidalgo et al. (eds.), *Deformation Models*, Lecture Notes in Computational Vision and Biomechanics 7, DOI: 10.1007/978-94-007-5446-1_5,

the error between deformations and we validate the error of several users (foot and last). Also the low error obtained guarantees the comfort of the foot that is a very important objective in this area of research.

1 Introduction

The footwear manufacturers from Europe, and particularly from Spain, need to bet on high performance and added value products, since they cannot compete in price against the low cost producing countries invading the markets. Personalization is an ideal strategy: It offers a high added value and links the client to the enterprise. Once the feet of the client are scanned and registered, the client could demand his personalized shoes from the catalogue or Internet. The system presented in this paper completes and follows one of the research and development lines which is being carried out at the INESCOP for several years, and is a clear example of collaboration between industry and university research group. Until now this project (European ERGOSHOE project: http://www.ergoshoe.inescop.es; also CEC-made shoe project: http://www.cec-made-shoe.com) led to the development of a low cost 3D foot scanner, a high precision 3D last scanner and software tools (Forma-3D and INFOHORMA) for last design which also allows to superimpose and compare different volumes (e.g. foot and last). These ideas will be addressed in Sect. 2. However there are still unsolved issues. With the general aim in mind of improving the automatic process of personalizing footwear. This paper address the issue of deforming the foot in order to predict his geometry once it is placed upon an elevated sole profile (e.g. high heels), considering the scan performed of the feet and the profile of the sole last. We need to be able to use the foot scan of the foot for multiple lasts, and most of the lasts have a certain amount of heel rise. The elevated sole profile is determined by the last.

The idea of "measuring" the shoe fit to the foot is not new. Lowe in 1927 obtained a patent for a fluoroscope intended to check the fitting of the shoes visually [11]. At that time the radiation was not taken very seriously and the idea was quite successful. Many shoe stores offered this service in Europe until the 1950s. But these fluoroscopes were removed in 1953. Letting radiation aside, this method was limited to observe the fit with the feet in an existing shoe. However, having the right technology, there has not been great progress since then. The process of digitizing a foot and changing the last shape from its numerical representation is something that is already being done at the theoretical and practical level [13, 15–17]. There have been some advances in foot volume parametrization from 3D coordinates of discrete points [12–14], and numerous mathematical models have appeared to describe the foot motion (see example [1, 2], although it does not offers any information about the volume of the foot). There are models that describe the shape of the foot, but these tend to be static [12]. Some of them describe the deformation of the foot depending on the load [5] and there are also attempts to record real-time full volume during walking (CEC made shoe project). But, in our concern, we have not found articles performing deformation models of the foot for applications with time restrictions.

Foot deformation on a high level position, as we can see in Fig. 6, is not so simple. Techniques such as free form deformation (FFD) (see [17]) suggest a technique of spatial deformation able to deform the scanned foot points. This technique associates some key points of the 3D model with the control points. The manipulation of these control points generates the deformation of the model. Techniques such as warping to the silhouette of the shoe last will not produce satisfactory results for such cases. Using an analytical model of foot bones that will lead to surface deformation could produce useful results [2], but has a high consuming time. Leon Kos et al. [8] propose that the scanned foot in the flat position should be matched with a similar foot from the database to obtain land mark similarities for the fitness analysis. The surfaces of the foot and the shoe last are scanned in 3D. Then they apply a series of deformation algorithms to evaluate the correspondence between the volumes. So you can have a shoe lasts database with which to compare foot for the analysis of the fitting of the shoes. Tang and Hui [22] presents an approach for modelling human foot tendons and determines their influence on the skin layer deformation. An anatomical foot model including skin, muscle, tendon and skeleton layers is adopted. Experimental results [22] showed that the axial deformation technique can model deformation of the foot tendons with satisfactory visual realism. In [23] they proposed an approach based on boundary element method (BEM) to model foot deformation and present its application in simulating interaction with footwear towards footwear design. Witana et al. in [27] developed a study on the feet of 16 participants showing that the interaction between the plantar surface of the foot and the load-bearing surface contributes to foot and surface deformations and hence to perceived comfort, discomfort or pain. Foot shape deformations were quantified using 3D laser scans. This study can have implications for the design and material selection of orthotics, insoles and footwear, and justify the importance of a suitable adjustment of foot to shoe last. Wang [26] develop a process for identifying the most suitable shoe last in a database for the human feet. Fuzzy techniques followed by an analytical hierarchy process (AHP) was applied to find the weighting factors for each girth to determine the fitness function among the shoe lasts and the feet.

Recently, Rupérez [19] (see also [20, 21]) conduce a similar research that in [23], but using the finite element method (FEM) instead BEM, to develop a software application for the simulation of the footwear deformation in gait. Artificial Neural Networks are successfully applied to predict the force exerted by a sphere that, simulating a bone, pushes a shoe upper material sample. Distribution of contact forces in the footwear is used for comfort analysis. The finite element method is applied to determine the pressure at the upper part of the shoe in a gait motion. In order to deform the foot model, the movement of 20 landmarks captured from a camera, is proposed to determine the foot motion. The deformation is calculated by minimizing the sum of the square of the distance between the landmarks. Alternatively, the foot model can be deformed with a geometric approach with 20 landmarks. On the other hand, creating foot deformation of different subjects requires an initial 3D foot model and captured motions of the landmarks of each subject. The method is time expensive

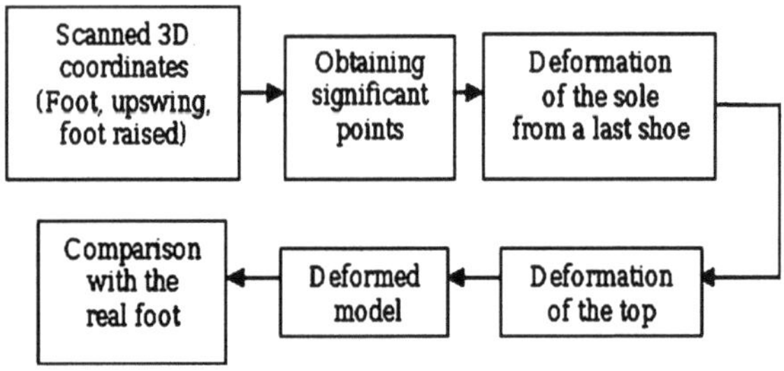

Fig. 1 Description of the deformation virtual process

for simulating foot motions and it is a drawback of the methods based on FEM or BEM, and are not suitable for applications with time constraints.

In [18] Rout et al. discuss the basic concepts and current methods being followed to convert foot to shoe-last, retrieve the best fitting shoe last based on the 3D foot scan of the customer, and to obtain customized shoe last. Leng et al. [9] propose to use a distance map to indicate how well the selected shoe last fits the specific foot shape and to guide the deformation, in this case, the idea is to deform existing last model to fit a consumer's foot shape with minimally affecting the original last style, this process is achieved by minimizing an equation. Only results using men's shoes with low profile are shown. As we can see a typical idea is to select from an existing shoe lasts database or deform an existing shoe last model into one that fits the scanned foot data (see [3, 7, 10, 15–17]).

In this paper we propose a system to predict and study the deformation of a consumer's foot when it adapts to different lasts, for in a subsequent process, design the best last that fits its foot.

The system uses as input a scanned 3D model of the foot obtained through the INESCOP 3D scanner (see [25] for other 3D surface scanning) and a model of the last bottom line. The foot is scanned over a flat plane and its orientation is irrelevant since the system is designed to automatically place the foot so it matches up with the sole last orientation. The process described in this paper is shown in Fig. 1. Moreover, the computations for each vertex of the 3D foot model are independently, therefore the whole process is easily parallelizable. This fact is important because you can compare several sole lasts with foot deformation. Therefore the computation time decreases.

The paper is organized as follows. Section 2 goes into the ideas of the customization of footwear identified at the beginning of the introduction. Section 3 is devoted to the description of the deformation process of the foot. Section 4 shows the interface created. The analysis of the results is carried out in Sect. 5. Finally, the paper ends with the conclusions and future work.

Table 1 Commercial tools in advanced shoe design

Company	Personalization of shoe	3D scan foot	Pressure platform	Comfort	Foot deformation
Nike	Yes	No	No	No	No
Todo para sus Pies	Yes	Yes	Yes	Yes	No
Shoemaster	Yes	Yes	Yes	Yes	No
Romans CAD	Yes	No	No	Yes	No
Crispin-Powershape	Yes	Yes	No	Yes	No
UIB-INESCOP	Yes	Yes	Yes	Yes	Yes[a]

[a] This deformation process is based on 3D scanned plane foot. So we can deformate the foot virtually for different last and compare

2 A General Framework in Shoes Customization

Actually the shoe last models are made with traditional methods with the help of CAD/CAM systems to design some specific features and to mill it. The shoe last is produced manually by expert's craftsmen. The shape of the shoe last is often arbitrary and it is based on expert's experience, shoe's design and fashion trends. The materials used in the fabrication as well as the type of machines, also influence the final product. Although the integration of technologies allows a rapid fabrication of prototypes, the craftsman still must retouch the final product. There are several initiatives that try to automate the process to create a shoe last. This implementation proposes new methods for development and research in this area.

From a commercial point view, the case NikeId[1] is a Nike service that allows customizing footwear. The customer can design the product. The user can choose the materials and colors. The system can choose 31 parts of the Nike's shoe for the personalization, 82 materials and options. This service is growing, developing new applications to give a better on-line service. Although it has many advantages, the system only changes the foot's appearance. It does not enter in the comfort of the shoe. Similar initiative have been set up by other brands such ADIDAS (MyADIDAS), MUNICH (Myway) and others. More technical methods are used by "The Spanish company Todo Para sus Pies S.L."[2], that has a customization system for footwear and insoles. It serves patients with diabetic foot or vascular problems. So, the customer gets the shoe model that better adapts him.

Table 1 shows a non-exhaustive comparative between software tools of several companies engaged in this sector.

So we see a need to improve the processes of creation and design of the shoe at all levels and possible applications (medical, sports and general public). Following the conventional criteria to improve the quality in the production process of shoes,

[1] http://nikeid.nike.com/nikeid/index.jsp

[2] http://www.todoparasuspies.es

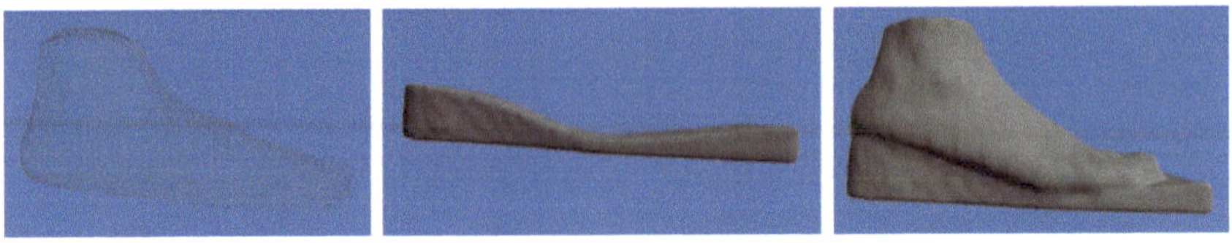

Fig. 2 Basic elements needed for simulation of a foot deformation. *Left* cloud of points of the scanned foot. *Center* scanned platform shoe. *Right* scanned of the foot and platform together

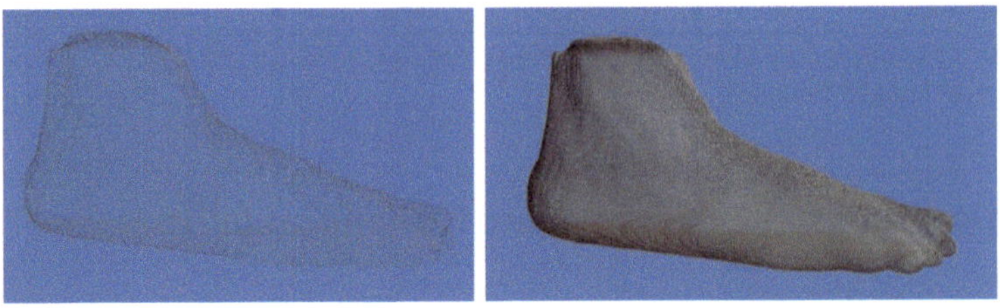

Fig. 3 Initial human foot. *Left* cloud of points. *Right* polygon mesh obtained by the DIGIPIE 3D

comfort is considered one of the most important. Our deformation process combines the features necessary to obtain an appropriate design to the biomechanical needs of each individual foot. This process is very fast and guarantees the comfort.

3 Deformation and Methodology

In this section, we describe the method used for the foot deformation. Through a 3D foot capture system we scan three basic elements: the foot, the platform shoe and both together. In Fig. 2 we show these basic elements. At left we can found the scanned points cloud corresponding to a consumer's foot, at center the scanned platform shoe, and at right the image of both together. Note that the goal is to deform the scanned foot and put it in the position determined by the platform shoe. The platform is used to obtain the line of deformation (see the red line of Fig. 6). To obtain it we cut the platform by a perpendicular plane and the points on the platform that intersect with this plane form the deformation line. The foot on the platform (see right part in Fig. 2) is necessary for the comparison between the deformed foot simulation and the real deformed foot in order to estimate the error and validate the method. So, we will obtain the final result.

In order to proceed to deform a foot, an initial foot model and a line deformation are required. The initial foot model is created using the DIGIPIE 3D, the INESCOP low cost foot scanner.[3] The scanned foot is represented by a cloud of points that is

[3] The DIGIPIE 3D and the high precision last scanner are developed by the INESCOP in the ERGOSHOE European project: http://ergoshoe.inescop.es/project.htm.

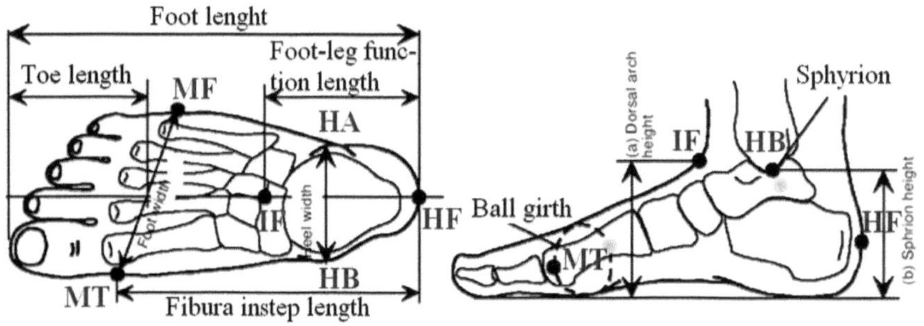

Fig. 4 Foot dimensions

model by a surface polygon mesh as shown in Fig. 3. The surface mesh of the foot model is composed of triangular elements with N vertices, and we have determined the 3D coordinates of each of them. The shoe platform model has been obtained using the INESCOP high precision scanner and the line deformation is modeled by a B-spline curve. The scanned foot and platform model are usually without noisy and are appropriate for conducting graphics applications.

In this paper, deformation of the foot model to adapt it to a platform is based on geometrical considerations. The proposed algorithm is divided in five steps:

1. Significant foot points.
2. Foot areas division.
3. Sole deformation.
4. Top deformation.
5. Fusion of both deformations.

In the following subsections will describe each of these steps.

3.1 Significant Foot Points

To obtain an accurate deformation we need to calculate several significant foot points. The most important points are IH, IF, MT, MF, HA and HB (see Fig. 4). The length of the foot is the x-coordinate, the width is the y-coordinate and the height is the z-coordinate. HF is the point with lowest x-coordinate and IF is the point with the lowest distance between the point HF and all the points of the top foot border line.

Now we divide the length of the foot in two parts: forefoot (from the middle of the foot to the toes) and hindfoot (from the middle of the foot to the heel). In the forefoot we find the points MF and MT. MF is the point with the lowest y-coordinate and MT is the point with the highest y-coordinate. We do the same in the hindfoot where HA is the point with the lowest y-coordinate and HB is the point with the highest y-coordinate.

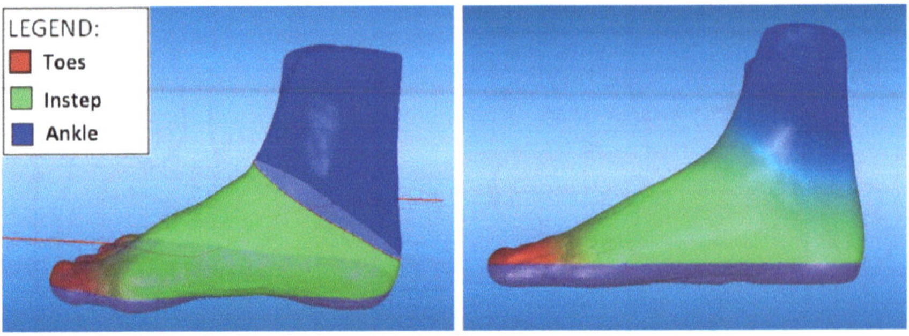

Fig. 5 *Left* Planes of the foot. *Right* areas in which we divide the foot

With the points MT and MF we create the plane that divides the toes from the instep and with the points IF and HF we create the plane that divides the instep from the ankle (see Fig. 5, left).

3.2 Foot Areas Division

On the basis of considerations made by footwear designers, we divide the foot into different areas which are affected differently by the deformation of the foot to adapt to the platform. We separate the foot into two principal parts (see Fig. 5, right): the sole (purple color) and the top that is divided in three areas (red, blue and green areas). The left side of the Fig. 5 shows the planes that define the different parts which we divides the foot. Recall that these planes are: the plane π_z of z-coordinate constant and containing the point MT, the plane π_{M} determined by the line joining the point MF with MT and, the plane π_{H} determined by the line connecting the points IF and HF. Each part will suffer a different type of deformation according to their characteristics.

- **The Sole**: The sole is composed of all the points below the plane π_z (Fig. 4).
- **The Top**: The top is the part of the foot that does not belong to the sole. This part is divided in three parts: the toes, the instep and the ankle, as we can see in Fig. 4. Region "toes" consists of all the points between the fingertips and the plane π_{M}. Region "instep" is composed of all points between the planes π_{M} and π_{H}. Finally, region "ankle" consists of all points above the plane π_{H}.

3.3 Sole Deformation

The deformation of the sole is the main step to obtain the deformation of the foot adapted to the platform. We must adapt with precision the sole to the deformation

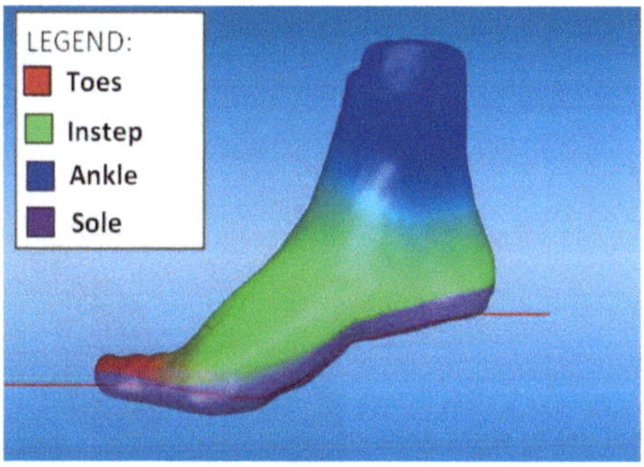

Fig. 6 Deformed foot and the final deformation of the sole in purple

line. In order to do that, we obtain the normal vector of every point of the line. This is necessary to correctly locate points on the deformed sole. The sole deformation varies according to deformation line. In Fig. 6, we show an example of the final deformation of the sole.

Sole of the foot is adapted to the deformation line starting from the heel. The process advances from the heel to the tiptoe, fitting the vertexes of the foot to the line deformation and taking into account the normal direction to the deformation line. The vertices to deform are at a distance d of the heel and height h. To obtain the deformed vertexes, the process goes over the deformation line until reach distance d, and moves it in normal direction h units until gets the exact deformed vertex.

3.4 Top Deformation

After the deformation of the sole must be addressed the deformation of the remaining regions of the foot. We must take into account that the deformation of these regions are subject to the restriction of the positions determined in the previous step for the points of the sole. Recall that divide the top of the foot in three parts: the toes area, the instep area and the ankle area. Each area is separated by a joint and it has a different rotation angle. The two principal joints are the metatarsal joint and the ankle joint.

- **Metatarsal Joint**: The metatarsal bones are responsible of the toes movement. This joint is essential in the virtual deformation because it allows the correct rotation of the forefoot. Its importance is due to is the most influential in the final shape of the foot. The metatarsal joint is obtained from the most prominent points of the forefoot (MT and MF). The axis of rotation passes through these two points (see Figs. 4 and 7).

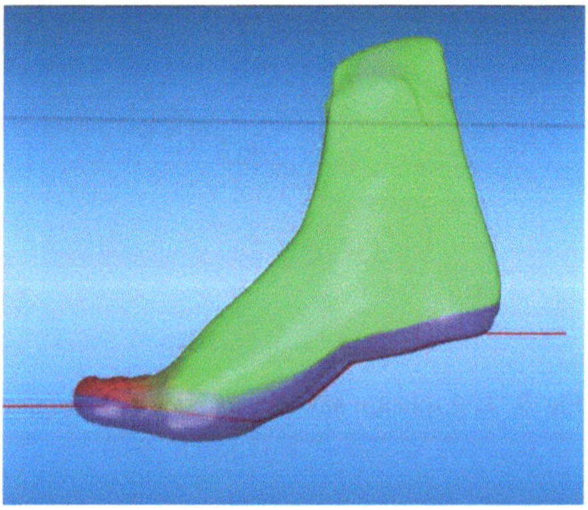

Fig. 10 Deformed foot without rotated ankle

As first solution we apply rotation to vertexes taking into account the area it belongs. The upper vertexes are catalogued as toes, instep or ankle. First we only rotate the toes and the instep areas (see Fig. 10). After that we deform the foot rotating the ankle (see Fig. 11). This solution produces discontinuities (see Figs. 11 and 12, middle image). To avoid this discontinuity we define a interpolation area. In this area the vertexes are rotated taking into account its position to get smoother deformation of upper vertexes (see Figs. 11 and 12, right image). To smooth the borders in the union of different areas of the foot we apply a interpolation and we obtain satisfactory results using two planes. We can appreciate the difference between the results obtained using one plane (see Fig. 11, right) or two planes (see Fig. 12, right) in order to compute the ankle area. Next section explains in detail the interpolation areas.

3.4.2 Interpolation Areas

The interpolation process is similar to the used by Kavan et al. in [6]. Then we need to delimit the interpolation zones. The following describes as we compute the interpolation zone for the union of the instep area with the ankle area. A similar process is carried out for the interpolation in the joining zone of toes area with instep area.

Let S be a surface in the original foot consisting of points belonging to the intersection of two planes with the foot, as we can see in Fig. 13. The the first plane π_1 is the plane parallel to the π_{H} plane (see Sect. 3.2, left) and passing through point HB. The second plane π_2 is the plane done by $z = \mathrm{HB}_z$. The interpolation zone is delimited by two surfaces obtained by the translation of S in the z-axis by two scalars t_{min} and t_{max}. Where $t_{min} = -\alpha \cdot c$ and $t_{max} = \alpha \cdot \tilde{c}$ and c, $\tilde{c}$ should be determined. Let S_m and S_M be the translated surfaces (see Fig. 13, right), the interpolation parameter t is done by

$$t = \frac{t'}{t_{max} - t_{min}}, \tag{5}$$

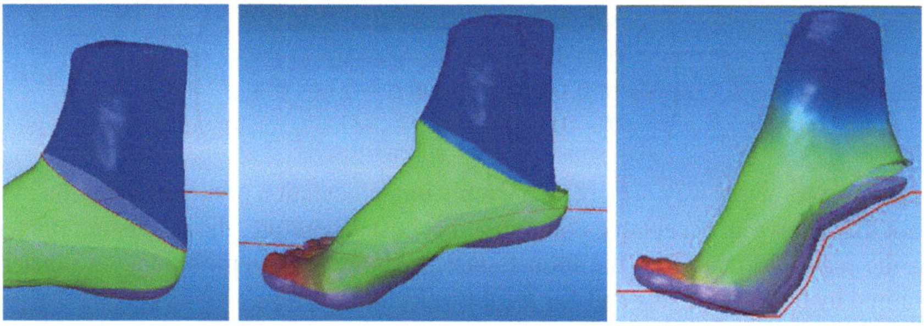

Fig. 11 *Left* unrotated ankle (one plane). *Middle* rotated ankle (one plane). *Right* interpolation with one plane

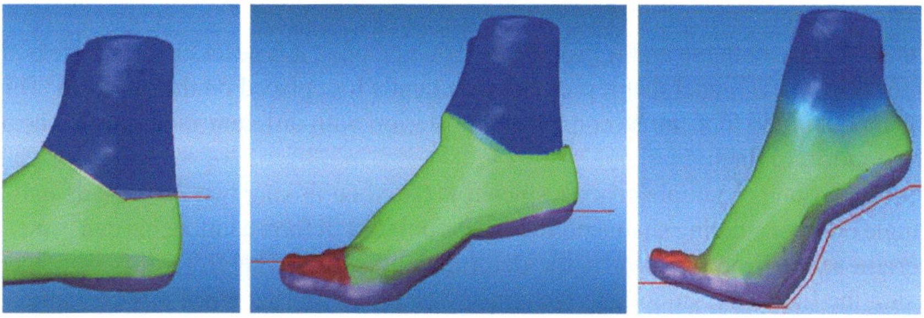

Fig. 12 *Left* unrotated ankle (two planes). *Middle* rotated ankle (two planes). *Right* interpolation with two planes

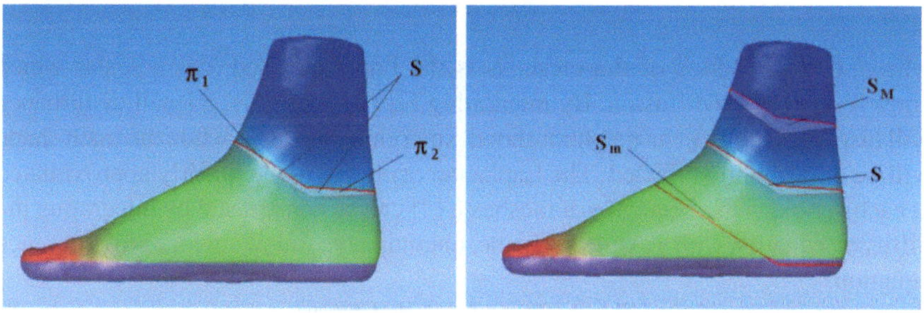

Fig. 13 Interpolation areas

where

$$t' = d(P, Proj_z(P, S_{min})),$$

d is the euclidean distance, P is a point on the interpolation zone and $Proj_z$ is the projection on z-axis direction of P in S_{min}.

Then, following [6] we apply an interpolation of angle β when we rotate all vertex of this zone. Where $\beta = \alpha \cdot t$ and α is the angle calculated for the ankle area. That is,

$$M = \begin{pmatrix} \cos\beta & 0 & \sin\beta & -x\cos\beta - z\sin\beta + x \\ 0 & 1 & 0 & 0 \\ -\sin\beta & 0 & \cos\beta & x\sin\beta - z\cos\beta + z \\ 0 & 0 & 0 & 1 \end{pmatrix}. \tag{6}$$

Observed in this way, we obtain a simple implementation of the deformation of a foot that allows an acceptable computational cost. Therefore, the results are tighter to the real deformation of the foot. Figure 13 shows the interpolation zone that is limited by the surfaces S_m and S_M for the ankle area.

3.5 Fusion of Both Deformations

Once we have both top and sole of the deformed model, it is necessary to merge both into a single 3D model that represents the client's foot placed on the platform. This 3D model of the foot can be compared a posteriori with different shoe lasts to check the comfort of them.

The fusion is a simple process since it is based on linking two 3D models into a single one, considering shared points. These points will form a union that can present certain anomalies produced by the deformation process. Therefore, it is smoothed using the Laplacian smoothing technique (see [4, 24]). This process is also known as diffusion. The equation of Laplacian smoothing is,

$$\frac{\partial X}{\partial t} = \lambda L(x), \tag{7}$$

where X is the vertex of the mesh, L is the Laplacian, and λ is a scalar which controls the rate of diffusion. By integrating Eq. (7) over time, a small disturbance will disperse rapidly in its neighbourhood, smoothing the high, while the main shape will be only slightly affected. The Laplacian operator can be linearly approximated at each vertex, then a sequence of meshes $\{X^n\}$ can be constructed by integrating the diffusion equation with a explicit scheme, obtaining the following forward difference equation,

$$X(n+1) = (I + \lambda\, dt\, L)X(n). \tag{8}$$

Obviously, the implementation of the previous explicit method, that have very nice properties such us linear time and linear memory size for each step (see [4]), is very straightforward. Unfortunately, if the mesh is large can be appear practical limitations on the time step, that can be avoid using an implicit scheme in order to integrate Eq. (7).

Note that the above process does not change the connectivity of the mesh. Each step changes the position of the vertices, but the mesh topology remains unchanged. The relaxation of a given vertex only requires information about its immediate neighbours. In Figs. 21 and 22 can be seen the final deformation obtained after the fusion process.

Fig. 14 Interface initial

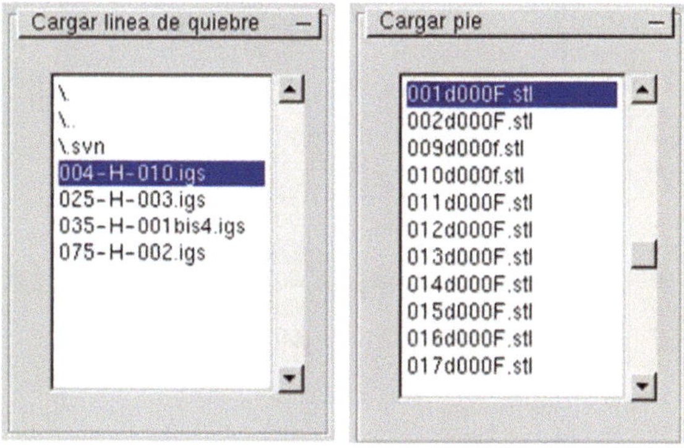

Fig. 15 Menu display. From *left* to *right* menu 1 and menu 2, respectively

4 Data Representation and Tool Description

As mentioned previously, the system needs two mandatory inputs for compute the position of a foot on a platform: the model of the foot and the deformation line corresponding to the selected platform. A third, optional input, concerning to the foot on the platform can be provided. This last is necessary if we want statistics of the errors. We have created a simple interface to load the three elements for the deformation. The initial interface is shown in Fig. 14.

The first component that is request by the program is the deformation line (see Fig. 15, menu 1). This line is stored in an igs file.

Once the deformation line is loaded, we will proceed to load the foot model (see Fig. 15, menu 2). The foot was previously scanned and stored as a polygon formed by tens of thousands of triangles in a stl file. Then, the program display the deformation line an the foot in its interface, and we can proceed with the deformation. To do this, we can click on the button generate deformation ("generar deformación"). The obtained result is the deformed foot over the deformation line (see Fig. 16).

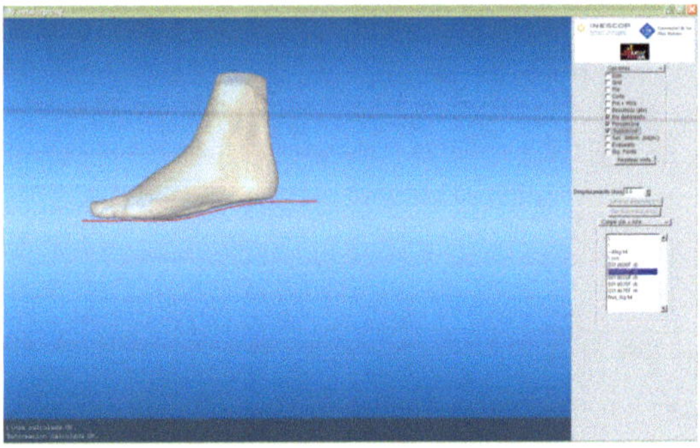

Fig. 16 Fina interface with the foot deformed

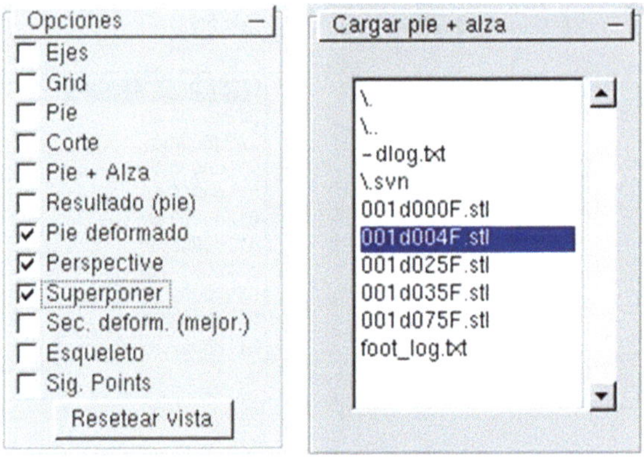

Fig. 17 Menu display. From *left* to *right* menu 1 and menu 2, respectively

Finally we can interact with both models easily (the foot without deforming and the deformed foot) with only a click:

- Left button: x-axis rotation (vertical moving) and z-axis rotation (horizontal moving).
- Middle button: z-axis translation (vertical moving) and x-axis translation (horizontal moving).
- Right button: y-axis translation.

Note that the y-axis is the depth and the z-axis is the height.

We also have several display options (see Fig. 17, menu 1):

- Ejes (axis): Displays the coordinate axis.
- Grid (grid): Displays a grid under the foot.
- Pie (foot): Displays the foot at rest.
- Corte (cut): Displays the sole at rest.
- Pie + alza: Displays the foot model placed on the platform shoe.
- Resultado (pie): Displays only the foot from the Pie + Alza model loaded.

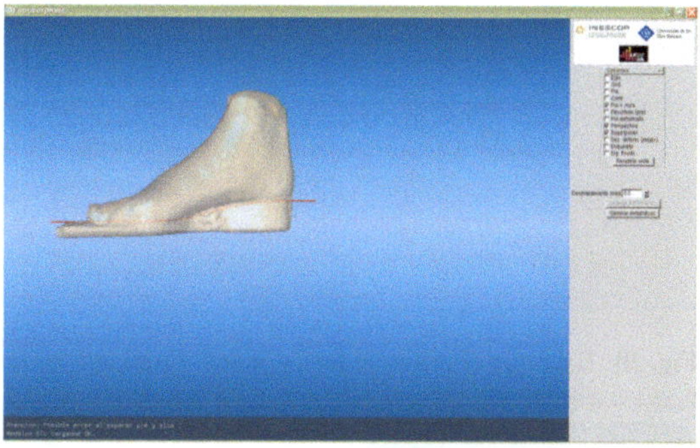

Fig. 18 The original foot with the platform shoe

Fig. 19 The colour map in the foot deformed

- Pie deformado (deformed foot): Displays the deformed foot.
- Perspective (perspective): Perspective or orthographic projection.
- Superponer (superimpose): Display an superimpose of the elements (deformation line, foot at rest and/or deformed foot).
- Sec. deform. (mejor.) (deformed section): Displays the deformed sole.
- Esqueleto (skeleton): Display of the deformed foot wireframed.
- Sig. Points (significant points): Displays some significant points and the legend of the areas in both foot models.

If we want to generate the numerical results associated with our approximation of foot position on the platform, we must load the model for the foot on the platform, see menu 2 in Fig. 17, this loads the "pie + alza" (see Fig. 18). Obviously, this model should be correspond to the platform from which we have loaded the deformation line.

Once the foot on the platform has been loaded we can generate statistics and calculate the errors in the approximation by clicking on the button "Generar estadísticas" (generate statistics). Thus, the program generates a color map in the foot deformed, based on the error on each face. The color changes depending on the amount of error of each face respect to the corresponding face in the model of the foot on the platform (see Fig. 19).

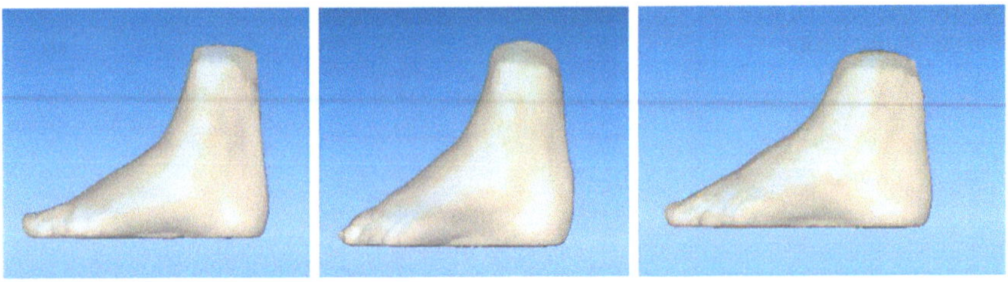

Fig. 20 Example of different feet scanned. From *left* to *right*: foot 1, foot 2 and foot 3

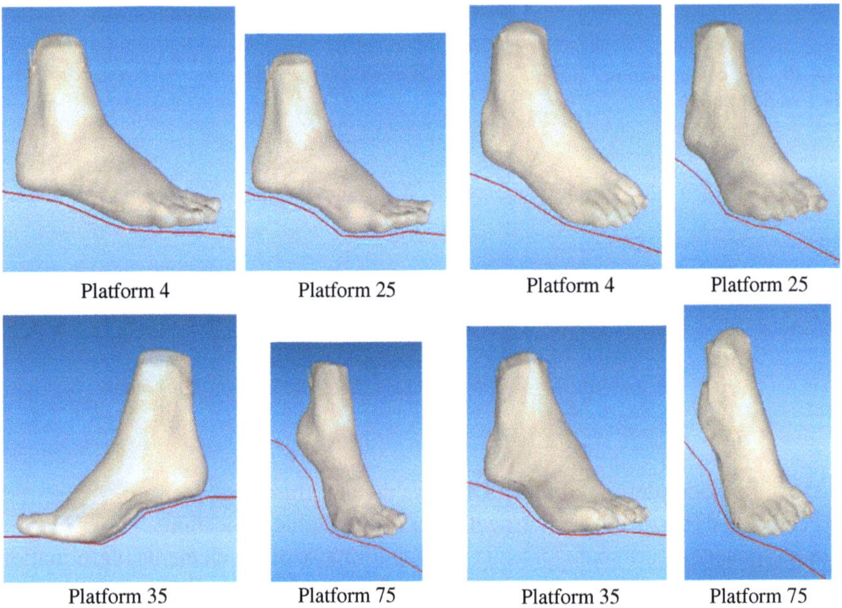

Fig. 21 From *left* to *right*: deformation and inclined platform adjustment of the foot 1 and foot 2 in Fig. 20. The deformation line is the *red line*

5 Results

In this section we describe the results obtained with the proposed system. To validate the results we need to perform a comprehensive study of the error between the deformed foot to adapt it to platform and the foot on the platform. Moreover, the results must be validated by footwear designers. In the first phase of the project we scanned nine women's feet with Europe size 37, considered a common size. Note that we scan the foot flat, the foot on the platform and the platform, and each foot for several platforms.

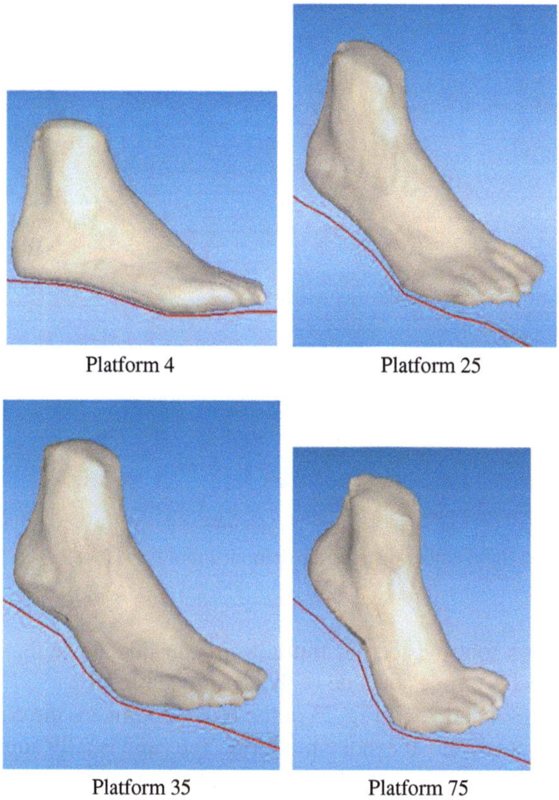

Fig. 22 Deformation and inclined platform adjustment of the foot 3 in Fig. 20. The deformation line is the *red line*

After obtaining the data (Fig. 20), it is filtered and processed with the deformation algorithm. The different deformations are shown in Figs. 21 and 22 with platform of 4, 25, 35, and 75 mm.

To estimate the error, the algorithm calculates the distances from deformed foot vertexes to the real foot. Real foot is computed from scanned foot over platform subtracting scanned platform. To get a more accurate error, distance is computed as Euclidean distance from vertexes to real foot in normal direction. The errors are shown in a color map, as we can see in Fig. 23, where green color means that the deformed foot is above the real model, red color the deformed foot is below the real model and grey the deformed foot coincides with the model.

Note that with Platform 4 of Fig. 23 of the foot 1, there are a lot of matches between the real model and the deformed model. It has more grey zones than zones in red and green. You can visually appreciate that this model has a lower error. With Platform 35 there are more red zones. That means that this person twisted her foot towards the inside above the platform. With Platform 75a of Fig. 23 there are more green zones.

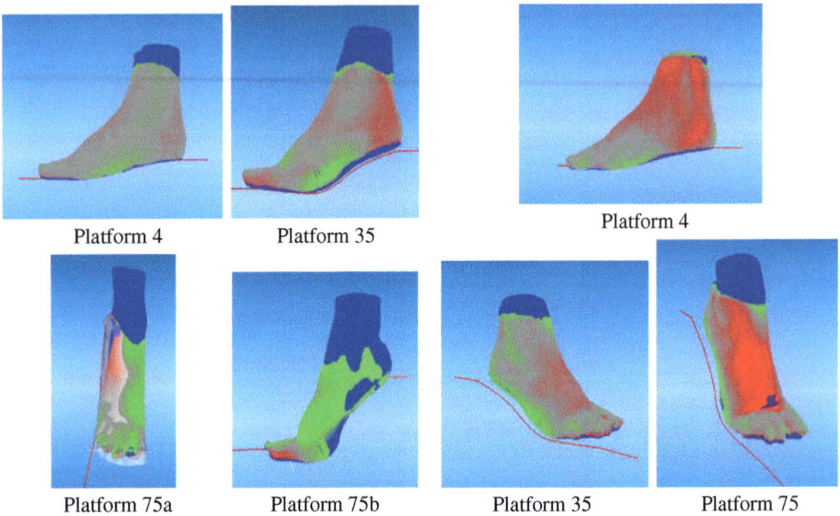

Fig. 23 From *left* to *right*: colour maps of the deformed foot 1 and foot 9

The woman of this foot twisted her foot towards outside above the platform. Note that the real model protrudes outside of the deformed model.

In the Platforms 4 and 75 of Fig. 23 of the foot 9 dominates the color red. So this person put her foot twisted above the platform. You can visually appreciate that she twisted her foot forwards the outside above the platform of 75 mm and inside above the platform of 4 mm. With Platform 35 the woman put her foot straighter than the others. Therefore, there is more grey zones in this figure.

We effect a comparison of the deformed foot and the real foot with the same rise. Comparing the real foot to the deformed foot with the same rise we gather the average errors of each foot with each platform in Table 2. Because of the foot positions during the scanning process some errors may seem high. In the Table 3 we show the total average error. We observe that the greater is the rise of the platform, more possibilities that the person twists the foot. In the Table 3 we show the total average error. We observe that the greater is the rise of the platform, more possibilities that the person twists the foot.

Finally the results were supervised by Footwear experts. They concluded that the results are more precise than initially estimated.

6 Conclusions and Future Work

In this research project, we have achieved a virtual deformation of a consumer's foot and adapt it to a previously selected platform. This way we will be able to determine the shoe that best fits the customer's foot. With the final goal in mind of a global

Table 2 Error

	Platform 04 (mm)	Platform 35 (mm)	Platform 75 (mm)
Foot 1	1.23059	2.13966	4.96364
Foot 2	4.14564	4.46409	4.18545
Foot 3	2.24506	4.07791	5.29221
Foot 4	2.73462	4.54069	4.00599
Foot 5	1.60728	3.39996	4.28592
Foot 6	2.06498	4.50866	5.28577
Foot 7	1.92616	2.79563	3.60909
Foot 8	2.82117	4.6632	4.54722
Foot 9	3.07567	2.25532	3.36666

Table 3 Average error

	Platform 04 (mm)	Platform 35 (mm)	Platform 75 (mm)
Maximum average error	4.14564	4.6632	5.29221
Minimum average error	1.23059	2.13966	3.36666
Total average error	2.688115	3.40143	4.329435

process of adaptation of the shoes to the feet of a customer, the speed of the process is essential. For this reason, we adopt in this simulation a deformation model based on geometry rather than on physics.

Up to our knowledge, the solution proposed is new because it includes a foot deformation procedure with conform criteria and a 3D scan foot that generates the personalization of shoe. We sacrifice physical realism for speed and efficiency but we guarantee the correctness of the deformation done. A client can't wait a long time to find out what type of shoe suits his/her feet better.

The system designed has a low computational cost and a low error rate. The error rate has been supervised by experts in footwear design and it is considered acceptable to guarantee the comfort of the foot. The control points needed to perform the deformation are detected automatically. A pleasant user interface, easy to use has been created. Using the system proposed we can obtain a better fit of the footwear to the customer's foot.

We have seen that the tests are satisfactory and that the system is easily parallelizable. In the section results a numerical and quantitative comparison is shown. However a more comprehensive study of the error will be performed and we expect to evolve the deformation to make it more realistic in the critical points. We will also extend the system proposed to special pathological feet that require a more specific deformations criterion.

Acknowledgments This work is subsidized by the national project DPI2009-14738-C02-01 of the MICIIN Spanish Government coordinated by INESCOP (Asoc. investigacin industrias del calzado y conexas) and developed in collaboration with UIB (University of Balearic Islands).

References

1. Arampatzis A, Brggemann GP, Klapsing GM (2002) A three-dimensional shank-foot model to determine the foot motion during landings. Med Sci Sports Exerc 34(1):130–138
2. Carson MC, Harrington ME, Thompson N, O'Connor JJ, Theologis TN (2001) Kinematic analysis of a multi-segment foot model for research and clinical applications: a repeatability analysis. J Biomech 34(10):1299–1307
3. Cheng FT, Peng DB (1999) A systematic approach for developing a foot size information system for shoe last design. Int J Ind Ergonomics 25(2):171–185. doi:10.1016/S0169-8141(98)00098-5
4. Desbrun M, Meyer M, Schröder P, Barr AH (1999) Implicit fairing of irregular meshes using diffusion and curvature flow. In: Proceedings of the 26th annual conference on computer graphics and interactive techniques, SIGGRAPH'99, pp 317–324. ACM Press/Addison-Wesley Publishing Co, New York. doi:10.1145/311535.311576
5. Houston VL, Luo G, Mason CP, Mussman M, Garbarini M, Beattie AC (2006) Changes in male foot shape and size with weightbearing. J Am Podiatr Med Assoc 96(4):330–343
6. Kavan L, Zara J (2003) Real-time skin deformation with bones blending. In: WSCG short papers proceedings
7. Kim SY, Lee K, Hwang T (2002) A grouping algorithm for custom-tailored products. J Mater Process Technol 130–131:618–625
8. Kos L, Duhovnik J (2002) A system for footwear fitting analysis. In: Proceedings of the international design conference design, pp 1187–1192
9. Leng J, Du R (2005) A deformation method for shoe last customization. Comput Aided Des Appl 2(1–4):11–18
10. Li G, Joneja A (2004) A morphing-based surface blending operator for footwear CAD. In: ASME-IMECE conference proceedings, pp 269–273. doi:10.1115/IMECE2004-60719
11. Lowe J (1927) Method and means for visually determining the fit of footwear. U.S. Pantent Publication. Washington DC. Publication No. US 1614988 A. Application No. US 275310 A
12. Luximon A, Goonetilleke RS (2004) Foot shape modeling. Hum Factors 46(2):304–315
13. Luximon A, Goonetilleke RR, Tsui KL (2003) Foot landmarking for footwear customization. Ergonomics 46(4):364–383
14. Luximon A, Goonetilleke RR, Tsui KL (2003) Footwear fit categorization. In: Tseng M, Piller F (eds) The customer centric enterprise: advances in mass customization and personalization, chap. 28. Springer, Heidelberg, pp 491–499
15. Luximon A, Goonetilleke RS, Tsui KL (2005) Foot landmarking for footwear customization. Comput Aided Des Appl 2(1):11–18
16. Luximon A, Goonetilleke R, Zhang M (2005) 3D foot shape generation from 2D information. Ergonomics 48(6):625–641
17. Mochimaru M, Kouchi M, Dohi M (2000) Analysis of 3-D human foot forms using the free form deformation method and its application in grading shoe lasts. Ergonomics 43(9):1301–1313
18. Rout N, Zhang YF, Khandual A, Luximon A (2010) 3D foot scan to custom shoe last. Int J Comput Commun Technol 1(2–3-4):14–18. Special Issue International Conference [ACCTA-2010]
19. Rupérez MJ (2011) Multidisciplinary techniques for the simulation of the contact between the foot and the shoe upper in gait: virtual reality, computational biomechanics, and artificial neural networks. Ph.D. thesis, Departamento de Ingienería Mecánica y Materiales, Universidad de Valencia
20. Rupérez MJ, Monserrat C, Alcañiz M (2008) Simulation of the deformation of materials in shoe uppers in gait. Force distribution using finite elements. Int J Interact Des Manuf 2:59–68
21. Rupérez MJ, Monserrat C, Alemany S, Juan M, Alcañiz M (2010) Contact model, fit process and, foot animation for the virtual simulator of the footwear comfort. Comput Aided Des 42(5):425–431
22. Tang Y, Hui K (2007) The effect of tendons on foot skin deformation. Comput Aided Des 39(7):583–597. doi:10.1016/j.cad.2007.01.013

23. Tang Y, Hui K (2011) Human foot modeling towards footwear design. Comput Aided Des 43(12):1841–1848. doi:10.1016/j.cad.2011.08.005
24. Taubin G (1995) A signal processing approach to fair surface design. In: Proceedings of the 22nd annual conference on computer graphics and interactive techniques, SIGGRAPH '95, pp 351–358. ACM, New York. doi:10.1145/218380.218473
25. Telfer S, Woodburn J (2010) The use of 3D surface scanning for the measurement and assessment of the human foot. J Foot Ankle Res 3:19
26. Wang CS (2010) An analysis and evaluation of fitness for shoe lasts and human feet. Comput Ind 61(6):532–540. doi:10.1016/j.compind.2010.03.003
27. Witana CP, Goonetilleke RS, Xiong S, Au EY (2009) Effects of surface characteristics on the plantar shape of feet and subjects perceived sensations. Appl Ergonomics 40(2):267–279. doi:10.1016/j.apergo.2008.04.014

Frame-Based Interactive Simulation of Complex Deformable Objects

Benjamin Gilles, François Faure, Guillaume Bousquet and Dinesh K. Pai

Abstract We present a new type of deformable model which combines the realism of physically based continuum mechanics models and the usability of frame-based skinning methods, allowing the interactive simulation of objects with heterogeneous material properties and complex geometries. The degrees of freedom are coordinate frames. In contrast with traditional skinning, frame positions are not scripted but move in reaction to internal body forces. The deformation gradient and its derivatives are computed at each sample point of a deformed object and used in the equations of Lagrangian mechanics to achieve physical realism. We introduce novel material-aware shape functions in place of the traditional radial basis functions used in meshless frameworks, allowing coarse deformation functions to efficiently resolve non-uniform stiffnesses. Complex models can thus be simulated at high frame rates using a small number of control nodes.

1 Introduction

Deformable models are essential in mechanical engineering, biomechanics and computer graphics, typically for simulating the behavior of soft objects. The classical approach is *physically based deformation*, typically using continuum mechanics,

B. Gilles (✉)
INRIA, LIRMM-CNRS, University of Montpellier, Montpellier, France
e-mail: benjamin.gilles@inria.fr

F. Faure · G. Bousquet
INRIA, LJK-CNRS, University of Grenoble, Grenoble, France
e-mail: francois.faure@inria.fr

G. Bousquet
e-mail: guillaume.bousquet@inria.fr

D. K. Pai
University of British Columbia, Vancouver, BC, Canada
e-mail: pai@cs.ubc.ca

M. González Hidalgo et al. (eds.), *Deformation Models*, Lecture Notes in Computational Vision and Biomechanics 7, DOI: 10.1007/978-94-007-5446-1_6,

This has the significant advantage of physical realism. Complex deformations are generated by numerical integration of discretized differential equations. However, these methods can be expensive and difficult to use. In the popular Finite Element Method (FEM) framework, the degrees of freedom of the discretized model are the vertices of a mesh, which must be constructed for each simulation object. A relatively fine mesh (i.e., a dense sampling of the deformation field) is required to capture common deformations such as torsion, leading to expensive simulations. Mesh adaptation can be difficult due to the topological constraints of the mesh. Particle-based meshless methods have been proposed to address these problems. While they obviate the need to maintain mesh topology, particles can not be placed arbitrarily. Therefore, these methods also need a dense cloud of particles not very different from the vertices of an FEM mesh.

Another approach, from the computer graphics community, is *skinning* (also known as vertex blending or skeletal subspace deformation). The deformation is kinematically generated by manipulating "bones," i.e., specific coordinate frames. This method is widely used, not only for its simplicity and efficiency, but because it provides natural and intuitive handles for controlling deformation. Skinning generates smooth deformations using a very sparse sampling of the deformation field. Adaptation is simple since frames can be inserted easily to control local features. These interesting features have made it the most widely used method for character animation. However, as a consequence of its purely kinematic nature (i.e., the frame positions need to be scripted), achieving physically realistic dynamic deformation is a major challenge with this approach.

We present a new approach that combines the advantages of both physically based deformation and skinning [13]. Instead of the vertices of a mesh, the degrees of freedom are a sparse set of coordinate frames. The equations of motion are derived for the moving frames by applying the principles of continuum mechanics across the volume of the deformed object, and solved using classical implicit time integration.

In addition, we show that it is possible to simulate complex heterogeneous objects with sparse sampling using new, material-aware shape functions [10]. So far, most of the work has focused on objects made of a single, homogeneous material. However, many real-world objects, including biological structures, are composed of heterogeneous material. The simulation of such complex objects using the currently available techniques requires a high resolution spatial discretization to resolve the variations of material parameters. However, dense sampling creates numerical conditioning problems, especially in the case of stiff material. Shape functions are geometrically designed to achieve a certain degree of locality and smoothness, independent of the material. The resulting deformations are rather homogeneous between the nodes. Consequently, the realistic simulation of such complex objects has remained impossible in interactive applications. Our approach is based on a simple observation: points connected by stiff material move more similarly than connected by compliant material. Given a deformable object to simulate and a number of control nodes corresponding to an expected computation time, optimization criteria can be used to compute, at initialization time, a discretization of the object and the associated shape functions, in order to achieve a good realism.

Our specific contributions are the following: (1) a new approach which unifies skinning and physically based deformation modeling. (2) material-aware shape functions using a novel distance function based on compliance; (3) a method to automatically model a complex object for this method, with an arbitrary number of sampling frames, based on surface meshes or volumetric data; (4) a system that implements the above methods and shows the ability to simulate complex deformation with a small number of dynamic degrees of freedom. The remainder of this chapter is organized as follows. We first briefly review in Sect. 2 relevant previous work, which allows us to motivate and sketch our approach with respect to the existing ones. In Sect. 3, we present the kinematic discretization using frames, and the interpolation function based on *skinning*. In Sect. 4, we derive the differential equation which governs the dynamics of the object, investigate precision issues and propose a strategy to optimize spatial integration. We then study in Sect. 5 the problem of material-aware shape functions starting in one dimension and extending to two or three dimensions, and propose a method to optimize the distribution of nodes. We finally present results and discuss future work.

2 Related Work

Physically based deformable models have attracted continuous attention in Computer Graphics, since the seminal work of Terzopoulos [39]. We refer the reader to the excellent survey of [31] on this topic. Here, we briefly review the main Lagrangian models of deformable objects.

Mesh-based methods: Early works on deformable models in Computer Graphics have focused on interconnected particles. In mass-spring systems [35], constraints on edge length are enforced to counter stretching. Bending and shear can be controlled using additional springs. More general constraints such as area or volume conservation can be enforced using appropriate energy functions [40]. To realistically model volumetric deformable objects, it is necessary to apply continuum mechanics. The spatial derivatives of the displacement field can be computed using finite differences on a regular grid [39]. Terzopoulos and Qin [38] studied the case of physically deformable NURBS surfaces for shape modeling. Finite elements [7, 8, 14, 33] allow irregular meshes, which are generally more convenient to sample objects with arbitrary shapes, but may be poorly conditioned. The spatial domain is subdivided into elements such as triangles, hexahedra or more frequently tetrahedra, in which the displacement field is interpolated using shape functions. At each point the strain can be computed using the spatial derivatives of the displacement field. Accurate material models have been implemented from rheological models relating stress and strain in hyperelastic, viscoelastic, inhomogeneous, transversely isotropic and/or quasi-incompressible media [41]. For simplicity, linearized strain has been applied assuming small displacements in rotated frames [27]. Precomputed deformations modes have been used to interactively deform large structures [6, 18, 22]. Using deformation modes rather than point-like nodes as DOFs allows

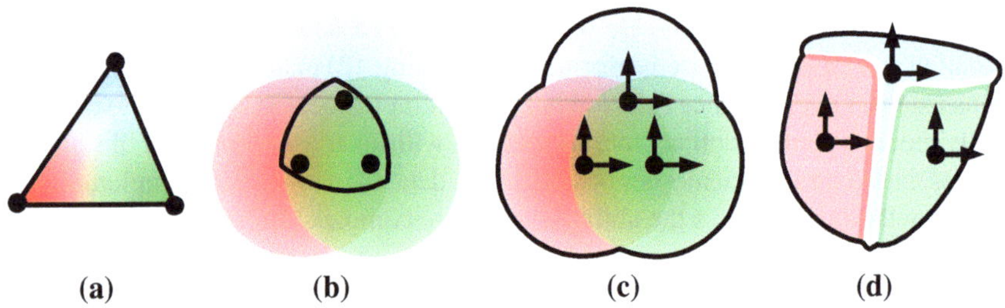

Fig. 1 Comparison of displacement functions. The *black line* encloses the area where the displacement function is defined, based on node positions (*black circles*) and associated functions (*colored areas*). **a** Finite element, **b** point-based, **c** frame-based with RBF kernels, **d** frame-based with our material-based kernels

to easily trade-off accuracy for speed. A layered model combining articulated body dynamics and a reduced basis of body deformation is presented in [12]. However, the deformation modes lack locality and pushing on one point may deform the whole object. Models based on Cosserat points have been proposed for large deformations in thin structures [34] and solids [30]. Since robustness problems such as inverted tetrahedra [17] or hourglass deformation modes in hexahedra [30] have been addressed, meshing remain the main issue in finite elements. To reduce computation time, embedding detailed objects in coarse meshes has become popular in computer graphics [27, 32, 37]. Multi-resolution approaches have been proposed [9, 15]. In recent work, disconnected or arbitrarily-shaped elements [19, 25] have been proposed to alleviate the meshing difficulties.

Meshless methods: Meshless methods do not use an underlying embedding structure but unstructured control points. In computer graphics, meshless methods have been first introduced for fluid simulation and then extended to solid mechanics [16, 28]. Besides continuum mechanics-based methods, fast algorithms have been developed for video games to simulate quasi-isometry [1, 29]. They are not able to model real materials, being based on geometry only. In meshless methods, each control node has a given influence that generally decreases with the distance to it. Standard approximation or interpolation methods have been investigated for physical simulation such as Shepard functions, radial basis functions and moving least squares (see [11] for an extensive review). Despite the added flexibility due to the absence of elements, sampling issues remain, since each interpolated point must lie in the range of at least four non-coplanar nodes, as illustrated in Fig. 1b, contrary to our method that explicitly use rotations in the degrees of freedom. A very interesting meshless approach using moving frames was recently proposed to alleviate this limitation [26], using the generalized moving least squares (GMLS) interpolation. In this method, even one single neighboring node is sufficient to compute a local displacement, as illustrated in Fig. 1c. Moreover, the authors introduce a new affine (first-degree) approximation of the strain, called elaston. In contrast with the plain (zero-degree) strain value traditionally used, this allows each integration point

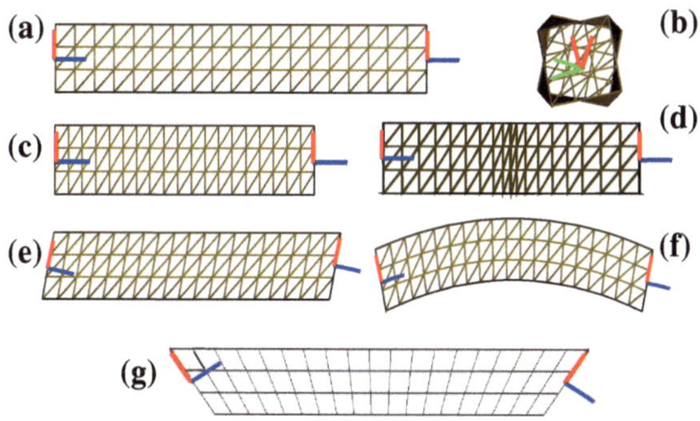

Fig. 2 Deformation modes obtained using two rigid frames. **a** Rest shape, **b** twisting, **c**, **d** compression with linear (resp. nonlinear) shape functions, **e** shear, **f** bending can be obtained using skinning, but **g** not using GMLS

to capture bending and twisting in addition to the usual stretch and shear modes. These improvements over previous methods remove all constraints on node neighborhood and allow the simulation of objects with arbitrary topology within a unified framework. However, a dense sampling of the objects is applied, leading to high computation times.

3 Frame-Based Deformation

In continuum media mechanics, it is necessary to numerically solve systems of differential equations (see Sect. 4). A general procedure is to smoothly approximate continuous functions in the solid from sought values at discrete sample locations. These values are the independent degrees of freedom (i.e., the DOFs q_i) which we will call *nodes*. In most simulation methods (Sect. 2), nodes are points and the deformation in the material is linearly interpolated from node displacements. In contrast, we consider rigid frames, affine frames and quadratic frames. Nodes are associated with *shape functions*, also called *weights*, which are combined to produce the displacement function of material points in the solid. To model deformable objects using a small number of control nodes, we need convenient, natural deformation functions. In character animation, the blending of frame displacements has been studied to deform a skin from an embedded articulated skeleton [21]. This method, called *skinning* or vertex blending or skeletal subspace deformation, is widely used, not only for its simplicity and efficiency, but because it provides natural and intuitive handles for controlling deformation. Skinning generates smooth deformations using a very sparse sampling of the deformation field. Here, we present two different blending techniques that we have explored for parameterizing a physically based deformable model, and how we measure the deformation.

Linear blend skinning: The simplest and most popular blending method is *linear blend skinning* [24] where the displacements of control nodes q_i are locally combined according to their shape function w_i. The following derivations hold for different

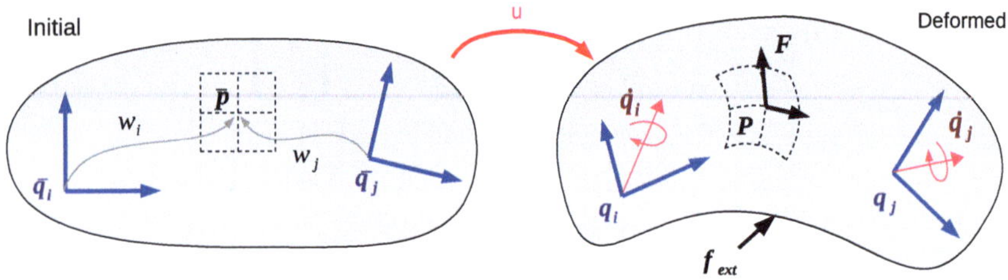

Fig. 3 The displacement u of a deformable object is discretized using nodes (*blue*). Strain is measured based on the deformation of local frames (*black arrow*) computed at integration points $\mathbf{p}$ in the material

types of control nodes: points, rigid frames, linearly deformable (affine) frames, and quadratic frames. Let $\bar{\mathbf{p}}$ and $\mathbf{p}$ be positions in the initial and deformed settings and $\mathbf{u} = (\mathbf{p} - \bar{\mathbf{p}})$ the corresponding displacement, expressed as: $\mathbf{u} = \sum_i w_i(\bar{\mathbf{p}})\mathbf{A}_i\bar{\mathbf{p}}^* - \bar{\mathbf{p}}$, where $(\bar{\mathbf{p}})^*$ denotes a vector of polynomials of dimension d in the coordinates of $\bar{\mathbf{p}}$. For point, affine or rigid, and quadratic primitives, we respectively use complete polynomial bases of order $n = 0$, $n = 1$, $n = 2$, noted as $(.)^n$. In 3D, we have $d = (n+1)(n+2)(n+3)/6$ and the three first bases are: $\mathbf{p}^0 = [1], \mathbf{p}^1 = [1, x, y, z]^T$, $\mathbf{p}^2 = [1, x, y, z, x^2, y^2, z^2, xy, yz, zx]^T$. The $3 \times d$ matrix $\mathbf{A}_i(q_i)$ represents the transformation of node i from its initial to its current position and is straightforwardly computed based on the independent DOFs q_i: for instance, the 12 DOFs of an affine primitive are directly pasted into a 3×4 matrix, while the 6 DOFs of a rigid primitive are converted to a matrix using Rodrigues' formula. $w_i(\bar{\mathbf{p}})$ is the shape function of node i evaluated at $\bar{\mathbf{p}}$. In linear blend skinning, weights need to constitute a partition of unity ($\sum w_i(\bar{\mathbf{p}}) = 1$). To impose Dirichlet boundary conditions, it is convenient to have interpolating functions at $\bar{\mathbf{x}}_i$, the initial position (frame origin) of node q_i in 3d space: $w_i(\bar{\mathbf{x}}_i) = 1$ and $w_j(\bar{\mathbf{x}}_i) = 0, \forall j \neq i$.

Dual quaternion skinning: Linear blend skinning suffers from well known volume loss artifacts when the relative displacement between nodes is large and non linear. To remedy this, extra nodes need to be inserted. Another solution, is to use a better blending function. For rigid frames, *dual quaternion blending* offers a good approximation of the linear interpolation of screws at a reasonable computational cost [20]. It provides a closed-form solution for more than two transforms contrary to screw interpolation that requires an iterative treatment. Here, the relative displacement of a rigid frame i is no more expressed using a 3×4 matrix $\mathbf{A}_i$, but using a 8d vector $\mathbf{a}_i = [\mathbf{a}_0^{iT}\ \mathbf{a}_\varepsilon^{iT}]^T$ where $\mathbf{a}_0^i$ (resp. $\mathbf{a}_\varepsilon^{iT}$) is a unit quaternion representing the rotation (resp. translation). Blended displacements are computed as normalized weighted sums of dual quaternions: $\mathbf{b}' = \sum w_i\mathbf{a}_i / \| \sum w_i\mathbf{a}_i \|$. Finally the blended dual quaternion is converted [20] into a 3×4 rigid transformation matrix $\mathbf{A}$ to transform material points: $\mathbf{u} = \mathbf{A}\bar{\mathbf{p}}^* - \bar{\mathbf{p}}$.

Strain measure: As shown in Fig. 3, the displacement is sampled at nodes, and is interpolated within the object based on nodal displacements. To apply the laws of continuum mechanics, we first need to measure the local deformation of the material. Consider a material whose undeformed positions $\bar{\mathbf{p}}(\theta)$ are parametrized by local, curvilinear coordinates θ (like texture coordinates). When the material undergoes a deformation, the points are displaced to new positions $\mathbf{p}(\theta) = \bar{\mathbf{p}}(\theta) + \mathbf{u}(\theta)$. At each material point, the derivatives of the position function $\mathbf{p}$ with respect to the coordinates θ are the vectors of a local basis called the deformation gradient $F = d\mathbf{p}/\mathbf{d}\theta$, with reference value $\bar{F} = d\bar{\mathbf{p}}/d\theta$, typically the identity. The local deformation of the material is the non-rigid part of the transformation $\bar{F}^{-1}F$ between the reference and current states (like the distortion of a checkerboard texture). The strain, ε, is a measure of this deformation. Different strain measures have been proposed, but all of them fit in our framework. For instance, the popular Green-Lagrange strain tensor, which is well suited for large displacements is computed as $(F^T \bar{F}^{-T} \bar{F}^{-1} F - \mathbf{I})/2$. Its six independent terms can be compactly stored in a $6d$ vector: $\varepsilon(\theta) = [\varepsilon_{xx}\, \varepsilon_{yy}\, \varepsilon_{zz}\, \varepsilon_{xy}\, \varepsilon_{yz}\, \varepsilon_{zx}]^T$.

4 The Dynamics of Frame-Based Continuum

This section explains how to set up the classical differential equation of dynamics for our models. An overview of the algorithm is given at the end of the Section. As shown in the following diagram, we apply a classical hyperelastic scheme: from the degrees of freedom $\mathbf{q}$, we interpolate a displacement field based on skinning, from which we compute the strain through spatial differentiation (Sect. 3). The elastic response $\sigma(\varepsilon)$ generates the elastic forces, and is a physical characteristic of the material. The elastic energy of a deformed object is the work done by the elastic forces from the undeformed state to the current state, integrated across the whole object (Sect. 4.4): $\mathscr{W} = \int_{\mathscr{V}} \int_0^{\varepsilon} \sigma^T d\varepsilon$. The associated elastic forces $\mathbf{f}$ are computed by differentiating the energy with respect to the DOFs (Sect. 4.1). After time integration (Sect. 4.3), we obtain the acceleration, velocity and the new position of each node.

$$
\begin{array}{ccccc}
\text{Position} & & \text{Strain} & & \text{Energy} \\
\mathbf{q} & \overset{\partial}{\rightarrow} & \varepsilon & \overset{\text{Material}}{\rightarrow} & \mathscr{W} \\
\int \uparrow & & & & \downarrow \partial \\
\dot{\mathbf{q}} & \overset{\int}{\leftarrow} & \ddot{\mathbf{q}} & \overset{\text{Mass}}{\leftarrow} & \mathbf{f} \\
\text{Velocity} & & \text{Acceleration} & & \text{Force}
\end{array}
$$

4.1 Elastic Force

The associated elastic forces $\mathbf{f}$ are computed by differentiating the energy with respect to the DOFs. Here, we explicitly introduce the deformation gradient F in the force computation:

Table 1 Polynomial degrees obtained with linear shape functions and linear blend skinning

Node	Strain measure	**u**	**M**	F	ε, σ	**f**
Affine/rigid	Corotational	2	4	1	1	2
Affine/rigid	Green-Lagrange	2	4	1	2	4
Quadratic	Corotational	3	6	2	2	4
Quadratic	Green-Lagrange	3	6	2	4	8

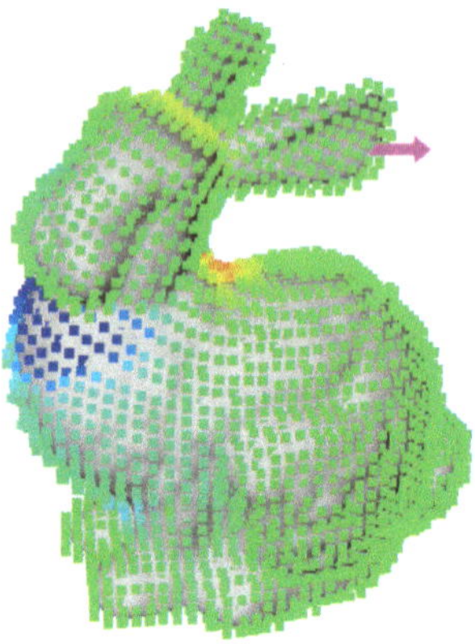

Fig. 4 In this example 15,000 integration samples are generated by rasterizing a bunny model, and a midpoint (zero order) integration scheme is used. *Colors* represent a hue mapping of the strain

as in [4], but it requires intensive static analysis at initialization time. Fortunately, displacements based on linear blend skinning can be easily differentiated and all quantities can be integrated explicitly in regions of linear weights. In a region $\mathscr{V}e$ centered on $\bar{\mathbf{p}}$, we express points as $\bar{\mathbf{p}} + \delta(\bar{\mathbf{p}})$. The integration of order n of a scalar quantity v in this region can be written as $\int_{\mathscr{V}_e} v = \mathbf{v}^T \int_{\mathscr{V}_e} \delta(\bar{\mathbf{p}})^n$ where $\mathbf{v}$ is a vector containing the quantity v and its spatial derivatives up to degree n, and $\int_{\mathscr{V}_e} \delta(\bar{\mathbf{p}})^n$ is the integrated polynomial basis of order n over the region. The last term can be accurately estimated at initialization time using a voxel grid, as the one shown in Fig. 4. This integration is exact if n is the polynomial degree of v. This formulation generalizes the concept of elastons [26] where quantities of order $n = 2$ are explicitly integrated in cuboid regions. Here, we consider arbitrary regions, and orders. Note that, using $n = 0$, the integration scheme is the classical midpoint rule: $\int_{\mathscr{V}_e} v \approx v\mathscr{V}e$.

The method presented in Sect. 5 generates as-linear-as possible shape functions. However, the gradients are discontinuous at the boundaries of the influence regions. We therefore partition the volume in regions influenced by the same set of nodes, and place one integration sample $\mathscr{V}e$ in each of them. To increase precision, we recursively subdivide the remaining regions up to the user-defined number of integration points. Our subdivision criterion is based on the error of a least squares fit of the voxel weights with a linear function.

Algorithm 4 Deformable model computations.

Data: Voxel map of material properties, number of control nodes
Initialization:

- Distribute the control nodes `// Sect. 5.4`
- Compute the shape functions `// Sect. 5.3`
- Compute the mass matrix `// Sect. 4.3`
- Generate the integration samples `// Sect. 4.4`
- Compute the weights of the surface vertices `// Sect. 4.2`

Loop:

- Accumulate force from each integration point: `// Eq. 1`
  - Compute F (and its spatial derivatives);
  - Compute ε (and derivatives) from F using a given strain measure;
  - Compute σ (and derivatives) from ε using a given material model;
  - Add integrated force to each influencing primitive;
- External forces and collision handling `// Eq. 3`
- At each solver iteration: `// Eq. 5`
  - Accumulate force change from each integration point `// Eq. 2`

5 Material-Aware Shape Functions

Building sparse frame-based physical models not only requires appropriate deformation functions as discussed in the previous section, but also anisotropic shape functions to resolve heterogeneous material as illustrated in Fig. 1d. In this section, we propose a method to automatically compute such functions.

5.1 Compliance Distance

Consider the deformation of a heterogeneous bar in one dimension, as shown in Fig. 5, where each point $\mathbf{p}$ is parameterized by one material coordinate x. Let the endpoints $\mathbf{p}_0$ and $\mathbf{p}_1$ be the sampling points of the displacement field. At any point, the displacement is a weighted sum of the displacements at the sampling points: $\mathbf{u}(x) = w_0(x)\mathbf{u}_0 + w_1(x)\mathbf{u}_1$. If the bar is heterogeneous, the deformation is not uniform and depends on the local stiffness, as illustrated in Fig. 5b. We call a shape function *ideal* if it encodes the exact displacement within the bar given the displacements of the endpoints, as computed by a static solution. Choosing the static solution as the reference is somehow arbitrary, since inertial effects play a role in dynamics simulation. However, the computation of interior positions based on boundary positions is an ill-posed problem in dynamics, since the solution depends on the velocities and on the time step. Moreover, for graphics, we believe that our perception of

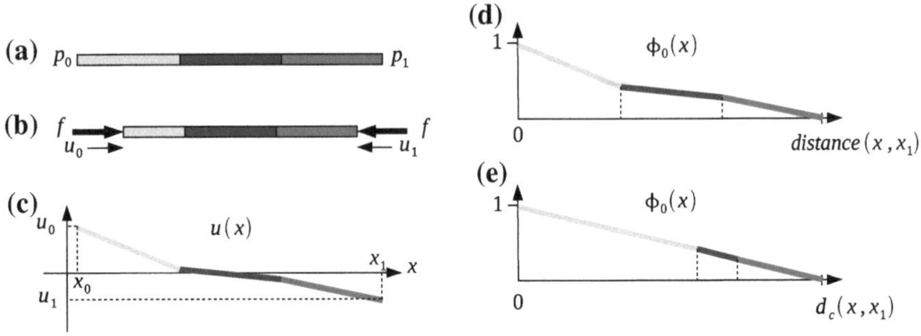

Fig. 5 Shape function based on compliance distance. **a** A bar made of 3 different materials, in rest state, with stiffness proportional to darkness. **b** The bar compressed by an external force. **c** The displacement across the bar. **d** The ideal w_0 shape function to encode the material stiffness. **e** The same, as a function of the compliance distance

realism is more accurate for static scenes than when the object is moving. Using the static solution as a shape function makes sense from this point of view, and encodes more information than a purely geometric shape function.

It is possible to derive the ideal shape functions by computing the static solution $\mathbf{u}(x)$ corresponding to a compression force $\mathbf{f}$ applied to the endpoints. Note that this precomputation is exact for linear materials only. For simplicity, we assume that the bar has a unit section. At any point the local compression is $\varepsilon = \frac{d\mathbf{u}}{dx} = \mathbf{f}/E = \mathbf{f}c$, where E is the Young's modulus, and its inverse c is the compliance of the material. Solving this differential equation provides us with: $\mathbf{u}(x) = \mathbf{u}(x_0) + \int_{x_0}^{x} \mathbf{f}c \; dx$, and since the force is constant across the bar, the shape function w_0 illustrated in Fig. 5d is exactly:

$$w_0(x) = \frac{\mathbf{u}(x) - \mathbf{u}(x_1)}{\mathbf{u}(x_0) - \mathbf{u}(x_1)} = \frac{\int_{x}^{x_1} c \; dx}{\int_{x_0}^{x_1} c \; dx} \tag{7}$$

Let us define the compliance distance between two points $\mathbf{a}$ and $\mathbf{b}$ as: $d_c(a, b) = \int_{x_a}^{x_b} c \; |dx|$. The slope of the ideal shape function is: $\frac{dw_0}{dx} = -c/d_c(\mathbf{p}_0, \mathbf{p}_1)$. It is proportional to the local compliance c and to the inverse of the compliance distance between the endpoints. Interestingly, the shape function is thus an affine function of the compliance distance, as illustrated in Fig. 5e, and it can be computed without solving an equation.

5.2 Extension to Two or Three Dimensions

We showed in the previous section that computing ideal shape functions in 1D objects, without performing compute-intensive static analyses as in [32], is straightforward based on compliance distance. Let n be the number of points where we want to compute exact displacements to encode in shape functions. In one dimension, both the static solution and the distance field can be computed in linear time. In two dimensions, computing the static solution for n independent points requires the solution of a $2n \times 2n$ equation system. The worst case time complexity of the solution is $O(n^3)$, and direct sparse solvers can achieve it with a degree between 1.5 and 2 in practice. In contrast, the computation of an approximate distance field in a voxel grid is $O(n \log n)$, which is much faster, but does not allow us to expect an exact solution like in one dimension. The reason is that there is an infinity of paths from one point to another to propagate forces across, thus the stress is not uniform and can not be factored out of the integrals and simplified like in Eq. 7. Another difference with the one-dimensional case is the number of deformation modes. Higher-dimensional objects exhibit several stretching and shearing modes, and we can not expect the ratio of displacement between two points to be the same in each mode. Since a single scalar value can not encode several different ratios, there is no ideal shape function in more than one dimension. Another limitation of this measure is that the compliance distance is the length (compliance) of the shortest (stiffest) path from one point to the other, independently of the other paths. Thus, two points connected by a stiff straight sliver are at the same compliance distance as if they were embedded in a compact block of the same material, even though they are more rigidly bound in the latter case. Moreover, material anisotropy is not modeled using a scalar stiffness value. Nonetheless, the compliance distance allows the computation of efficient shape functions, as shown in the following.

5.3 Voronoi Kernel Functions

In meshless frameworks, each node is associated with a kernel function which defines its influence in space, as presented in Sect. 3. A wide variety of kernel functions have been proposed in the literature, most often based on spherical, ellipsoidal or parallelepipedal supports. Our design departs from this, and is guided by a set of properties that we consider desirable for the simulation of sparse deformable models. To correctly handle the example shown in Fig. 1d, we need to restrict kernel overlap, to prevent the influence of the left node from unrealistically crossing the bone and reaching the flesh on the right. We thus need to constrain the kernel values, while keeping them as smooth as possible. In particular, we favor as-linear-as possible shape functions with respect to the compliance distance, in order to reproduce the theoretical solution in pure extension. A kernel value should not vanish before reaching the neighboring nodes, otherwise there would be a rigid layer around each node. With a sufficient number of radial basis functions, all the boundary conditions could be met [36]. Unfortunately, shape functions computed with RBFs are generally global and can increase with distance, producing unrealistic deformations. Local RBFs have isotropic compact support, and are thus only approximating.

Since there is no general analytical solution that can satisfy all the desired properties, we numerically compute a discrete approximate solution on the voxelized material property map. Solving a Laplace or heat equation on the grid would require

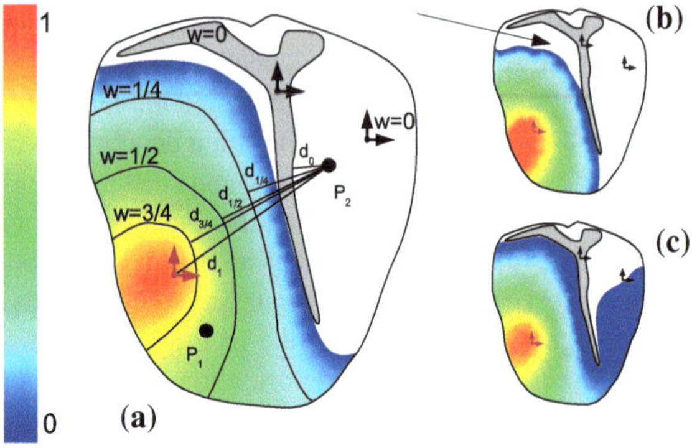

Fig. 6 Color map of the normalized shape function corresponding to the *red node*, computed using two Voronoi subdivisions (*left*), one subdivision (*top right*) and five subdivisions (*bottom right*)

the solution of a large equation system, and would compute nonlinear weight functions. A Voronoi partition of the volume allows us to easily compute kernels with compact supports, imposed values and linear decrease. This can be efficiently implemented in voxelized materials using Dijkstra's shortest path algorithm. Since a point on a Voronoi frontier is at equal distance from two nodes, we set the two kernel values to 0.5 at this point, and scale the distances accordingly inside each cell. To extend the distance function outside a cell, we generate the isosurface of kernel value $1/4$ by computing a new Voronoi surface between the $1/2$ isosurface and the other nodes. We can then recursively subdivide the intervals to generate a desired number of isosurfaces. The kernel values can straightforwardly be interpolated between the isosurfaces: for instance, the value at $\mathbf{P}_1$ in Fig. 6a is $(\frac{1}{2}d_{3/4} + \frac{3}{4}d_{1/2})/(d_{3/4} + d_{1/2})$, where d_i is the distance to isosurface of kernel value i, on Dijkstra's shortest path the point belongs to. To compute values between the last isosurface and 0 (the neighboring nodes), we apply a particular scheme since points beyond the neighbors, such as $\mathbf{P}_2$ in the figure, should not be influenced by the node. In this case, we linearly extrapolate the kernel function: $(-\frac{1}{2}d_{1/4} + \frac{1}{4}d_{1/2})/(d_{1/4} + d_{1/2})$. This technique is easily generalized to compliance distance and all the desired properties are met: it correctly generates interpolating, smooth, linear and decreasing functions between nodes. The corresponding cell shapes are not necessarily convex in Euclidean space, which allows them to resolve complex material distributions as illustrated by the compliance distance field in Fig. 10d. Since points can be in the range of more than two kernels, a normalization is necessary to obtain a partition of unity and the linearity of the shape functions is not perfectly achieved, however the functions are often close to linear as shown in the accompanying video. When using a small number of Voronoi subdivisions, we are not guaranteed to reach all the expected regions due to inaccurate extrapolation (see arrow tip in Fig. 6b, where $d_1 < 2d_{1/2}$). Increasing the number of isosurfaces reduces this artifact, but can lead to unrealistically large influence regions as shown in Fig. 6c, where the right part is influenced by the red node due to the linear interpolation between the right and left nodes. In practice, a small number of subdivisions are sufficient to remove noticeable artifacts while maintaining realistic bounds. The design of more realistic kernels in the extreme case of a very sparse discretization, large material inhomogeneities and complex geometry, is deferred to future work.

5.4 Node Distribution

The Voronoi computations provide us with a natural way to uniformly distribute nodes in the space of compliance-scaled distances. We apply a standard farthest point sampling followed by a Lloyd relaxation (iterative repositioning of nodes in the center of their Voronoi regions) as done in [1, 26]. The uniform sampling using the compliance distance results in higher node density in more compliant regions, allowing more deformation in soft regions. Since a whole rigid object corresponds to a single point in the compliance distance metric, all its points in Cartesian space have the same shape function values. Interestingly, it thus undergoes a rigid displacement, even if it is not associated to a single node. However, due to the well-known artifacts of linear blend skinning, it may actually undergo compression in case of large deformations. This artifact can be easily avoided by initializing nodes in the rigid parts, and keeping them fixed during the Lloyd relaxation. Another solution would be to replace linear blend skinning with dual quaternion skinning [20], at the price of more complex mechanical computations due to normalization.

6 Results

6.1 Validation

We implemented our method within the SOFA framework [2] to exploit its implicit and static solvers, as well as its GPU collision detection and response [3]. To encourage its use, our software will be freely available in the upcoming release. We measured the displacement of the centerline of a 10×4 thin plate, as shown in Fig. 7. The left side is fixed, while a uniform traction is applied to the right side. As expected, we obtain similar results using FEM and our method, with the same material parameters. A slight over-extension occurs in dense frame distributions, probably due to numerical issues in the voxel-wise integration of the deformation energy. We have also compared the simulations of cantilever beams, as illustrated in Fig. 8. We used regularly spaced nodes along the axis, with piecewise linear weight functions. We apply an extension force to the beam and verify that the force-extension law precisely matches the theoretical St. Venant-Kirchhoff model $f = \varepsilon + 3\varepsilon^2/2 + \varepsilon/2$, independently of the number of frames and the volume sample densities. Bending is more complex because it simultaneously involves extension–compression and shear,

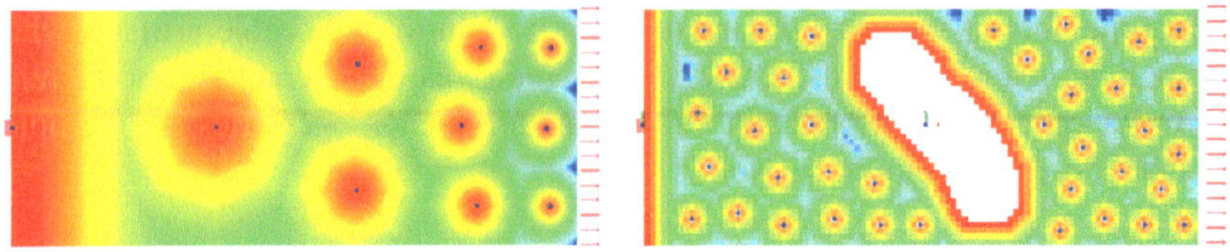

Fig. 7 Comparison with FEM on the extension of a plate. *Left* with a stiffness gradient. *Right* uniform stiffness and rigid part. *Top* compliance distance field within each Voronoi cell, *left* 500 × 200 grid, *right* 100 × 40 grid

Fig. 8 Comparison of our models (*solid colors*) with FEM (wireframe). *Blue* with two affine frames. *Green* three affine frames. *Red* five affine frames. *Yellow* nine affine frames

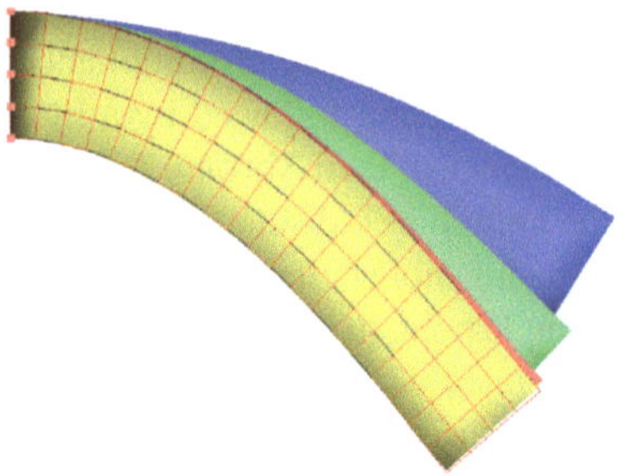

especially with large displacements as shown in the example in Fig. 8. This confirms that accurate continuum mechanics can be performed using our model. The behaviors converge as we increase the number of nodes. As usual, fewer degrees of freedom result in more stiffness.

6.2 Performance

Computation times are difficult to compare rigorously because we use an iterative solver based on the conjugate gradient algorithm. In the test on heterogeneous material shown in Fig. 7, we measured the total number of CG iterations applied to reach their final state with less than 1 % of precision. The frame-based models converged from one to three orders of magnitude faster, thanks to the reduced number of DOFs. We used a regular FEM mesh. A more sophisticated meshing strategy taking the stiffness into account would certainly be more efficient, unfortunately implementations of these are not easily available. Carefully designed meshes can greatly enhance the speed of the FEM method. However, resolving geometrical details requires fine meshes with a large number of DOFs, and in case of large variations of stiffness, numerical issues considerably slow down the convergence, even using preconditioning. The ability of our method to encode the stiffness in the shape functions not only reduces the number of necessary DOFs, but also seems to reduce the conditioning problems. The pre-computation times range from less than one second for 10 frames in a 100 × 40 voxel grid to 10 min for 200 frames in a 500 × 200 grid.

Table 2 Timings

Model	# frames	# samples	# vertices (K)	# voxels	t_{ini} (s)	FPS
Steak	3	10	5	67 K	3	500
Steak	10	53	5	67 K	8	100
Steak	20	140	5	67 K	12	40
Dragon	3	6	20	7 M	100	300
Dragon	10	41	20	7 M	220	150
Dragon	20	158	20	7 M	360	27
Ribbon	5	9	4	60 K	4	200
Knee	10	200	35	500 K	11	10
Rat	30	230	600	1.5 M	90	8

Our implementation is straightforward and there is plenty of room for optimization and parallelization.

Table 2 presents frame rates achieved on a common PC (2.67 Hz processor, 8 GB, Nvidia 295GTx). They include all the computations, including rendering and collision detection. The dragon and the ribbon demos are shown in the video. The computation times strongly depend on the number of integration points, which suggests that a GPU implementation of the force computations may dramatically increase the speed. A faster node relaxation [23] would speed up the precomputations in fine grids.

Corotational strain is about 1.5 times faster than Green-Lagrange strain due to the lower degree integration. However, in our implementation, it is not as robust because the rotation part of F is not differentiated to compute forces (Eq. 2). Their accuracy on static solutions is comparable in our tests. Rigid and affine primitives exhibit similar computational time. Quadratic primitives are about 15 % slower with the same number of integration points of the same degree. We believe that the best compromise between accuracy and performance is achieved using affine primitives: they have more DOFs than rigid frames so can capture more deformation modes, and they require significantly fewer integration points than quadratic primitives if we limit the expansion of the deformation gradient to the first order, and the integration degree to 4 (= 30 polynomial terms). In theory, quadratic primitives would need a second order expansion of F and an eighth order integration (=165 polynomial terms), which would be more costly. Affine and rigid frames require only one integration point per region with linear shape function, providing the main deformation modes of a rod using only two frames and one integration point (see Fig. 2). In our implementation, linear blend skinning of rigid frames is about five times faster than dual quaternion skinning [13] with the same number of integration points. Dual quaternion skinning is more accurate in large bending (no volume loss) but requires more integration points due to the non-linear blending function and is significantly more complex to implement.

We found that sampling integration points in the overlapping influence regions was a suitable strategy, since it allowed a good linear approximation of the fine grained

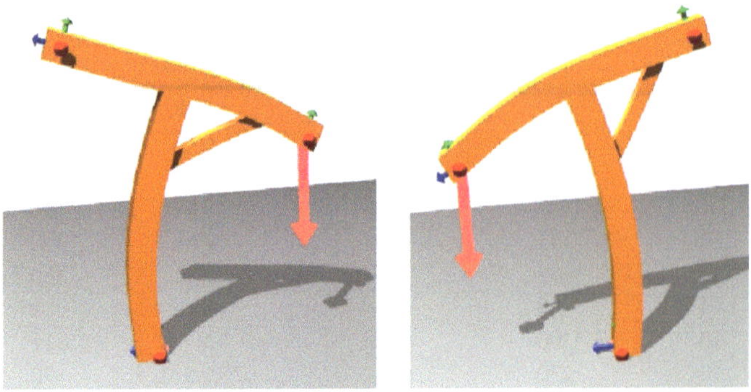

Fig. 9 An object with asymmetric stiffness automatically computed based on its asymmetric shape

shape function defined in the voxel grid: in our test, the average difference was 0.05 (the shape function being defined between 0 and 1, making the approximation error less than 5 %). This result also shows that the normalization of the kernel function does not significantly change the linearity. Uniformly distributed sample points unrealistically increase the stiffness because soft parts are assigned with a high stiffness due to averaging in the sample region.

6.3 Simulations

The most appealing feature of our method is probably its ability to easily model deformable objects using a reduced number of control nodes. The T-shaped rubber object shown in Fig. 9 (Young's modulus $E = 200\,\text{kPa}$, Poisson's ratio $\nu = 0.3$) exhibits compression, shear, bending and torsion using only two frames, corresponding to a total of 12 DOF. The same number of DOF only allows to model a single linear tetrahedron in FEM, which can not exhibit torsion and bending! The object automatically exhibits an asymmetric stiffness reflecting its asymmetric shape (Fig. 9).

Figure 11 shows a close-up of the high speed simulation presented in Fig. 10, which runs at haptic rates. The fat undergoes more deformation than the flesh because it is more compliant, even though they are interpolated between the two same control frames. The method of [32] also realistically resolves heterogeneous materials, but a rigid bone across several elements would result in high stiffnesses and generate numerical problems. In contrast, our method handles the rigid parts straightforwardly, independently of their shape.

Our method allows the interactive simulation of complex biological systems such as the knee joint shown in Fig. 12. In this model, we have integrated four different tissues: bones, muscles, fat and ligaments. With only 10 nodes, we are able to realistically simulate flexion and fine movements such as the motion of the patella (knee cap) at ten frames per second, without any prior knowledge of the kinematic

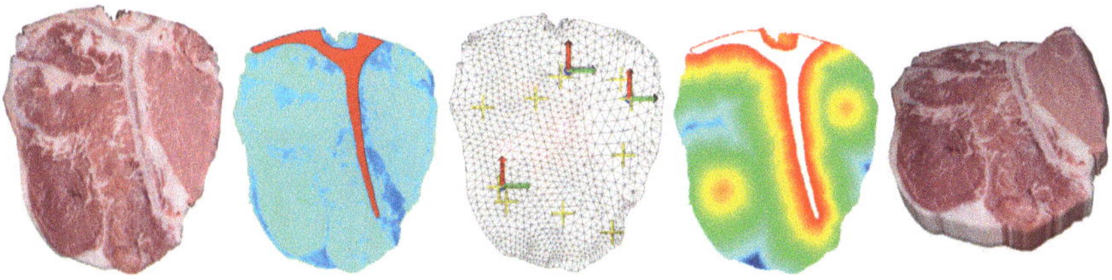

(a) T-Bone Steak **(b)** Stiffness **(c)** Discretization **(d)** Distance map **(e)** Deformation

Fig. 10 The T-bone steak (**a**) has a rigid bone and softer muscle and fat, as seen in the volumetric stiffness map (**b**). Our method can simulate it using only three moving frames and ten integration points (**c**), running at 500 Hz on an ordinary PC. The frame placement is automatically generated using a novel compliance-scaled distance (**d**). Observe that when one side of the meat is pulled (**e**), the bone remains rigid and the two meaty parts are correctly decoupled

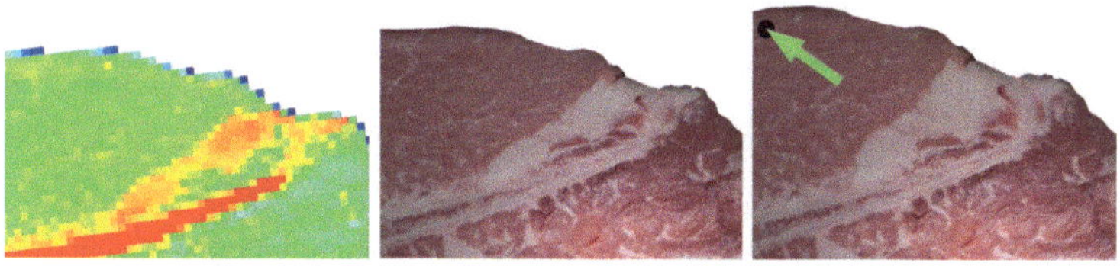

Fig. 11 The flesh and the fat, although interpolated between the same two control frames, pulled at the *black* point, exhibit different strains due to different stiffnesses

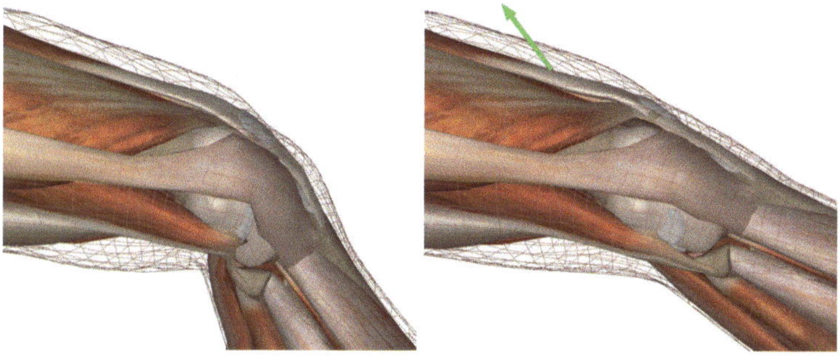

Fig. 12 Interactive knee simulation using 10 nodes. Pulling the quadriceps lifts the tibia

skeleton. Forces are transmitted from the quadriceps to the tibia suggesting that accurate dynamic models of the anatomy, taking into account muscle actuation, could be built. A few modifications in the voxelization and material modules would allow motion discontinuity between tissues in contact and a more accurate simulation of the highly anisotropic non-linear fibrous biological tissues.

Adding mechanical degrees of freedom during the simulation by inserting new frames with custom radial-basis shape functions is dramatically simpler than editing the mesh of an FEM model. In Fig. 13, we show that a dynamically inserted frame at the contact point with an object can be used to generate a local deformation. The

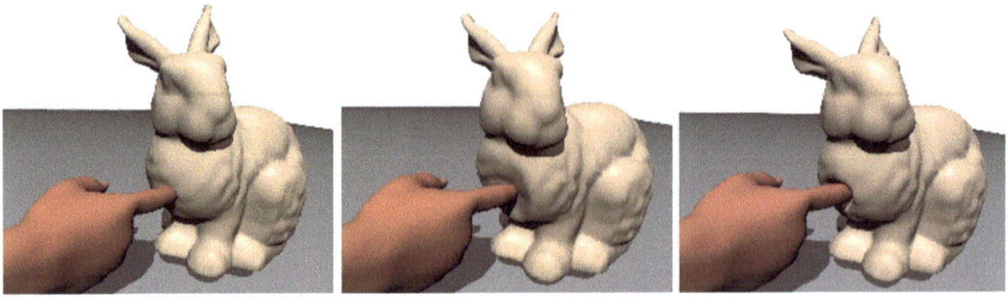

Fig. 13 More or less global deformation produced by dynamically inserting a frame with two different weight functions

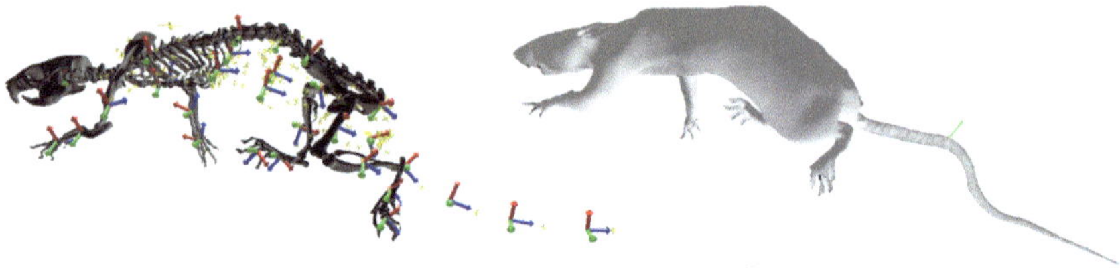

Fig. 14 Mix of animation and simulation. *Left* model subset animated using motion capture and added physical nodes. *Right* result, with flesh and tail animated using physics

range of the local deformation can be tuned using the shape function of the inserted frame. Such a high level of adaptivity in a physical model is straightforward with our model, while it is difficult to implement using previous methods.

Our method can also be used to easily simulate physically-based secondary motions from skeleton-based animation or motion capture, as illustrated in Fig. 14. The skin and the complete skeleton of a rat were acquired from micro-CT data. We applied our method to automatically sample the intermediate soft tissues and the tail with additional nodes. Optical motion capture was used to capture the movements of the limbs, head and three bones on the back.

7 Conclusion

In this chapter, we have presented a new type of deformable model using continuum mechanics applied to objects undergoing skinning deformation fields. Our approach allows the creation of sparse meshless models with arbitrary constitutive laws, and we have demonstrated it using St. Venant-Kirchhoff materials. Moreover, we have introduced novel, anisotropic kernel functions using a new definition of distance based on compliance, which allow the encoding of detailed stiffness maps in coarse meshless models. We have shown that the behavior of heterogeneous objects with complex materials and geometries can be simulated using a small number of control nodes and small computation times. The models are robust to large displacements and deformations.

In contrast with classical FEM and with the methods using geometrical shape functions, our approach decouples the resolution of the material from the resolution of the displacement function. The ability of setting an arbitrarily low number of frames, combined with a compliance-based distribution strategy, allows fast models to capture the most relevant deformation modes. Sampling is easier than with traditional particle-based meshless methods because there is no constraint on the number and on the placement of the nodes. Compared with FEM, adaptivity is easier because no volumetric mesh is used. However, due to computational time issues, the shape functions of the dynamically inserted nodes are currently limited to analytic radial-basis functions with local support. A faster computation of material-aware shape functions and hardware implementations are currently under investigation.

Acknowledgments We would like to thank Florent Falipou, Michaël Adam, Laurence Boissieux, François Jourdes, Estelle Duveau and Lionel Revéret for models and data. This work is partly funded by European project *PASSPORT for Liver Surgery* (ICT-2007.5.3 223894) and French ANR project *SoHuSim*. Thanks to the support of the Canada Research Chairs Program, NSERC, CIHR, Human Frontier Science Program, and Peter Wall Institute for Advanced Studies.

References

1. Adams B, Ovsjanikov M, Wand M, Seidel H-P, Guibas LJ (2008) Meshless modeling of deformable shapes and their motion. In: Symposium on computer animation, pp 77–86, 2008
2. Allard J, Cotin S, Faure F, Bensoussan P-J, Poyer F, Duriez C, Delingette H, Grisoni L (2007) SOFA–an open source framework for medical simulation. In: Medicine meets virtual reality, MMVR 15, pp 1–6, Long Beach, California, Etats-Unis, 2007
3. Allard J, Faure F, Courtecuisse H, Falipou F, Duriez C, Kry P (2010) Volume contact constraints at arbitrary resolution. ACM Trans Graph 29(3):205–223
4. An SS, Kim T, James DL (2008) Optimizing cubature for efficient integration of subspace deformations. ACM Trans Graph 27(5):1–10
5. Baraff D, Witkin A (1998) Large steps in cloth simulation. SIGGRAPH Comput Graph 32: 106–117
6. Barbič J, James DL (2005) Real-time subspace integration for St. Venant-Kirchhoff deformable models. ACM Trans Graph (SIGGRAPH 2005) 24(3):982–990
7. Bathe K (1996) Finite element procedures. Prentice Hall, Englewood Cliffs
8. Cotin S, Delingette H, Ayache N (1999) Real-time elastic deformations of soft tissues for surgery simulation. IEEE TVCG 5:62–73
9. Debunne G, Desbrun M, Cani M-P, Barr A (2001) Dynamic real-time deformations using space and time adaptive sampling. In: SIGGRAPH, computer graphics, pp 31–36, 2001
10. Faure F, Gilles B, Bousquet G, Pai DK (2011) Sparse meshless models of complex deformable solids. ACM Trans Graph 30(4):73
11. Fries T-P, Matthies HG (2003) Classification and overview of meshfree methods. Technical report, TU Brunswick, Germany
12. Galoppo N, Otaduy MA, Moss W, Sewall J, Curtis S, Lin MC (2009) Controlling deformable material with dynamic morph targets. In: Proceedings of the ACM SIGGRAPH symposium on interactive 3D graphics and games, Feb 2009
13. Gilles B, Bousquet G, Faure F, Pai DK (2011) Frame-based elastic models. ACM Trans Graph 30(2):15

14. Gourret J-P, Thalmann NM, Thalmann D (1989) Simulation of object and human skin formations in a grasping task. SIGGRAPH Comput Graph 23(3):21–30
15. Grinspun E, Krysl P, Schröder P (2002) Charms: a simple framework for adaptive simulation. In: SIGGRAPH computer graphics, pp 281–290, 2002
16. Gross M, Pfister H (2007) Point-based graphics. Morgan Kaufmann, San Francisco
17. Irving G, Teran J, Fedkiw R (2006) Tetrahedral and hexahedral invertible finite elements. Graph Models 68(2):66–89
18. James DL, Pai DK (2003) Multiresolution green's function methods for interactive simulation of large-scale elastostatic objects. ACM Trans Graph 22:47–82
19. Kaufmann P, Martin S, Botsch M, Gross M (2008) Flexible simulation of deformable models using discontinuous Galerkin fem. In: Symposium on computer animation, 2008
20. Kavan L, Collins S, Zara J, O'Sullivan C (2007) Skinning with dual quaternions. In: Symposium on interactive 3D graphics and games, pp 39–46, 2007
21. Kavan L, Collins S, Zara J, O'Sullivan C (2008) Geometric skinning with approximate dual quaternion blending, vol 27. ACM Press, New York
22. Kim T, James DL (2009) Skipping steps in deformable simulation with online model reduction. ACM Trans Graph 28:123:1–123:9
23. Liu Y, Wang W, Lévy B, Sun F, Yan D-M, Yang C (2009) On centroidal voronoi tessellation–energy smoothness and fast computation. ACM Trans Graph 28:08
24. Magnenat-Thalmann N, Laperrière R, Thalmann D (1988) Joint dependent local deformations for hand animation and object grasping. In: Graphics interface, pp 26–33, 1988
25. Martin S, Kaufmann P, Botsch M, Wicke M, Gross M (2008) Polyhedral finite elements using harmonic basis functions. Comput Graph Forum 27(5):1521–1529
26. Martin S, Kaufmann P, Botsch M, Grinspun E, Gross M (2010) Unified simulation of elastic rods, shells, and solids. SIGGRAPH Comput Graph 29(4):39
27. Müller M, Gross M (2004) Interactive virtual materials. In: Graphics interface, 2004
28. Müller M, Keiser R, Nealen A, Pauly M, Gross M, Alexa M (2004) Point based animation of elastic, plastic and melting objects. In: Symposium on computer animation, pp 141–151, 2004
29. Müller M, Heidelberger B, Teschner M, Gross M (2005) Meshless deformations based on shape matching. ACM Trans Graph 24(3):471–478
30. Nadler B, Rubin MB (2003) A new 3-d finite element for nonlinear elasticity using the theory of a cosserat point. Int J Solids Struct 40:4585–4614
31. Nealen A, Müller M, Keiser R, Boxerman E, Carlson M (2005) Physically based deformable models in computer graphics. Comput Graph Forum 25(4):809–836
32. Nesme M, Kry P, Jerabkova L, Faure F (2009) Preserving topology and elasticity for embedded deformable models. In: SIGGRAPH computer graphics, 2009
33. O'Brien J, Hodgins J (1999) Graphical models and animation of brittle fracture. In: SIGGRAPH computer Graphics, pp 137–146, 1999
34. Pai DK (2002) Strands: interactive simulation of thin solids using Cosserat models computer graphics forum. Int J Eurograph Assoc 21(3):347–352
35. Platt SM, Badler NI (1981) Animating facial expressions. In: SIGGRAPH computer graphics, pp 245–252, 1981
36. Powell MJD (1990) The theory of radial basis function approximation. University numerical analysis report
37. Sifakis E (2007) Der KG, Fedkiw R (2007) Arbitrary cutting of deformable tetrahedralized objects. In: Symposium on computer animation, 2007
38. Terzopoulos D, Qin H (1994) Dynamic nurbs with geometric constraints for interactive sculpting. ACM Trans Graph 13(2):103–136
39. Terzopoulos D, Platt J, Barr A, Fleischer K (1987) Elastically deformable models. AMC SIGGRAPH comput graph 24(4): 205–214
40. Teschner M, Heidelberger B, Muller M, Gross M (2004) A versatile and robust model for geometrically complex deformable solids. In: CGI, 2004
41. Weiss JA, Maker BN, Govindjee S (1996) Finite element implementation of incompressible, transversely isotropic hyperelasticity. Comput Methods Appl Mech Eng 135:107–128

Part II
Tracking and Computer Vision Applications

Robust Deformable Models for 2D and 3D Shape Estimation

Jorge S. Marques, Jacinto C. Nascimento and Carlos Santiago

Abstract Deformable models are useful tools to extract shape information from images and video sequences. However, the model has to be initialized in the vicinity of the object boundary, in order to foster convergence towards the desired features. This chapter describes four methods which alleviate this restriction. Despite their differences, they share three common features: (i) they use middle level features (edge segments) instead of low level ones; (ii) they explicitly assume that the measured features contain outliers and assign confidence degrees to the detected features and (iii) they adopt robust model updates, taking the confidence degrees into account. These four methods are reviewed and their performance is illustrated with selected examples.

1 Introduction

The automatic estimation of object boundaries in images and video sequences is a key operation in many computer vision systems. However, several factors make this task difficult namely, the high variety of objects shape and appearance, partial occlusion and the presence of clutter in the background image. Deformable models are popular techniques for dealing with this kind of problems [3, 11, 12]. They assume that object boundaries can be approximated by elastic curves or surfaces. The model is

J. S. Marques (✉) · J. C. Nascimento · C. Santiago
Institute for Systems and Robotics, Instituto Superior Tecnico,
av. Rovisco Pais, 1049-001 Lisboa, Portugal
e-mail: jsm@isr.ist.utl.pt; jsr@isr.ist.utl.pt

J. C. Nascimento
e-mail: jan@isr.ist.utl.pt

C. Santiago
e-mail: carlos.santiago@ist.utl.pt

M. González Hidalgo et al. (eds.), *Deformation Models*, Lecture Notes in Computational Vision and Biomechanics 7, DOI: 10.1007/978-94-007-5446-1_7,

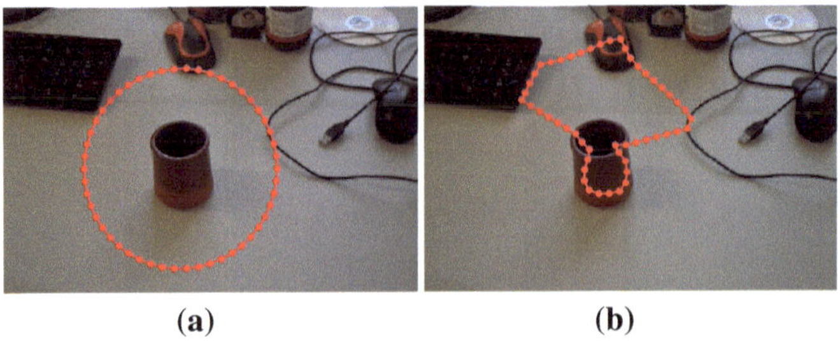

Fig. 1 Convergence difficulties in the estimation of the mug boundary with poor initialization: **a** initial contour and **b** final contour estimate obtained with the Snake algorithm

initialized close to the object boundary and it is automatically attracted towards the boundary by features detected in the image.

Despite the simplicity of this approach, model adaptation is not an easy task due to the following difficulties: (i) boundary cues (e.g., edge lines) are often subtle and incomplete, (ii) they are mixed with cues produced by other objects and by the cluttered background and (iii) the object is often partially occluded. Therefore, image noise, occlusion and cluttered background degrade the performance of deformable models. Figure 1 shows typical difficulties found in practice.

Classic deformable models deal with such difficulties by initializing the deformable contour in the vicinity the object boundary and by reducing its attraction range, to decrease the influence of outliers. This is often considered as a myopic behavior, since in such approaches, the model only "sees" the features located in a small neighborhood of the deformable contour [10].

Several attempts have been made to improve classic methods by using more informative features such as color [15], motion [6, 19] or gradient vector field [23]. Other approaches use prior information about the object shape by characterizing their statistics e.g., mean shape and main deformation modes [5]. The main problem remains unsolved, however: these algorithms are unable to separate valid features from outliers. They require a careful contour initialization, close to the object boundary, and myopic estimation algorithms.

The estimation of models in the presence of invalid data (outliers) is an old challenge in Computer Vision and has been addressed using robust estimation methods [14]. A popular approach (RANSAC) is based on the generation of multiple hypotheses, each of them based on different subset of data, considered as valid [7]. The model is estimated from the subset of data considered as valid and applied to all the data. The model which achieves the best fit is selected. This approach performs well if the number of parameters to estimate is small (<10) but it cannot be extended to flexible models that depend on tens or hundreds of parameters.

This chapter reviews four methods developed by the authors (Adaptive Snakes, S-PDAF, MMDAT, 3DS-PDAF) which allow a robust estimation of deformable curves and surfaces from image data, corrupted by outliers [8, 13, 16–18]. Instead of assuming that all the features are valid, we assume that some of them are outliers and

Table 1 Structure of robust adaptation methods

initialization: curve or surface initialization in the vicinity of the object boundary;
cycle
    feature extraction
    weights calculation
    model update
end

assign a confidence degree to each detected feature. Model update is accomplished by taking the confidence degrees into account. The second main idea concerns feature extraction: the proposed methods extract middle level features (curve segments or surface patches) from the image instead of low level ones (e.g., edge points). Both strategies contribute to improve the robustness of the algorithms. Table 1 summarizes the main steps of the robust model update algorithms. The methods discussed in this paper fit into this structure with one exception. In Adaptive Snakes, feature extraction is done once for all at the beginning and it is not repeated in each iteration.

The paper is organized as follows. Section 2 reviews two classic algorithms which fail in the presence of outliers. Section 3 describes Adaptive Snakes for the estimation of static objects. Section 4 describes the S-PDAF method for object estimation and tracking. Sections 5 and 6 extend S-PDAF to cope with multiple dynamics and 3D images. Section 7 concludes the paper.

2 Two Classic Algorithms

The Snake algorithm proposed in [11] estimates a deformable curve, called snake, $x(s) : [0, 1] \rightarrow \mathbb{R}^2$ by minimizing an energy functional $E(x) = E_{int}(x) + E_{img}(x)$, where $E_{int}(x)$ is an internal energy which prevents the snake from taking unusual configurations, and $E_{img}(x)$ is an image energy that attracts the model towards the object boundary. The image energy is defined by

$$E_{img}(x) = \int P(x(s))ds \tag{1}$$

where $P(x) : \mathbb{R}^2 \rightarrow \mathbb{R}$ stands for an *image potential* function, chosen in such a way that some of its valleys are located on the object boundary.

Several image potential functions have been proposed e.g., $P(x) = \|\nabla I(x)\|$ where $I(x)$ denotes the image intensity at point x and $\nabla I(x)$ is the image gradient [11]. Another alternative is the edge-based potential [4]

$$P(x) = -\sum_k G(x - e_k) \tag{2}$$

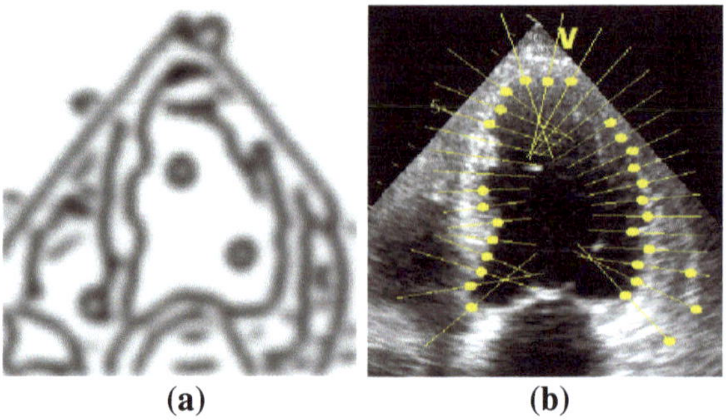

Fig. 2 *Bottom-up* and *top-down* image analysis: **a** image potential and **b** model guided feature extraction

where $G(x)$ is a Gaussian kernel and e_k denotes the kth edge point detected in the image. This potential function attracts the contour towards regions with high concentration of edge points. An example of an edge-based potential is shown Fig. 2a. The Snake algorithm equipped with this potential function performs well, if the edge points are located at the object contour and there are no other edges in the vicinity of the object. However, it performs poorly otherwise.

The Snake algorithm is slow and tends to get trapped in textured regions. In order to speed up convergence, top down algorithms were proposed [3]. Instead of computing a potential function, they extract feature points in the vicinity of the deformable curve i.e., the model is used to extract information from the image in a top-down way. Typically, the deformable contour is sampled at equally spaced points. Features points are detected by searching intensity transitions along lines orthogonal to the contour (see Fig. 2b). In tracking applications, the contour model and the set of measurements are characterized by vectors $x \in \mathbb{R}^n$, $y \in \mathbb{R}^m$, respectively, whose evolution is described by a linear dynamical system

$$x(t) = Ax(t-1) + w(t) \tag{3}$$

$$y(t) = Cx(t) + v(t) \tag{4}$$

where t denotes the frame number, $A \in \mathbb{R}^{n \times n}$, $C \in \mathbb{R}^{m \times n}$ characterize the dynamical behavior of the curve and the relationship between the curve parameters and the observations; $w(t) \sim N(0, Q)$, $v(t) \sim N(0, R)$ are uncorrelated random perturbations, with zero mean and covariance matrices Q, R. The estimation of the curve parameters from the observed measurements $y(t)$ can be performed by Kalman filtering and this method is called Kalman Snakes [21]. This approach is fast and well suited to real-time tracking applications. However the robustness difficulties remain unsolved: after a few frames, the deformable contour gets stuck in the outlier features.

It should be mentioning that several models are used to represent deformable curves. In some works, the deformable curve is represented by a sequence of 2D points (curve samples), $(x_1, \ldots, x_N)$, or by a spline curve, $x(s)$, whose shape is modified by a set of control points. Some works, however assume that the observed shape, $x(s)$, is obtained from a known reference shape, $x_r(s)$, modified by a global transformation T_θ (e.g., affine transform) and local deformation [3]. Therefore,

$$x(s) = T_\theta[x_r(s)] + x_d(s)\,, \tag{5}$$

where θ is the set of global parameters and $x_d(s)$ is the deformation spline. Both the global parameters, θ, and the deformation parameters have to be estimated from the input images.

3 Adaptive Snakes

As discussed in Sect. 2, the classic Snake algorithm gets easily trapped in wrong image features i.e., features associated to other objects or to the textured background, which are considered as *outliers*. Ideally, we would like to assign a binary label (valid/invalid) to each feature (edge point) and let the model be attracted by valid features only. This strategy inspired the Adaptive Snake algorithm proposed in [18]. It must be stressed, however, that some difficulties must be addressed as the feature labels are unknown.

The Adaptive Snakes algorithm follows a bottom-up approach to shape estimation. It starts by detecting edge points in the image using a standard edge detection algorithm. The edge points are then organized in connected line segments, called strokes in this paper. The set of strokes is denoted by $E = (e^1, \ldots, e^M)$ where the jth stroke $e^j = (e^j_1, \ldots, e^j_{L^j})$ is a sequence of edge points $e^j_p \in \mathbb{R}^2$ and L^j is the stroke length. Let $K = (k^1, \ldots, k^M)$ be a sequence of binary labels associated to the strokes in which $k^j \in \{0, 1\}$ is the label associated with the jth stroke e^j.

The ideal potential function is the sum of valid and invalid stroke potentials. We will assume that the potential function associated to valid strokes is given by (2) while the potential associated to the outliers is constant C. Therefore,

$$P(x, E, K) = -\sum_j \left(k^j \sum_p G(x - e^j_p) + (1 - k^j) L^j\, C \right). \tag{6}$$

Unfortunately, we do not know the labels k^j and cannot use this ideal potential function to directly estimate the deformable contour. To overcome this difficulty, the problem will be stated in a probabilistic framework.

Let $X = (x_1, \ldots, x_N)$, $x_i \in \mathbb{R}^2$, be N samples of the deformable contour. We will assume that the strokes and labels, E, K, given the contour samples X, follow a Gibbs distribution,

$$p(E, K|X) = \alpha \prod_{i=1}^{N} e^{-P(x_i, E, K)}. \tag{7}$$

We want to estimate X given the observed strokes E. This is a statistical inference problem which can be tackled by the Maximum Likelihood method. However, the binary labels are not observed and make this problem difficult since we must resort to $p(K \mid X)$ and a marginalization of the joint distribution $p(E, K \mid X)$ is required.

This difficulty can be circumvented by using the Expectation-Maximization (EM) method. The expressions for the E and M steps are derived in [18] and will not be repeated here. The E-step computes the confidence degrees associated to each stroke

$$w^j = Pr(k^j = 1|e^j, \hat{X}) \tag{8}$$

where $\hat{X}$ denotes the most recent deformable model estimate and the M-step minimizes the snake energy

$$E(X) = E_{int}(X) + \sum_{i=1}^{N} P_a(x_i) \tag{9}$$

where

$$P_a(x) = -\sum_j w^j \sum_p G(e_p^j - x), \tag{10}$$

changes in every iteration of the EM method and is called an *adaptive potential function*. The EM method leads to a recursive estimation algorithm similar to the Classic Snakes. However, the stroke potentials are multiplied by their confidence degrees, updated in each new iteration. All the strokes contribute to the estimation of the object contour. Although, the strokes with low confidence degrees have a negligible influence on the final contour estimates.

Figures 3, 4, illustrate the performance of Classic and Adaptive Snakes in the estimation of the mug boundary, using a deformable model represented by a sequence of 2D points. While in Classic Snakes, the potential function remains invariant during the convergence process and the contour is attracted towards outlier valleys, in Adaptive Snakes the potential changes as the contour deforms. After 20 iterations (see Fig. 4c), only the potential valleys associated to the mug receive a high confidence degree and the model converges towards the mug boundary.

Adaptive Snakes have been successfully used in several problems, e.g., in the segmentation of skin lesions displayed in dermoscopic images. A comparison among several segmentation algorithms was carried out in [20], using a database of dermoscopic images annotated by a medical expert. The Adaptive Snakes were selected as the best method (ex-aequo) in this study. Figure 5 shows an example of the Adaptive Snakes performance in the segmentation of a melanocytic skin lesion.

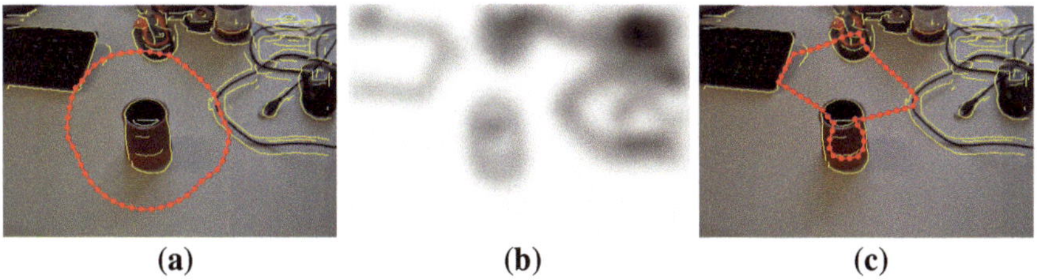

Fig. 3 Classic Snakes performance: **a** edge points and initial contour, **b** edge potential and **c** final snake configuration

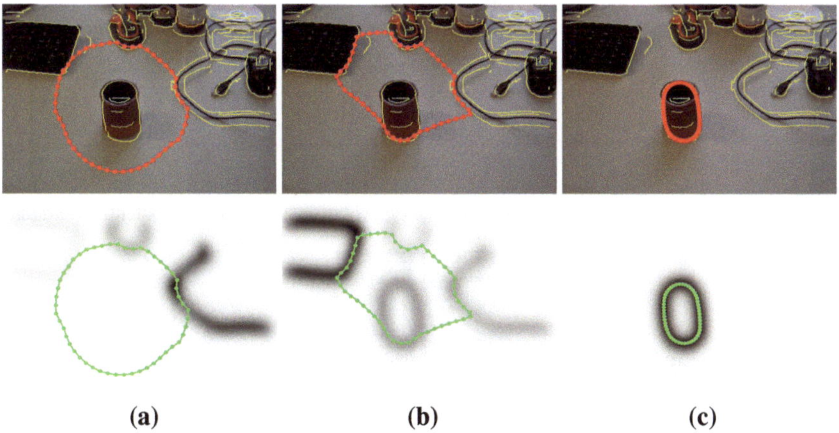

Fig. 4 Convergence of Adaptive Snakes; Snake configuration at iterations 1, 8 and 20 (*1st row*) and corresponding adaptive potential functions (*2nd row*)

4 Robust Tracking with S-PDAF

As mentioned before (Sect. 2), Kalman Snakes extract image features in the vicinity of the current contour estimate using directional search [3]. This strategy is very fast and tailored to object tracking in video sequences. However, some of the features detected by directional search are outliers and jeopardize the contour estimates after a few frames. Figure 2b illustrates feature extraction guided by the model estimate.

The method described in this section tries to alleviate this difficulty by using two complementary strategies. First, it replaces feature points by strokes i.e., sequences of points detected by directional search at consecutive samples of the deformable contour and such that their distances to the model boundary change smoothly (see Fig. 6a). Strokes are more informative and reliable than isolated points. Second, we will explicitly assume that some of the strokes are outliers (we do not know which) and should not be considered by the tracker.

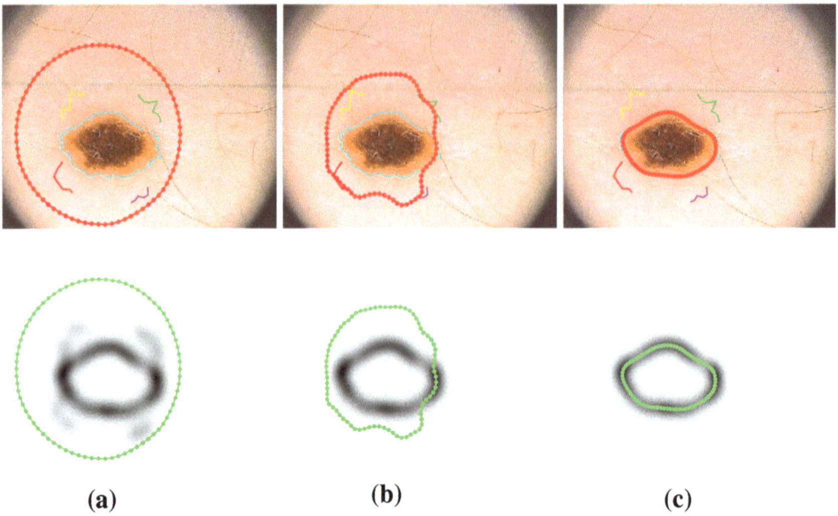

Fig. 5 Convergence of Adaptive Snakes in skin lesion segmentation: image features and evolution of the elastic curve in iterations 1, 8 and 20 (*1st line*); evolution of the adaptive potential (*2nd line*)

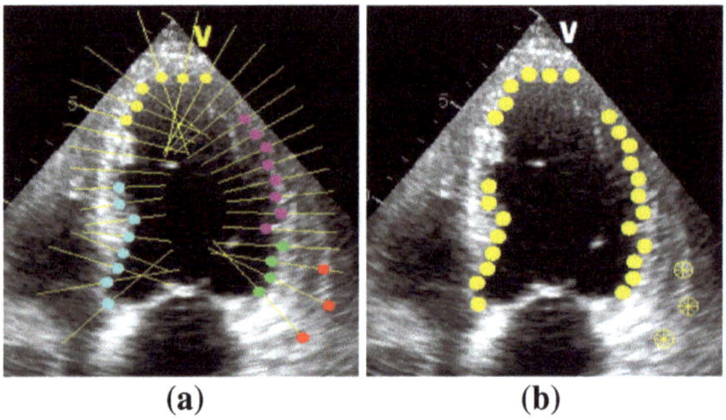

Fig. 6 **a** Strokes detected in an ultrasound image of the left ventricle (each color represents a different stroke). In this example 32 interpretations are possible. One of these interpretations is illustrated in **b** where strokes with the *filled dot lines* are considered as valid and the *dashed* stroke as invalid

We will assume that there is a binary label associated to each stroke (valid/invalid) and consider all the admissible label sequences. Each binary sequence will be denoted as a *data interpretation*. Thus, the ith data interpretation is defined by $I_i = (I_i^1, \ldots, I_i^M)$ where $I_i^j \in \{0, 1\}$ is the label of the jth stroke in the ith interpretation.

If all the observations were valid, we could use the dynamical model (3, 4) to describe shape and observations evolution. However, this model does not hold in the case of invalid observations. Specifically, if the ith interpretation, I_i, is true, only a subset of the observations, y_i, is valid and all the other measurements are considered as outliers. Therefore, the model associated to the ith interpretation is

$$x(t) = Ax(t-1) + w(t) \tag{11}$$
$$y_i(t) = C_i x(t) + v_i(t), \tag{12}$$

where matrix C_i is obtained from matrix C by removing the rows associated to the invalid observations. This makes the inference problem more difficult since we cannot apply Kalman filtering based on the correct interpretation of the data, as we do not know which interpretation is true. There are multiple observation models and we do not know which is the correct one at time t.

The *a posteriori* distribution of the state vector $x(t)$, given all the observations until the time instant t, $Y^t = \{y(1), \ldots, y(t)\}$, is a mixture of Gaussians. Unfortunately the number of Gaussians exponentially increases with t and exact inference is unfeasible.

A similar problem occurs in the tracking of moving point targets from Radar measurements. Although the radar system detects multiple echoes at each instant of time, only one of them, at most, corresponds to the target to be tracked. All the others echoes are produced by clutter or by other targets. In this case, exact inference is also unfeasible, since the number of hypotheses exponentially grows with the operation time and the number of detected echoes in each frame. This difficulty has been addressed by using approximate methods. Remarkable results have been obtained by the probabilistic data association filter (PDAF) [1, 2], used to compute the *a posteriori* distribution of the state vector $x(t)$, given a sequence of observations, most of them being outliers. This filter approximates the distribution of the state vector $x(t)$, given past observations $Y^{t-1} = (y(1), \ldots, y(t-1))$, by a normal distribution

$$p(x(t)|Y^{t-1}) = N(x(t); \hat{x}(t|t-1), P(t|t-1)) \tag{13}$$

where $\hat{x}(t|t-1)$, $P(t|t-1)$ are the mean vector and covariance matrix of $x(t)$ given past observations until time $t-1$, Y^{t-1} [1].

The PDAF can be modified to cope with multiple observation models in the context of robust shape tracking and the presence of multiple sensor models [17]. Using (13), it can be shown that the *a posteriori* distribution of $x(t)$, given Y^t, with multiple observation models is still Gaussian, i.e., $p(x(t)\,|Y^t) = N(x(t); \hat{x}(t\,|\,t), P(t\,|\,t))$, with mean vector being updated by

$$\hat{x}(t|t) = \sum_{i=1}^{M} \alpha_i(t)\hat{x}_i(t|t) \tag{14}$$

where $\hat{x}_i(t|t)$ is the mean vector obtained by Kalman filtering tailored to the ith interpretation of data, and $\alpha_i = p(I_i(t)\,|\,Y^t)$ is the corresponding data association probability. Similarly, the covariance matrix is updated by combining the Kalman filters associated to each interpretation of the data

$$P(t|t) = \left[I - \sum_{i=1}^{M} \alpha_i(t) K_i(t) C_i \right] P(t|t-1) + \sum_{i=1}^{M} \alpha_i(t) \hat{x}_i(t|t) \hat{x}_i(t|t)^T - \hat{x}(t|t)\hat{x}(t|t)^T \quad (15)$$

where K_i is the Kalman gain associated to the ith interpretation. These equations show that the contour estimate considers all data interpretations, each of them weighted by the corresponding association probability.

The association probabilities play an important role in this algorithm and they are computed using the Bayes law,

$$\alpha_i(t) = \beta p(y(t)|I_i(t), Y^{t-1}) p(I_i(t)) , \quad (16)$$

where β is a normalization constant. Equation (16) requires the distribution of the data associated to the ith interpretation, given previous observations, and the prior distribution for the interpretations. The first factor is computed as follows. Assuming that the observed features $y(t)$ are conditionally independent, we obtain

$$p(y(t), |I_i(t), Y^{t-1}) = \prod_{j=1}^{M} \prod_{p=1}^{L^j} p(y_p^j | I_i(t), Y^{t-1}) . \quad (17)$$

where $p(y_p^j | I_i(t), Y^{t-1})$ is a Gaussian density function, if the observation is valid, and a uniform density function, if the observation is invalid [17]. The prior, $p(I_i(t))$, is defined according to two main assumptions: (i) interpretations with long valid strokes are more probable than interpretations with small valid strokes, and (ii) if two valid strokes overlap, the interpretation has zero probability.

Figure 7 shows the operation of the S-PDAF tracker in the analysis of facial expressions (lips and eyebrows). A large number of outliers are detected since the search window along each direction is large. The tracker successfully manages to discard the outliers and to correctly update the deformable contours in this example.

5 Dealing with Multiple Motion Models

Sometimes, the object shape deforms according to different motion regimes (e.g., vehicles moving in streets, heart deformation in systole and diastole phases) and a single dynamical system (3, 4) is not enough to the describe the evolution of an object shape in the video sequence. In this case we should resort to multiple motion models

$$x(t) = A_{k(t)} x(t-1) + w(t) \quad (18)$$

$$y(t) = C x(t) + v(t) \quad (19)$$

where $k(t) \in \{1, \ldots, O\}$ is the label of the active model at time t, matrices A_1, $\ldots$, $A_O \in \mathbb{R}^{n \times n}$ define O dynamical behaviors of the state vector $x(t) \in \mathbb{R}^n$

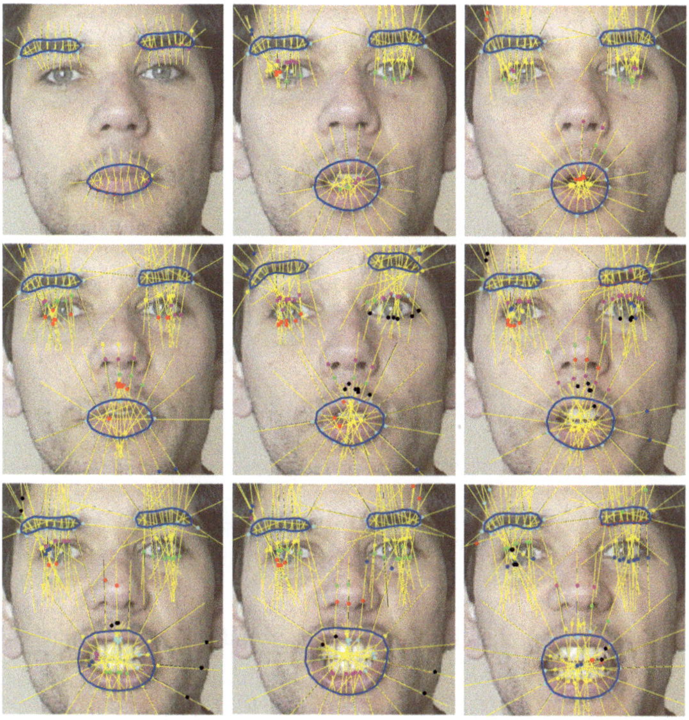

Fig. 7 Robust model tracking with S-PDAF. The tracker manages to discard the influence of outliers

and matrix $C \in \mathbb{R}^{m \times n}$ characterizes the relationship between the state vector and the observations $y(t) \in \mathbb{R}^m$; $w(t), v(t)$ are uncorrelated random sequences with normal distributions, $w(t) \sim N(0, Q_{k(t)}), v(t) \sim N(0, R)$. Furthermore, it is assumed that the label sequence $k(t)$ is a first order Markov process with transition probabilities

$$P(k(t) = j | k(t-1) = i) = T_{ij} . \tag{20}$$

Equations (18–20) define a switched dynamical model with a hybrid state $(x(t), k(t))$.

The most probable estimate of the hybrid state given the observations until time t, is obtained by solving the optimization problem

$$(\hat{x}(t), \hat{k}(t)) = \arg \max_{x(t),k(t)} p(x(t), k(t) | Y^t) . \tag{21}$$

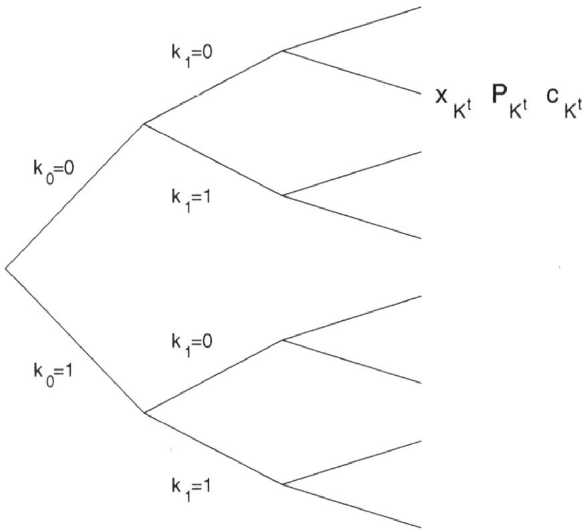

Fig. 8 Multi-model tracking: tree of Kalman filters

Using the law of total probabilities, the *a posteriori* distribution becomes

$$p(x(t), k(t)|Y^t) = \sum_{K^{t-1}} p(x(t), K^t|Y^t) \tag{22}$$

$$= \sum_{K^{t-1}} p(x(t)|K^t, Y^t)p(K^t|Y^t) \tag{23}$$

$$= \sum_{K^{t-1}} c_{K^t} p(x(t)|K^t, Y^t) \tag{24}$$

where $c_{K^t} = p(K^t|Y^t)$ is one mixture coefficient and $p(x(t)|K^t, Y^t)$ is a normal density function whose parameters can be obtained by Kalman filtering. Equation (24) shows that the *a posteriori* distribution of the hybrid state is a mixture of Gaussians each of them associated to a different label sequence K^t. Although, each Gaussian can be computed by Kalman filtering on a tree structure (see Fig. 8), the number of hypothesis, K^t exponentially increases, making a direct computation unfeasible.

This difficulty can be overcome by mode merging and elimination. If the probability c_{K^t}, for a specific label sequence K^t, becomes too small, the associated mode will be eliminated. In addition, if a pair of Gaussians have similar parameters (the similarity being measured by the Kullback-Leibler divergence) the two modes are merged and replaced by a single mode [13]. After performing mode merging and elimination, the estimation of the best shape estimate is determined by maximizing the *a posteriori* distribution, according to (21). This involves finding the maximum value of a multivariate mixture of Gaussians.

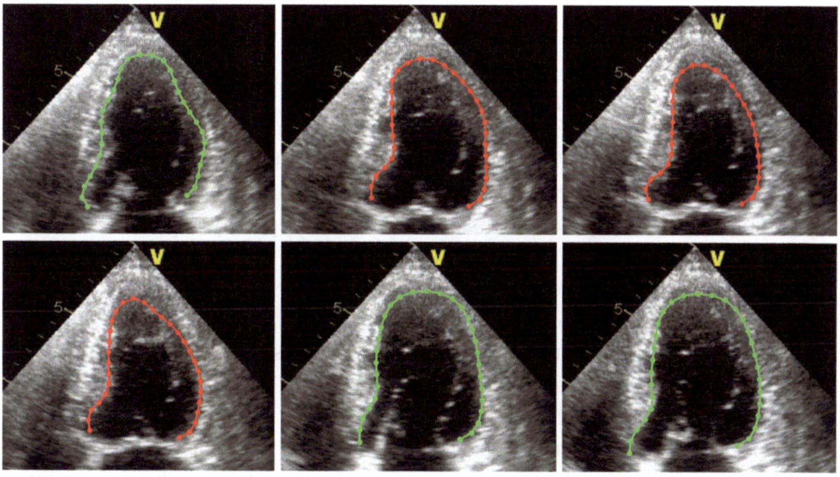

Fig. 9 Tracking of the *left ventricle* using the MMDAT, during systole and diastole phases. The contour color displays the label of the selected model: contraction (*red line*) or expansion (*green line*) models. Notice that whenever the mitral valve is closed (open) the cardiac cycle is in systole (diastole) phase

This algorithm can be used to track moving objects under different motion regimes and will be denoted as Multi-model tracker (MMT). However, the MMT is not robust since the state estimates are strongly affected by outlier measurements. To deal with this problem, we will extend the robust S-PDAF tracker to this multi-model context. We assume that the observations are organized in strokes and some of them may be outliers. A binary label will be associated to each stroke as we did in Sect. 4. The sequence of all binary labels, $I_i = (I_i^1, \ldots, I_i^M)$, is called a data interpretation.

The dynamical model associated to the model label $k(t)$ and data interpretation I_i is given by

$$x(t) = A_{k(t)}x(t-1) + w(t) \tag{25}$$

$$y_i(t) = C_i x(t) + v_i(t) \tag{26}$$

where $y_i(t)$ denotes the vector of valid observations according to interpretation I_i. The true data interpretation is unknown, however, and the estimation of the hybrid state based on this model can be done in a similar way to the one used in the MMT. However, the tree of Kalman filters is replaced by a tree of robust S-PDAF filters. This multi-model tracker with robust state estimate will be denoted Multi-model Data Association Tracker (MMDAT).

Figure 9 shows the application of the MMDAT in a sequence of ultrasound images of the heart. Two dynamical models are used to represent the evolution of the left ventricle during systole (contraction) and diastole (expansion) phases. Figure 9 shows the contour estimates obtained with this algorithm and the contour color represents the label of the active model, $k(t)$, which was automatically selected by the algorithm in each frame (see (21)).

6 From 2D to 3D

This section considers the segmentation of 3D objects in 3D images, $I : [0, 1]^3 \rightarrow \mathbb{R}$ e.g., ultrasound, MRI or PET images. The previous methods can be used in this context, specially the S-PDAF method which combines fast feature extraction, low memory requirements and robust model update. Memory and computational time play important roles in this context since the models typically depend on hundreds or thousands of parameters.

Four issues must be considered in order to extend the S-PDAF method to 3D shape estimation: (i) the surface model, (ii) model initialization, (iii) feature extraction and (iv) robust model update. Each of these issues is discussed in the sequel.

Surface model: we wish to approximate the object boundary by a deformable surface in 3D space. The choice of the surface model is an important issue since there is a tradeoff between model accuracy and number of free parameters. Popular solutions range from parametric models, such as superquadrics, with a small number of parameters and limited representation capability, to non-parametric models with many hundreds of parameters [22]. We adopted a non-parametric simplex mesh [9] which consist of a set of nodes, associated to 3D points, and edges among neighboring nodes. The model is constructed in such a way that each node has 3 neighbors which define a tangent plane to the surface. The normal vector is therefore available for each node.

Model initialization: this is an important step since we wish to create a simplex mesh close to the object boundary as a starting point for adaptation. We adopt a space carving approach. We discretize the region of interest (data volume) and intersect the region of interest by several inspection planes (e.g., three orthogonal planes as shown in Fig. 10). We then define a 2D contour in each plane using a graphical editor. All the voxels in the data volume are projected onto the inspection plane and those outside the contour are zeroed. At the end, we obtain a binary 3D mask which is then approximated by a simplex mesh. The simplex mesh is initialized with a spherical shape and deformed until it fits the binary mask. This is an easy task for the deformable simplex mesh since we are dealing with a binary object without clutter or noise.

Patch extraction: this operation is done in two steps. First, we detect transitions along straight lines orthogonal to the surface model. This task is repeated for each node and may lead to multiple detections in each direction (see Fig. 11a). Then the nodes are grouped in surface patches. Neighboring nodes receive the same label if their distances to the surface are similar. Patch construction depends on the order by which nodes are visited. Figure 11b shows an example of two patches.

Robust model update: model update is done using the S-PDAF Eqs. (14, 15) but there are two important differences. First the dimension of the state space is much higher. Instead of a few tens of parameters needed to represent a deformable curve, we typically need a few thousands of coordinates to represent the nodes positions. The update iteration is therefore much more demanding and slow. Second, the middle level features (surface patches) also have a large number of observations (sometimes

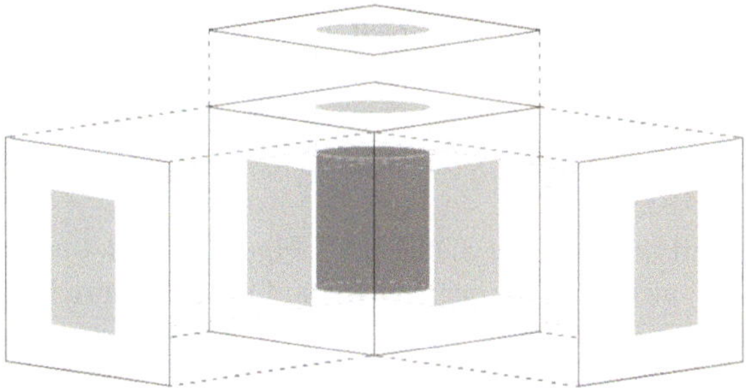

Fig. 10 Model initialization by space carving: image data is observed along three or more planes and the object boundary is approximated by a 2D contour in each plane. All voxels are projected onto the planes and those who project outside the contour are zeroed

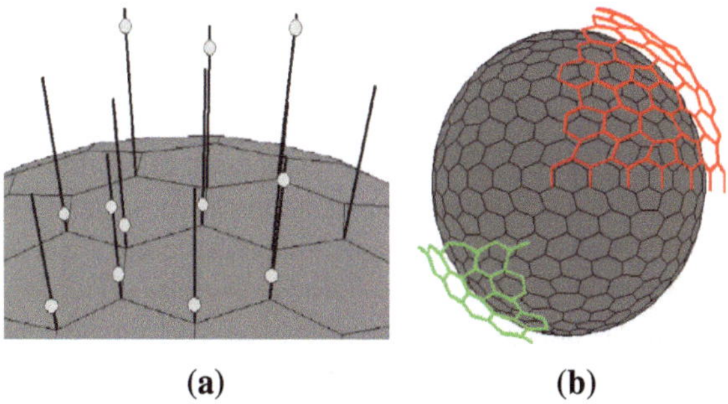

Fig. 11 Feature extraction: **a** directional search; **b** patch creation by region growing

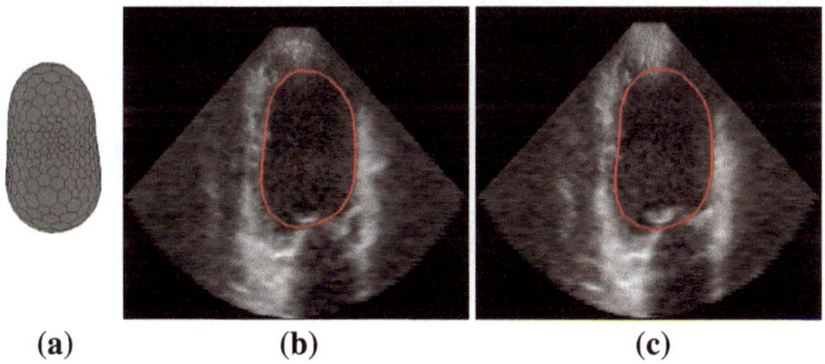

Fig. 12 Segmentation of the *left ventricle* in an ultrasound 3D volume: **a** estimated 3D model using 3DS-PDAF; **b, c** pair of observed images and model cross sections obtained by intersecting the deformable mesh by the inspection planes

hundreds of transitions are detected in the 3D volume of data). It is no longer possible to assume that all the observations in the same patch are statistically independent. The confidence degrees under these hypotheses are almost binary: patches closest to the surface have confidence degrees equal to 1, while other patches have 0 confidence degree. We therefore adopted a different probabilistic model which takes into account the average distance of the valid features to the surface along the normal direction, d_{av}. The association probabilities are given by (16) as before but with

$$p(y(t)|I_i(t), Y^{t-1}) = Ce^{-\gamma d_{av}^2}. \tag{27}$$

This algorithm will be called the 3DS-PDAF. Figure 12 shows the results obtained by the 3DS-PDAF method in the segmentation of the left ventricle in an ultrasound volume of data. This is a difficult problem due to the low signal-to-noise ratio, the presence of speckle noise and lack of visual boundary information (edge dropout) in some regions of the heart boundary. Figure 12a shows an example of a deformable surface estimated by the 3DS-PDAF method. Figure 12b, c shows cross sections of the 3D volume of data and of the estimated model.

7 Conclusions

This chapter presents a set of robust shape estimation algorithms based on robust feature extraction and robust model update techniques. These methods can be applied in a variety of problems ranging from static shape estimation, to object tracking in 2D and 3D applications.

The proposed algorithms exhibit an improved performance when compared with the classical deformable model techniques. The key ideas explored in these approaches are: (i) the use of more reliable features (line segments or surface patches) and (ii) explicit modeling of outlier observations by assigning a binary label to each of them. Formulated in this way, the estimation of the object shape is an inference problem with unobserved variables (labels) which can be tackled by the Expectation-Maximization method or by Data Association filtering algorithms.

Acknowledgments Part of the work on multi-model tracking was done in collaboration with Profs. Gilles Celeux (Paris-Sud University) and João M. Lemos (IST). The dermoscopic image was kindly provided by Dr. Jorge Rozeira of Hospital Pedro Hispano. This work was partially supported by FCT under projects PTDC/EEA-CRO/098550/2008, PTDC/SAU-BEB/103471/2008 and PTDC/EEA-CRO/103462/2008.

References

1. Bar-Shalom Y (1987) Tracking and data association. Academic Press, Boston
2. Bar-Shalom Y, Daum F, Huang J (2009) The probabilistic data association filter. IEEE Control Syst 29:82–100
3. Blake A, Michael I (1998) Active contours. Springer, London
4. Cohen LD, Cohen I (1993) Finite element methods for active contour models and balloons for 2-D and 3-D images. IEEE Trans Pattern Anal Mach Intell 15:1131–1147
5. Cootes TF, Cooper D, Taylor CJ, Graham J (1995) Active shape models—their training and application. Comput Vis Image Underst 61(1)38–59

6. DeCarlo D, Metaxas D (1996) The integration of optical flow and deformable models: applications to human face shape and motion estimation. IEEE computer vision and pattern recognition, pp 231–238
7. Fischler MA, Bolles RC (1981) Random sample consensus: a paradigm for model fitting with applications to image analysis and automated cartography. Commun ACM 24(1981)
8. Gilles C, Marques JS, Nascimento JC (2004) Learning switching dynamic models for objects tracking. Pattern Recognit 37(9):1835–1840
9. Hervé D (1999) General object reconstruction based on simplex meshes. Int J Comput Vision 32:111–146
10. Jain AK, Zhong Y, Marie-Pierre Dubuisson-Jolly (1998) Deformable template models: a review. Signal Process 71:109–129
11. Kass M, Witkin A, Terzopoulos D (1988) Snakes: active contour models. Int J Comput Vis 1(4):321–331
12. Lei H, Zhigang P, Bryan E, Xun W, Chia YH, Kenneth LW, William GW (2008) Comparative study of deformable contour methods on medical image segmentation. Image Vis Comput 26:141–163
13. Marques JS, Lemos JM (2001) Optimal and Suboptimal Shape Tracking Based on Multiple Switched Dynamic Models, Image and Vision Comput 19:539–550
14. Meer P, Mintz D, Rosenfeld A, Kim DY (1991) Robust regression methods for computer vision: a review. Int J Comput Vision 6:59–70
15. Moreno-Noguer F, Sanfeliu A, Samaras D (2007) Integration of deformable contours and a multiple hypotheses Fisher color model for robust tracking in varying illuminant environments. Image Vis Comput 25:285296
16. Nascimento JC, Marques JS (2002) Improving the robustness of parametric shape tracking with switched multiple models. Pattern Recognit 35:2711–2718
17. Nascimento JC, Marques JS (2004) Robust shape tracking in the presence of cluttered background. IEEE Trans Multimedia 6(6):852–861
18. Nascimento JC, Marques JS (2005) Adaptive snakes using the EM algorithm. IEEE Trans Image Process 14(11):1678–1686
19. Nilanjan R, Acton ST (2004) Flow motion gradient vector: an external force for tracking rolling leukocytes with shape and size constrained active contour. IEEE Trans Med Imaging 23:1466–1477
20. Silveira M, Nascimento JC, Marques JS, Marçal AR, Mendonça T, Yamauchi S, Maeda J, Rozeira J (2009) Comparison of segmentation methods for melanoma diagnosis in dermoscopy images. IEEE J Sel Top Sign Proces 3:35–45
21. Terzopoulos D, Szeliski R (1992) Tracking with kalman snakes. In: Blake A, Yuille AL (eds) Active vision. MIT Press, Cambridge, pp 3–20
22. Xu C, Pham DL, Prince JL (2000) Medical image segmentation using deformable models. In: Handbook of medical imaging, vol 2. SPIE Press, Bellingham, pp 129–174
23. Xu C, Prince J (1998) Snakes, shapes, and gradient vector flow. IEEE Trans Image Process 7:359–369

Deformable Face Alignment via Local Measurements and Global Constraints

Jason M. Saragih

Abstract This chapter will review a particular approach to deformable face alignment coined constrained local models (CLM). The approach leverages the excellent generalisation properties of local appearance representations of parts and the strong global constraints imposed by the geometrical relationships between part locations. We begin by posing CLM in the general context of deformable face alignment, highlighting its similarities and differences with other approaches and motivating its benefits. An overview of the approach is then presented, explicating its various components and touching briefly on the interrelated issues of optimisation, feature representation and geometry regularisation. The following three sections discuss each of these three components in detail. The chapter concludes with a general discussion and directions of future work.

1 Introduction

Deformable face alignment has had a long history, starting from the seminal works of Cootes and Taylor with their active shape model (ASM) [5] and active appearance model (AAM) [7]. A particular feature of ASM and AAM that distinguishes them from earlier work is the use of annotated data to build statistical models of the face's appearance and geometry rather than human-defined or biologically inspired attributes. Since then, a tremendous amount of work has been invested in improving this basic paradigm and applications of its underlying techniques have been seen in fields as diverse as human-computer interaction [22, 47, 49], medical image analysis [17, 36, 53], industrial vision [6, 13, 30] and graphics [19, 43, 45].

The ASM/AAM paradigm is to statistically model all sources of variation that the face exhibits, namely that of geometry and appearance. ASM and AAM share a

J. M. Saragih (✉)
1 Technology Crt, CSIRO, Pullenvale, QLD4096, Australia
e-mail: jason.saragih@csiro.au

M. González Hidalgo et al. (eds.), *Deformation Models*, Lecture Notes in Computational Vision and Biomechanics 7, DOI: 10.1007/978-94-007-5446-1_8,

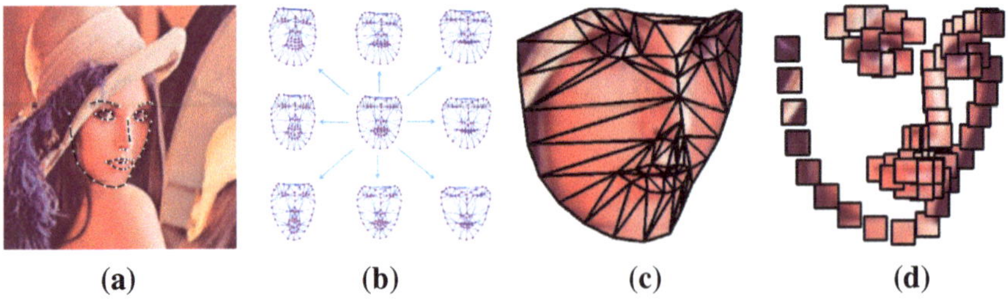

Fig. 1 AAM and ASM both represent a face in an image (**a**) using a statistical shape model (**b**) but AAMs represent appearance holistically (**c**) and ASM in parts (**d**)

common representation of facial geometry as a statistical model of joint variations in facial feature locations. A 2D linear model is most commonly used for this purpose and its utility for modelling dense 3D facial shapes has also been demonstrated in 3D morphable models (3DMM) [2]. Extensions that employ mixture models [21] and kernel based methods [37] have also been proposed for handling large pose variations in a 2D representation. The core difference between ASM and AAM is in how the appearance of the face is modelled. In AAMs, the appearance of the whole face is modelled jointly, whereas in ASM, each facial feature is modelled independently of all others (see Fig. 1). As such, AAMs can potentially capture a more faithful representation of the underlying statistics of facial appearance. However, in practice, due to the large space of appearance variability of the face and its high dimensionality, a compact representation that generalise well can only be afforded in highly restricted settings such as in the person-specific case [18, 34].

The primary use of ASM/AAM is to infer the locations of facial features in an image. Thus, discourse on ASM/AAM techniques in the literature often couples their representation with the particular inference strategy employed, herein referred to as deformable face alignment. Major advancements on this aspect of ASM/AAM have been proposed since the original works of Cootes and Taylor, with notable improvements in accuracy, computational complexity and generalisation. These inference algorithms can be broadly categorised into two groups; regression-based and optimisation-based. Regression-based approaches learn a regression model that predicts the location of facial features in an image from a set of features extracted from it. Improvements to the original regression approach in [8] include the use of more sophisticated feature representations [11, 14, 52], averaging multiple predictions [33, 50] and leveraging successive regression iterations [39, 40, 46]. The main advantage of the regression-based approach is its simplicity and efficiency; one simply extracts features from the image and applies the regression model to them. The main drawback of this approach is that the relationship between the image features and the location of facial features in an image is often highly nonlinear, requiring high-capacity regression models that are difficult to train and often generalise poorly. In contrast, optimisation-based approaches design/learn an objective function that encodes the degree of misalignment between the model and face in the image [21, 28, 25]. To reduce sensitivity to local minima, one can either learn an objective function that exhibits fewer local minima [26, 32] or simplify components of the problem such that the objective is amendable to exact inference methods [16, 48]. The main drawback of these provisions against local minima is that they often come at the cost of perturbing the global solution away from the true configuration in the image.

Despite substantial progress in all aspects of the ASM/AAM paradigm, the problem of real-time deformable face alignment in the presence of pose, identity, expression and lighting variations as well as image noise, resolution and partial occlusions, remains unresolved. Nonetheless, the space of variations that state-of-the-art methods can handle is such that their application in real-world settings has started to be realised.

In this chapter, one such method will be discussed in detail. It shares a common representation with the original ASM, but is referred to as a constrained local model (CLM) to emphasise its adoption of the various lessons learned about deformable face alignment since ASM was first proposed over two decades ago.[1] The remainder of this chapter is structured as follows; an overview of CLM is presented in Sect. 2, where motivations for its form are given and the three interrelated components of feature detection, regularisation and optimisation are identified. In Sect. 3, issues pertaining to the choice of facial feature detectors, their training and employment are discussed. The role of regularising geometry is discussed in Sect. 4, where a method for a highly compact image-specific regulariser is presented. In Sect. 5 an efficient and accurate optimisation procedure that leverages the forms of the feature detectors and regularisation described in the preceding sections is discussed in detail. This chapter concludes in Sect. 6 with a discussion and mention of promising directions for future work.

2 Overview

Posing deformable face alignment probabilistically, all existing algorithms strive to maximise the posterior over the location of facial features in an image:

$$p(\mathbf{x}|I) \propto p(\mathbf{x})\ p(I|\mathbf{x}); \quad \mathbf{x} = [\mathbf{x}_1; \ldots; \mathbf{x}_n], \quad \mathbf{x}_i \in \Re^2. \tag{1}$$

Here, I denotes the image and $\mathbf{x}$ is a vector of concatenated 2D-coordinates of the facial features. The term $p(I|\mathbf{x})$ denotes the likelihood that the face in the image has its features configured according to $\mathbf{x}$ and $p(\mathbf{x})$ denotes the prior distribution of plausible facial shapes.

A distinguishing feature of CLM, compared to other approaches, is that the likelihood of a particular location in the image corresponding to a particular facial feature

[1] The term CLM was first coined in [10].

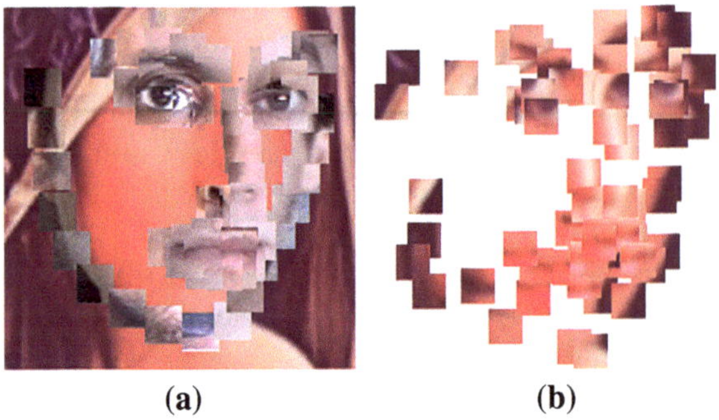

Fig. 2 A geometrically plausible composition of facial parts of different individuals in (a) constitutes a valid face, whereas a facial parts from the same individual arranged in a geometrically unlikely configuration in (b) does not

is assumed to be conditionally independent of all other facial features:

$$p(I|\mathbf{x}) = \prod_i p(I_i|\mathbf{x}_i). \tag{2}$$

Here, I_i denotes a spatially-coherent region in the image centred around $\mathbf{x}_i$ (i.e. an image patch). An illustration of the effects of this assumption on the alignment model is illustrated in Fig. 2. Although this assumption may appear restrictive, the advantage of this parameterisation is two-fold. First, modelling each facial part independently of all others is far easier; it requires less data to generalise well compared to holistic representations that model the appearance of all parts jointly together. The second advantage of this form is that the likelihood of each part over all locations in the image can be computed independently, admitting an efficient implementation through parallelisation. Having a lookup-table of the likelihood of part locations in the image is also advantageous as it allows posing alignment as a generic graph inference problem.

Maximising the posterior in Eq. (1) is equivalent to minimising its negative logarithm, which leads to an objective of the following form:

$$\min_{\mathbf{x}} \underbrace{\mathscr{R}(\mathbf{x})}_{\text{regulariser}} + \sum_i \underbrace{\mathscr{D}_i(\mathbf{x}_i)}_{\text{data terms}} . \tag{3}$$

The form of the objective in Eq. (3) is not unique to CLM. In fact, its use in vision can be traced back to the seminal work on optical-flow by Horn and Schunck [23]. Differences between problems that utilise an objective of this form stem primarily from how the regularisation and data terms are modelled, which often restricts the types of optimisation strategies one can employ. The original Horn-Schunck method for optical flow uses the Lucas-Kanade appearence-constancy objective [27] as the data term along with a Laplacian regulariser, which leads to a sparse-linear system

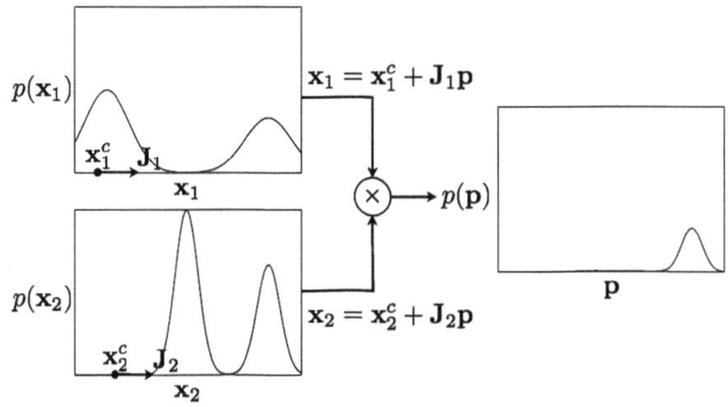

Fig. 3 The two boxes on the *left* illustrate the likelihood functions for two facial feature locations (i.e. $\mathbf{x}_1$ and $\mathbf{x}_2$). By constraining the configuration of the features via a low-dimensional subspace (i.e. $\mathbf{x} = \mathbf{x}^c + \mathbf{J}\mathbf{p}$), spurious detections that are not geometrically consistent are attenuated

that can be solved efficiently. More recently, pictorial structure models [16], which were popularised in the context of body alignment and pose estimation, use a tree-structured gaussian markov random field as a regulariser, which admits a global solution through dynamic programming. For deformable face alignment, the regulariser of choice is still the linear deformation model of ASM. Despite its simplicity, it has been shown to adequately span the space of human faces while affording a substantial constraint over the space of plausible solutions, removing the ambiguity in part detections to a large extent (an illustration of this is shown in Fig. 3). In the sections that follow, the specific forms of the data- and regularisation-terms used in CLM face alignment will be discussed.

3 Facial Feature Detectors

Modelling the local appearance of facial features independently of the rest of the face often exhibits better generalisation properties than holistic representations largely because the dimensionality of the data is much lower and some degree of invariance towards lighting variation can be obtained by a simple power normalisation [48]. One of the main problems with patch based methods is that the appearance of facial features can vary greatly between people, pose, lighting and expression. Thus, to account for these variations, it may be necessary to use high-capacity models, which tend to have poorer generalisation and higher evaluation costs. The second, and perhaps, more pressing issue, is the aperture problem; the local appearance of some facial features are inherently ambiguous. By observing only a small patch around a facial feature, it can be very difficult to pin-point where within an image a facial feature is located as many locations can share a similar local appearance (see Fig. 4a). This is complicated further when one has to account for inter-personal variabilities. As such, even the use of highly sophisticated models that account for inter-personal variations may not help resolve these ambiguities.

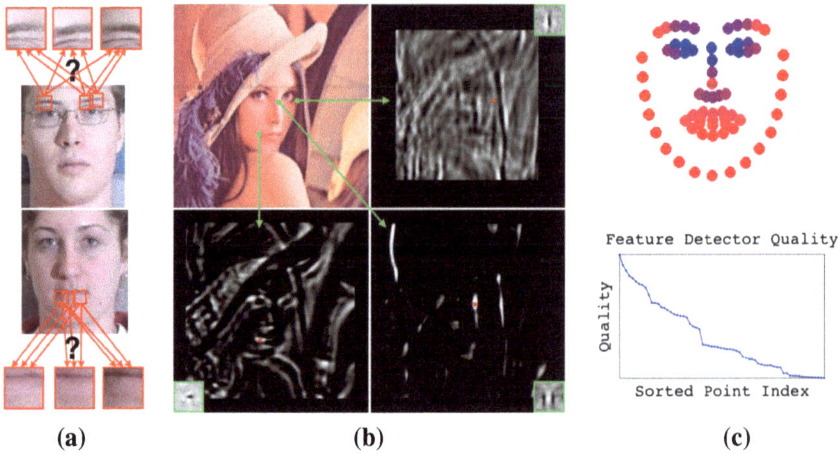

Fig. 4 **a** Examples of the aperture problem for points on the *eyebrows* and *lips* in two images from [20]. **b** The likelihood of facial features at all locations in the image (i.e. response-maps) for three different facial features. The scaled SVM template (i.e. $\mathbf{w}_i$ in Eq. (4)) for each feature is shown at the corners of their respective response map. The three features exhibit different degrees of detection ambiguities. **c** Visualisation of the relative quality of facial feature detectors. In the *top* image, the quality of each facial feature detector is colour coded (*blue* is high and *red* is low). A plot of the relative quality of each detector is shown in the graph below, ordered from highest to lowest quality. As one would expect, facial features that exhibit small inter/extra-personal variations, such as the eyes and nose, admit better feature detectors than those with large variations, such as those on the mouth and the periphery of the face

Faced with these difficulties, a pragmatic approach is to use a simple feature detector that is easy to train and efficient to employ, relying instead on global constraints over geometry and the inference strategy to find the best solution over their locations. Perhaps the simplest of all feature detectors is the linear-SVM [9]:

$$\mathscr{C}(\mathbf{d}) : \Re^D \to \mathbb{Z} = \begin{cases} 1 & \text{if } \mathbf{w}^T\mathbf{d} + b \geq 0 \\ 0 & \text{otherwise} \end{cases} . \tag{4}$$

Here, $\mathbf{d}$ denotes the features extracted from an image patch centred around the hypothesised facial feature location,[2] $\mathbf{w}$ is the SVM gain vector and b is the bias. Linear-SVMs have the advantage that the likelihood of feature locations in an image can be evaluated extremely rapidly using efficient convolution operations.

In order to use a linear-SVM for the feature detector in CLM alignment, a few issues must first be addressed. The first issue is how to choose the negative data for training the SVM. Unlike conventional detection tasks, the role of the SVM in CLM alignment is to generate the likelihood scores in Eq. (2) at each location in the image. As such, the negative data should consist of features extracted from the

[2] In [48], choosing $\mathbf{d}$ as the mean- and power-normalised raw pixel values was shown to work well for facial features. For body parts in articulated pose estimation, HoG features [12] were shown in [16] to give good results.

image at locations other than that corresponding to the facial feature. A question then naturally arises: how close to a feature location in an image can a negative sample be? Choosing negative samples too close the feature location can limit the accuracy of the classifier, while choosing only negative samples that are far away may limit its spatial precision. A practical approach to resolve this dilemma is through cross-validation, whereby a number of settings for the minimum proximity to the positive sample in each image are used to generate the negative set, and the efficacy of each choice is evaluated by using the detectors, in conjunction with the global constraints over geometry, in the inference process (i.e. solving Eq. (3)).

Another issue with using a linear-SVM to model the likelihood function is the great imbalance between positive and negative samples. For each image, there is only a single positive sample (i.e. that centred at the facial feature of interest), where as there are potentially an infinite number of negative samples that can be generated from anywhere outside the *positive-proximity* region. One way to address this problem is to use a bootstrapping procedure [12] to retain only the *difficult* negative samples. The idea here is that since the decision hyperplane of an SVM is described entirely by its support-vectors, the solution is not changed by training only with data that consists entirely of samples pertaining to the support vectors. In practice, this bootstrapping process can be approximated efficiently by simply choosing the negative samples which are most correlated with the positive one.

Finally, there is the issue of how to choose the best regularisation constant used for training the SVM. Unlike conventional applications of SVMs, where one desires a binary classification score, here, the linear-SVM is used to generate the likelihood that a location in an image corresponds to a particular facial feature. Specifically, the likelihood of a particular location in the image in Eq. (2) can be obtained by passing the classification score through a sigmoid function:

$$p(I_i \mid \mathbf{x}_i) = \frac{1}{1 + \exp\{\alpha(\mathbf{w}^T \mathbf{d}_{\mathbf{x}_i} + b) + \beta\}}, \tag{5}$$

where (α, β) are learned through a calibration process [35]. Ideally, the likelihood at the feature location should be higher than at all other locations. If it is assumed that the facial feature is located *somewhere* in the image, then the probabilities in Eq. (5) must sum to one. Thus, the entity of interest is not the SVM's binary classification score (or even its regressed value), but rather, the relative scores and their distribution in the image (i.e. a large likelihood close to the true feature location is better than one farther away). This is not reflected in the SVM objective as it penalises all miss-classifications equally. A heuristic to overcome this problem is to evaluate the efficacy of an SVM by a weighted sum of feature likelihoods over the whole image, where the weights are chosen to decay as one moves away from the true facial feature location in the image, using, for example, a gaussian distribution. Assuming isotropic Gaussian noise over the facial feature annotations, this is equivalent to measuring the marginal likelihood over the entire image (see discussion on optimisation in Sect. 5). One can use this measure in cross-validation to choose the best setting for the regularisation constant in SVM training.

Some examples of typical response maps for three facial features using a linear SVM are shown in Fig. 4b. It is evident that the simple linear-SVM can not adequately discriminate between the true feature locations and many other locations in the image. In the ideal case, there would be a single large response at the true location and small values everywhere else. Instead, a multimodal distribution of responses is evident. This reflects the difficulty in accounting for inter-personal variabilities as well the aperture problem stemming from local appearance ambiguities. Nonetheless, each of the response-maps contains some information about the true location of the feature and the ambiguities are often non-complementary. Thus, the simple linear-SVM detector can be sufficient for CLM alignment when coupled with an optimisation strategies that finds a solution through consensus between the different response maps and a strong global constraint over facial geometry.

4 Global Shape Constraints

The regularisation term in Eq. (3) serves to restrict the space over which consensus between facial feature detections is to be found through optimisation. Most deformable face alignment methods employ a linear approximation to how the shape of facial features deform, coined the point distribution model (PDM) by [5]:

$$\mathbf{x} = \bar{\mathbf{x}} + \Phi \mathbf{p}. \tag{6}$$

Here, $\bar{\mathbf{x}}$ denotes the mean shape, Φ denotes the basis of deformation and $\mathbf{p}$ denotes the deformation parameters. A common choice for the PDM is that learned by applying principal component analysis (PCA) to manually annotated facial feature locations in a set of training images. This model is both simple and efficient, and has been shown to adequately span the deformation space of objects such as the human face [5] and organs in medical image analysis [51]. Although PCA affords a significant compaction of the space of plausible facial shapes, the dimensionality of the deformation basis that accounts for a major portion (i.e. 95–98 %) of variations is rarely sufficient to eliminate enough spurious configurations. That is, the employment of the PCA prior over facial shapes in Eq. (3) does not eliminate a sufficient number of local minima such that local optimisation strategies converge to the global solution at acceptable rates.

In this section, we will review a method for multivariate regression that recognises the limitations of the conventional approach in building a deformation basis for face alignment. Through a method named principal regression analysis (PRA) [38], a basis is learned that spans the space of sample-specific regressors. In a departure from the standard notion of regression, this model does not, in itself, generate a specific prediction for a given input. Rather, it predicts a low dimensional subspace thought to contain the solution. As such, it is not devised to replace optimisation based alignment, but to enhance it by administering a strong prior over the space of plausible solutions.

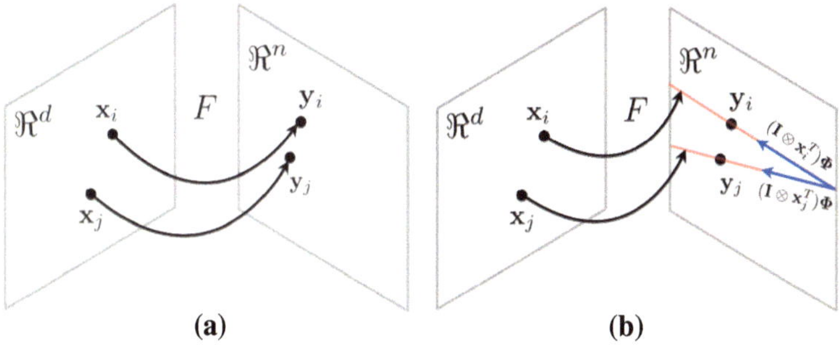

Fig. 5 Illustration of regression versus principal regression. **a** Regression maps points in input space to points in target space. **b** Principal regression maps points in input space to a specialised subspace containing the target

4.1 Principal Regression Analysis

Given a training set of input/output pairs[3] $\{\mathbf{x}_i, \mathbf{y}_i\}_{i=1}^N$, where $\mathbf{x}_i \in \Re^d$ and $\mathbf{y}_i \in \Re^n$, regression learns a function F that maps input variables to their target output. Conventionally, F is a static function that enables the prediction of responses for new input. The main idea in PRA is to use a predictor of the following form:

$$F : \Re^d \times \Re^h \rightarrow \Re^n = \sum_{i=1}^{h} p_i F_i(\mathbf{x}), \tag{7}$$

where $F_i : \Re^d \rightarrow \Re^n$ and $\mathbf{p} \in \Re^h$ is an additional variable that serves to linearly combine the predictions from $\{F_i\}_{i=1}^h$. In general, the value of $\mathbf{p}$ for a given $\mathbf{x}$ is typically unknown. As such, rather than mapping the input to a point in $\Re^n$ as in conventional regression, the form in Eq. (7) maps the input to a linear subspace containing the target output (see Fig. 5 for an illustration). The coordinates of the target output within this subspace are then found by other means. The idea is that if $h \ll n$, then the space of plausible target configurations is reduced considerably, simplifying the task of finding the correct one.

In the case where F_i is linear:

$$F_i(\mathbf{x}) = \mathbf{W}_i^T \mathbf{x}, \tag{8}$$

Equation (7) reduces to a bilinear function:

$$F(\mathbf{x}; \mathbf{p}) = (\mathbf{I} \otimes \mathbf{x}^T)\Phi\mathbf{p}; \;\; \Phi = [\text{vec}\{\mathbf{W}_i\} \ldots \text{vec}\{\mathbf{W}_h\}], \tag{9}$$

---

[3] In this section, we follow the usual convention in the regression literature and denote $\mathbf{x}$ as the input variable, rather than the facial feature locations as in preceding sections.

and for a given input $\mathbf{x}$ we have:

$$\mathbf{y} \in \text{span}\{(\mathbf{I} \otimes \mathbf{x}^T)\Phi\}. \tag{10}$$

Learning, then, involves minimising the regularised out-of-subspace component of the training targets:

$$\min_{\mathbf{p}_i, \Phi} \sum_{i=1}^{N} \|\mathbf{y}_i - (\mathbf{I} \otimes \mathbf{x}_i^T)\Phi\mathbf{p}_i\|^2 + \beta\|\Phi\mathbf{p}_i\|^2. \tag{11}$$

Here, $\Phi\mathbf{p}_i$ can be thought of as a conventional linear regressor specialised to the ith training example. As such, the second term in Eq. (11) is simply the Tikhonov regulariser applied separately to the predictors of each sample. The user defined parameter β trades off prediction error against each sample's regressor complexity. The affine ambiguity between Φ and $\mathbf{p}_i$ can be removed by, for example, enforcing the orthonormality constraint: $\Phi^T\Phi = \mathbf{I}$.

Drawing parallels with PCA, F_i are called principal regressors since they define a subspace over regressors. Similarly, the process of learning the principal regressors (i.e. solving Eq. (11)) is denoted principle regression analysis.

4.2 Kernel PRA

A common 'trick' in machine learning is to kernelise problems that involve only inner products between their input variables. This has the advantage of expanding modelling capacity whilst preserving the algorithmic properties and training complexity of linear models.

PRA belongs to a class of methods that admit a kernelisation via their adherence to Mercer's theorem [29]. That is, for $\Phi = [\Phi_1; \ldots; \Phi_n]$, each column of Φ_j is spanned by the columns of $\mathbf{X} = [\mathbf{x}_1 \ldots \mathbf{x}_N]$. To see this, consider the objective in Eq. (11), where the basis is replaced by its components within, and orthogonal to, the column space of $\mathbf{X}$: $\Phi = \Phi_{\mathbf{X}} + \Phi_{\perp\mathbf{X}}$. Since Φ influences the first term in Eq. (11) only through inner products with $\mathbf{x}_i$, $\Phi_{\perp\mathbf{X}}$ has no effect on it. Therefore, a solution will have components outside the span of $\mathbf{X}$ only if the second term is reduced as a result. However, since for any $\mathbf{p}$ we have: $\|(\Phi_{\mathbf{X}} + \Phi_{\perp\mathbf{X}})\mathbf{p}\|^2 \geq \|\Phi_{\mathbf{X}}\mathbf{p}\|^2$, without loss of generality the optimal basis admits the following decomposition:

$$\Phi = (\mathbf{I} \otimes \mathbf{X})\mathbf{B}. \tag{12}$$

Letting $\mathbf{K} = \mathbf{X}^T\mathbf{X}$, Eq. (11) can be written:

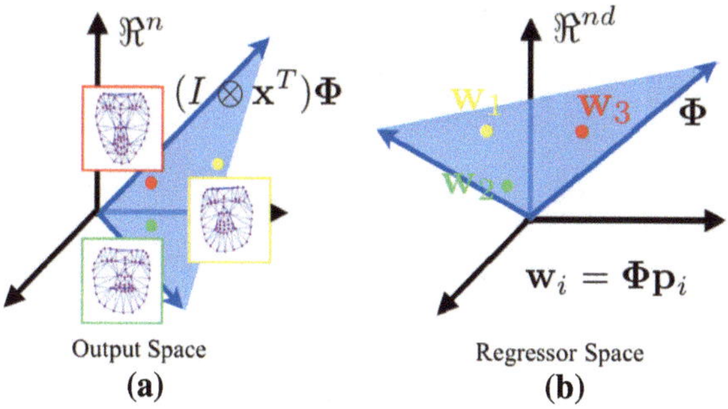

Fig. 6 When considering the use of PRA as a prior for face alignment, it can be interpreted in one of two ways. **a** PRA predicts an image-specific subspace containing the target facial shape. **b** PRA describes the subspace that spans all the image-specific regressors

$$\min_{\mathbf{B},\mathbf{p}_i} \sum_{i=1}^{N} \left\| \mathbf{y}_i - \left(\mathbf{I} \otimes \mathbf{K}_{(i)}^T\right) \mathbf{B}\mathbf{p}_i \right\|^2 + \beta \, \|\mathbf{B}\mathbf{p}_i\|^2_{(\mathbf{I}\otimes\mathbf{K})}. \tag{13}$$

For $\mathbf{B} = [\mathbf{B}_1; \ldots; \mathbf{B}_n]$, the prediction for the jth output dimension given a new input takes the form:

$$f_j(\mathbf{x}) = \mathbf{p}^T \mathbf{B}_j^T [k(\mathbf{x}; \mathbf{x}_1); \ldots; k(\mathbf{x}, \mathbf{x}_N)], \tag{14}$$

where k is the kernel function. For a particular choice of $\mathbf{p}$, the form reduces to that of conventional kernel regression.

4.3 PRA Regularisation for Face Alignment

The effect of the regularisation term in Eq. (11) is that the output data does not typically lie entirely within the predicted subspaces. As such, restricting search to within the predicted subspace may be suboptimal. Instead, it may be more beneficial to treat the predicted subspace as a soft constraint during search. One way to do this is by introducing it as an image-specific prior over the space of facial shapes for that image.

The variable $\mathbf{p}$ in Eq. (7) has a dual interpretation. For a given $\mathbf{x}$, $\mathbf{p}$ denotes coordinates of $\mathbf{y}$ within the image-specific (non-orthonormal) subspace in Eq. (10) (see Fig. 6a). It can also be interpreted as the coordinates of regressors within Φ (see Fig. 6b). Analogous to how the distribution of coordinates in PCA space is often modelled as a Gaussian distribution [31], one can similarly model the distribution of regressors in Φ:

$$\mathbf{p} \sim N(\mu_p, \Sigma_p). \tag{15}$$

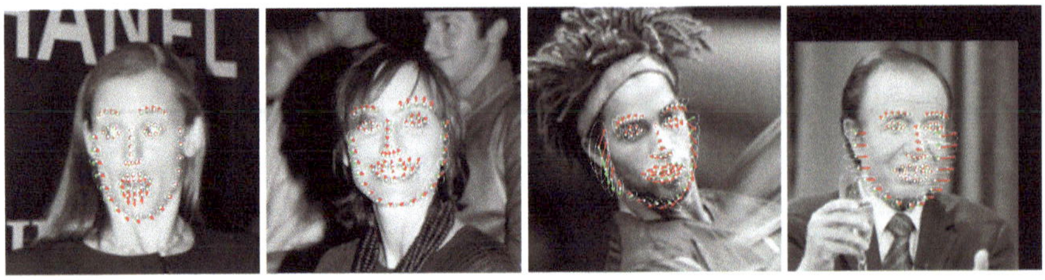

Fig. 7 Examples of the subspaces for regularising face alignment predicted using PRA with $h = 2$ on images from the Faces in the Wild dataset [24]. The red and green arrows denote the first and second direction of variation respectively. The white ellipses denote the (2×2)-block diagonal components of the covariance of the PRA-prior in Eq. (16)

For use as a prior over the coordinates of $\mathbf{y} \in \Re^n$, the regression basis Φ and the current observation $\mathbf{x}$ can be used to transform the regressor prior to an image-specific prior $\mathbf{y} \sim N(\mu_y, \Sigma_y)$, with:

$$\mu_y = \hat{\mathbf{x}}\Phi\mu_p \quad \text{and} \quad \Sigma_y = \hat{\mathbf{x}}\Phi\Sigma_p\Phi^T\hat{\mathbf{x}}^T + \sigma^2\mathbf{I}. \tag{16}$$

Here, $\hat{\mathbf{x}} = \left(\mathbf{I} \otimes \mathbf{x}^T\right)$ and σ^2 is the variance of the estimation error (i.e. the first term in Eq. (11)).

Incorporating the prior in Eq. (16) into a search procedure entails setting the regularisation term in Eq. (3) as the negative log of the Gaussian distribution:

$$\mathscr{R}(\mathbf{y}) = \frac{1}{2}\|\mathbf{y} - \mu_y\|^2_{\Sigma_y^{-1}}. \tag{17}$$

Since this regulariser is quadratic in $\mathbf{y}$, it preserves the properties of most search objectives, allowing the same search strategy to be used with or without the prior. The application of PRA in this way is closely related to regularisation with a PCA prior. The main difference is that in PCA, the prior is static as it constitutes the marginal likelihood of the output. In contrast, PRA prescribes a prior specialised to the current input that is much more compact (i.e. close to low rank when σ^2 is small).

Some examples of the predicted subspace constraints using PRA are shown in Fig. 7. The first thing to notice is that the priors are different for each image and are, in a way, specialised to the face they contain. For example, in the first two images, the predicted subspaces appear to capture the fact that the subjects are displaying facial expressions, where the directions of variability pertain to intensifying/attenuating those specific expressions (i.e. surprise and smile). Similarly, for the other two images, the subspaces appear to capture the fact that the faces are not in a frontal pose and, thus, encode uncertainty about pose of the face.

5 Optimisation

With the descriptions of the feature detectors in Sect. 3 and the PRA induced prior over facial geometry in Sect. 4, the components of the objective in Eq. (3) can now be fully described.

5.1 Formulation

As described in Sect. 4, the way in which the locations of facial features in an image can vary is governed by a linear deformation model (PDM):

$$\mathbf{x} = \bar{\mathbf{x}} + \Phi\mathbf{p}; \quad \mathbf{x} = [\mathbf{x}_1; \ldots; \mathbf{x}_n]. \tag{18}$$

The deformation basis, Φ, can be found by applying a low-rank decomposition to the sample covariance matrix. In the case where PRA is used, the mean and covariance of this deformation space is described by Eq. (16).[4] In CLM alignment, the true facial feature locations are assumed to be at one of the locations where the detector was evaluated:

$$\mathbf{x}_i \in \Omega_i; \quad \mathscr{D}_i(\mathbf{x}_i) = -\log\{p(I_i|\mathbf{x}_i)\}. \tag{19}$$

Due to the truncation used in PCA, or the prediction error in PRA, the deformation model can not perfectly reconstruct the true feature locations in an image. That is, there may be no combination of $\mathbf{x} = [\mathbf{x}_1; \ldots; \mathbf{x}_n]$ that is spanned by the PDM. To account for this, it is often assumed that the reconstruction error over facial feature locations is homoscedastic isotropic Gaussian distributed:

$$\mathbf{x}_i = (\bar{\mathbf{x}}_i + \Phi_i\mathbf{p}) + \varepsilon_i; \quad \varepsilon_i \sim N(\mathbf{0}, \sigma^2\mathbf{I}). \tag{20}$$

Here, σ denotes the standard deviation of the noise on feature locations. Assuming the training set of facial feature locations were all located at integer pixel locations, and that Ω_i denotes all integer pixel locations in the image, σ can be inferred from the covariance matrix as follows [31]:

$$\sigma^2 = \frac{1}{2n-h}\sum_{i=h+1}^{2n}\lambda_i; \quad \Sigma = [\Phi \mid \Psi]\,\mathrm{diag}\{\lambda\}\,[\Phi \mid \Psi]^T, \tag{21}$$

which is simply the arithmetic average of the eigenvalues in Ψ; the subspace orthogonal to Φ.

[4] Note that in this section we revert back to the convention of denoting by $\mathbf{x}$ the concatenated coordinates of facial features, which corresponds to $\mathbf{y}$ in Sect. 4, and Φ again refers to the deformation basis rather than the regression basis as in Sect. 4.

Describing the relationship between the PDM and the feature detections in this way leads to a decomposition of the regulariser in Eq. (3) into two terms:

$$\mathscr{R}(\mathbf{x}) = \frac{1}{2\sigma^2}\|\mathbf{x} - (\bar{\mathbf{x}} + \Phi\mathbf{p})\|^2 + \frac{1}{2}\|\mathbf{p}\|^2_{\Lambda^{-1}}; \quad \Lambda = \text{diag}\{[\lambda_1; \ldots; \lambda_h]\}, \tag{22}$$

where λ_i denotes the eigenvalue of the ith mode of deformation. When assuming a non-informative (uniform) prior over the PDM parameters (i.e. $\lambda_i \rightarrow \infty$), the formulation in Eq. (3) leads to a Maximum Likelihood (ML) estimate [41], otherwise it leads to a Maximum *a-posterior* (MAP) estimate.

The final objective for CLM alignment takes the form:

$$\min_{\mathbf{p}\in\Re^h, \mathbf{x}_i\in\Omega_i} \frac{1}{2\sigma^2}\|\mathbf{x} - (\bar{\mathbf{x}} + \Phi\mathbf{p})\|^2 + \frac{1}{2}\|\mathbf{p}\|^2_{\Lambda^{-1}} - \sum_i \log\{\pi_{\mathbf{x}_i}\}; \quad \pi_{\mathbf{x}_i} = p(I_i \mid \mathbf{x}_i). \tag{23}$$

This is a mixed integer programming problem which is well known to be intractable. Common strategies for solving this problem include parameterising the detector responses using simpler forms [48] or relaxing the prior over facial geometry to one that admits exact inference [16]. The main drawback of these approaches is that, the optimal solution they afford does not necessarily correspond to the global solution of the original problem. Recently, an exact method for solving Eq. (23) was proposed in [1]. However, as it is an instance of the branch-and-bound (BnB) approach to optimisation, it relies on a sparse set of candidate facial feature locations (i.e. using non-maximum suppression etc.) in order for the BnB strategy to terminate rapidly. As such, it is difficult to incorporate characteristics of the response maps that reflect the aperture effect.[5]

Face alignment usually plays the role as a front end process for higher level inference process. Thus, the computational complexity of its inference must be kept to a minimum. Given inherent intractability of the problem in Eq. (23) one must rely on local optimisation strategies in order to afford rapid alignment. The main problem with most local optimisation strategies is their sensitivity towards local minima of the objective. In the following section, a method that aims to reduce the effects of local minima whilst preserving computational efficiency will be discussed.

5.2 Subspace Constrained Mean-Shifts

A common approach for reducing sensitivity of local optimisation strategies against local minima is the so called coarse-to-fine method (also called the continuation method). This approach involves minimising a successively less smoothed version

[5] The specific application of the method in [1] was for initialising a 3DMM. As such, a highly precise solution was of less importance.

of the original problem, where the solution at the coarser level is used as an initial estimate at the finer scale. One way to smooth the CLM objective is to marginalise over the facial feature locations in the image:

$$p(\mathbf{p}|I) \propto p(\mathbf{p}) \prod_i \sum_{\mathbf{x}_i \in \Omega_i} p(I_i|\mathbf{x}_i)\, p(\mathbf{x}_i|\mathbf{p}) \tag{24}$$

$$= p(\mathbf{p}) \prod_i \sum_{\mathbf{x}_i \in \Omega_i} \pi_{\mathbf{x}_i} N(\hat{\mathbf{x}}_i; \mathbf{x}_i, \sigma^2 \mathbf{I}); \quad \hat{\mathbf{x}}_i = \bar{\mathbf{x}}_i + \Phi_i \mathbf{p}. \tag{25}$$

This can be interpreted as making a non-parametric estimate of the response map in the form of a homoscedastic isotropic Gaussian kernel density estimate (KDE) [44].

The quality of this nonparametric estimate of the response map depends largely on the choice of candidate feature location sets $\{\Omega_i\}_{i=1}^n$. If the candidates are sampled too sparsely, they may not adequately cover the space of variations and σ, which is learned from training data using Eq. (21), will be underestimated. However, since the space of $\mathbf{x}_i$ is the 2D image plane, it is computationally tractable to compute the likelihood of a dense set of candidates locally around the current PDM estimate through an exhaustive local search over all integer pixel locations.

In [42], it was shown that Eq. (25) can be maximised using the EM algorithm. Treating the true facial feature locations $\{\mathbf{x}_i\}_{i=1}^n$ as hidden variables, in the E-step the *posterior* over the candidates are evaluated:

$$\omega_{\mathbf{x}_i} = p(\hat{\mathbf{x}}_i \mid \mathbf{x}_i, I_i) = \frac{\pi_{\mathbf{x}_i}\, N(\hat{\mathbf{x}}_i; \mathbf{x}_i, \sigma^2 \mathbf{I})}{\sum_{\mathbf{z}_i} \in \Omega_i \pi_{\mathbf{z}_i}\, N(\hat{\mathbf{x}}_i; \mathbf{z}_i, \sigma^2 \mathbf{I})} \tag{26}$$

Then, the M-step involves minimising:

$$Q(\mathbf{p}) = E\left[-\log \left\{ p(\mathbf{p}) \prod_i p(\hat{\mathbf{x}}_i \mathbf{x}_i, I_i) \right\} \right] \tag{27}$$

$$\propto \|\mathbf{p}\|^2_{\Lambda^{-1}} + \sum_i \sum_{\mathbf{x}_i \in \Omega_i} \frac{\omega_{\mathbf{x}_i}}{\sigma^2} \left\| \mathbf{x}_i - \hat{\mathbf{x}}_i \right\|^2. \tag{28}$$

Using the relationship: $\sum_{\mathbf{x}_i \in \Omega_i} \omega_{\mathbf{x}_i} = 1$, the solution for the linear deformation model can be written:

$$\Delta \mathbf{p} = \left(\sigma^2 \Lambda^{-1} + \Phi^T \Phi \right)^{-1} \left(\Phi^T \mathbf{v} - \sigma^2 \Lambda^{-1} \alpha \right); \quad \mathbf{p} \leftarrow \mathbf{p} + \Delta \mathbf{p}, \tag{29}$$

where $\mathbf{v} = [\mathbf{v}_1; \ldots; \mathbf{v}_n]$ is the concatenation of the mean-shift vectors for each facial feature:

$$\mathbf{v}_i = \left(\sum_{\mathbf{x}_i \in \Omega_i} \omega_{\mathbf{x}_i} \mathbf{x}_i \right) - \hat{\mathbf{x}}_i \tag{30}$$

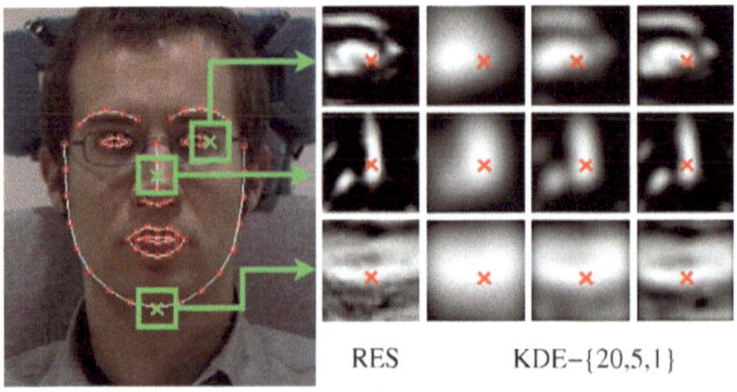

Fig. 8 Visualisation of the marginalised objective for three response maps locally around three facial features in an image from [20]. The *first row* denotes the raw response maps computed using the facial feature detector in Eq. (5), and the other three are their non-parametric estimates using different settings for σ^2

Notice that in the case that a ML solution is desired (i.e. $p(\mathbf{p})$ is non-informative), the solution for the parameter update $\Delta\mathbf{p}$ is simply the projection of the mean-shift vectors onto the deformation subspace Φ. Thus, this strategy was coined the name subspace constrained mean-shift (SCMS). In either case, the form in Eq. (29) suggests a simple and efficient implementation, consisting of an alternation between computing the mean shift-vectors and their regularisation by the deformation model.

In [15] it was previously shown that mean-shift is a bound optimisation. This was later extended in [3] by showing that for Gaussian kernels, mean-shift is equivalent to employing the EM algorithm as an optimisation strategy. With the derivation of presented in [42], it was show that such an interpretation can be generalised further to problems with conditionally independent likelihoods. As such, the desirable properties of the EM algorithm, namely provably convergent and improving, are adopted by this optimisation strategy.

5.3 Smoothing, Subspaces and Local Minima

SCMS provides a convenient way to implement a coarse-to-fine fitting strategy over the objective in Eq. (23). Specifically, the variance of the kernel, σ^2, used in the non-parametric estimate of the response maps defines the degree of smoothing one applies over the response maps (see Fig. 8). The guiding principle here is similar to that of optimising on a Gaussian pyramid. It can be shown that when using Gaussian kernels, there exists a $\sigma < \infty$ such that the KDE is unimodal, regardless of the distribution of samples [4]. As σ is reduced, modes divide and smoothness of the objective's terrain decreases. However, it is likely that the optimum of the objective at a larger ρ is closest to the desired mode of the objective with a smaller σ, promoting its convergence to the correct mode. As such, the policy under which σ is reduced acts to guide optimisation towards the global optimum of the true objective.

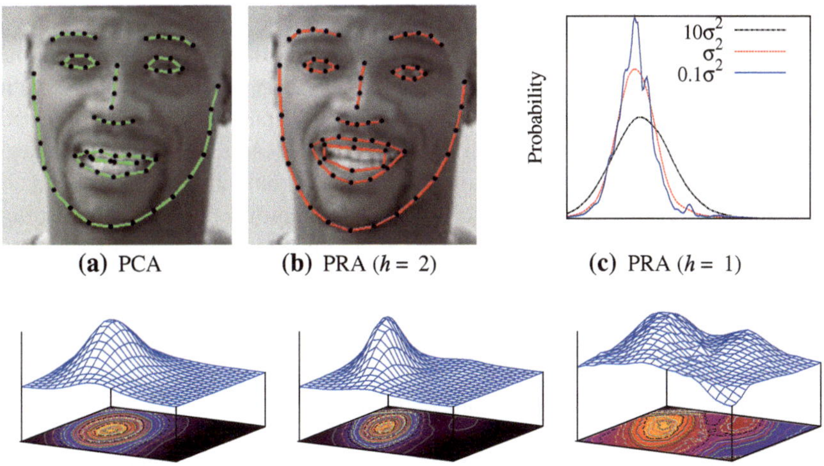

(a) PCA **(b)** PRA (h = 2) **(c)** PRA (h = 1)

(d) PRA (h = 2, ρ = $10\sigma^2$) **(e)** PRA (h = 2, ρ = σ^2) **(f)** PRA (h = 2, ρ = $0.1\sigma^2$)

Fig. 9 Illustration of the effects of smoothing on the posterior likelihood in the predicted PRA subspace. **a** SCMS alignment result using a PCA subspace ($h = 20$). **b** SCMS alignment result using a PRA subspace ($h = 2$). **c** Posterior likelihood within the PRA predicted subspace ($h = 1$) between $\pm 3\lambda$ for different settings of the smoothing parameter ρ. **d–f** Posterior likelihood within the PRA predicted subspace ($h = 2$) between $\pm 3\lambda_i$ for different settings of the smoothing parametre ρ

It should be noted that in formulating SCMS, the Gaussian kernel is in fact a particular incarnation of the likelihood of selecting a feature candidate given the PDM's feature estimate. As such, given sufficient granularity in the local search (i.e. the choice of $\{\Omega_i\}_{i=1}^n$), then the variance of the kernel, σ, that best represents the likelihood is given by Eq. (21). However, since PCA retains the majority of total variance of the shape, typically in the order of 95–98 %, σ will generally be quite small, which results in a highly multimodal objective terrain. When initialisation is far from the global maximum, optimisation over this terrain is susceptible to terminating in local maxima. The variance tightening policy described above essentially replaces the optimal objective terrain with a smoothed estimate of itself, for which local maxima are reduced, but the global optimum is perturbed, the magnitude of which is directly related to the choice of σ.

One aspect of optimisation to consider here is wether smoothing is necessary given the substantial reduction of the space of plausible solutions afforded by the PRA predicted subspace. As discussed previously, restricting the solution to a low dimensional subspace can substantially reduce the likelihood of the optimisation strategy terminating in a local minimum since the space in which consensus between strong facial feature detections is possible is greatly reduced. In Fig. 9, the terrain of the posterior likelihood in Eq. (25) over the PDM parameters for one- and two-dimensional PRA basis are shown at different settings for the smoothing parameter ρ. One notices that for $\rho = \sigma^2$, calculated using Eq. (21), the terrain is smooth with a single dominant mode. In fact, when $h = 1$, the posterior exhibits only a single stationary point. However, for $h = 2$, the terrain exhibits a (shallow) spurious mode. Thus, when poorly initialised, a local optimisation strategy may converge to that sub-optimal solution. When using $\rho = 10\,\sigma^2$, this spurious mode is eliminated. Thus, even with the substantial constraint afforded by PRA, a coarse-to-fine strategy is still beneficial. Although coarse-to-fine strategies is only a heuristic that can not guarantee convergence to the optimal solution, the removal of many of the *strong* spurious modes by the PRA subspace constraint greatly increases its efficacy. As a comparison, Fig. 9a, b shows the solutions found by applying the SCMS algorithm to the same image using the PCA and PRA priors respectively. The reason why PCA fails in this case is because there are still many strong spurious modes within the PCA subspace. Many of these have been eliminated in the much smaller space predicted by PRA, promoting convergence to the optimal solution.

Finally, for extremely small dimensional PRA (i.e. $h = 1$ or 2), it may be possible to perform an exhaustive search for the best solution within the predicted subspace. However, questions still remain about how to best tessellate the space within this subspace. Furthermore, despite being computationally tractable, the computational complexity of such an approach is typically much higher than a the local optimisation strategy described in this section. Furthermore, results from experiments in [38] suggest that for general pseudo-frontal face alignment, PRA performs the best for $h = 3$, which can be computationally expensive to explore exhaustively. Thus, for practical applications that require an efficient face alignment algorithm, the combination of PRA and SCMS constitutes a good balance between computational complexity and accuracy.

6 Conclusion

In this chapter we have reviewed a method for deformable face alignment that leverages the excellent generalisation properties of local appearance representations of parts and the strong global constraints imposed by the geometrical relationships between part locations. The utility of simple linear support vector machines for facial feature detectors was motivated primarily from the perspective that high-capacity models are unlikely to be capable of distinguishing regions with similar local appearance in an image. In order to eliminate a large portion of *strong* spurious candidate configurations, the employment of a highly compact prior over possible face shapes was shown to be necessary. For this, the method of principal regression analysis was shown to perform well, capable of reducing the search space to a few directions of uncertainty. Finally, an efficient local optimisation algorithm that admits an elegant coarse-to-fine strategy called subspace constrained mean-shifts was shown to be capable of steering away from the spurious configurations remained within the PRA predicted subspace. In summary, the combination of simple feature detectors, a strong global prior over allowable facial shapes, and a robust and efficient optimisation has been shown to be a combination that work well together, with each component complementing the weaknesses of the others.

Directions of future work will involve extending the applications of these methods to more general settings, including large pose variations, extreme lighting conditions and partial occlusions. Given the difficulty of this problem, it is unlikely that a single solution will solve the problem in all cases. A more promising direction would be to leverage the strengths of various disparate methods in a complementary fashion.

References

1. Amberg B, Vetter T (2011) Optimal landmark detection using shape models and branch and bound. In: Proceedings ICCV'11: international conference on computer vision
2. Blanz V, Vetter T (1999) A morphable model for the synthesis of 3D-faces. In: SIGGRAPH'99
3. Carreira-Perpinan M (2007) Gaussian mean-shift is an EM algorithm. PAMI 29(5):767–776
4. Carreira-Perpinan M, Williams C (2003) On the number of modes of a gaussian mixture. Lect Notes Comput Sci 2695:625–640
5. Cootes T, Taylor C (1992) Active shape models–'smart snakes'. In: British achine vision conference (BMVC'92), pp 266–275
6. Cootes TF, Cooper D, Taylor CJ, Graham J (1995) Active shape models–their training and application. Comput Vis Image Underst 61:38–59
7. Cootes T, Edwards G, Taylor C (1998) Active appearance models. In: ECCV'98, pp 484–498, 1998
8. Cootes TF, Edwards GJ, Taylor CJ (1998) A comparative evaluation of active appearance model algorithms. In: BMVC'98: proceedings of the 9th British machine vision conference, vol 2, pp 680–689
9. Cortes C (1995) Vapnik V (1995) Support-vector networks. Mach Learn 20:273–297
10. Cristinacce D, Cootes T (2004) Feature detection and tracking with constrained local models. In: EMCV'04, pp 929–938
11. Cristinacce D, Cootes T (2007) Boosted active shape models. In: BMVC'07: proceedings of the 18th British machine vision conference, vol 2, pp 880–889
12. Dalal N, Triggs B (2005) Histograms of oriented gradients for human detection. In: ICCV, vol 2, pp 886–893
13. di Mauro EC, Cootes TF, Page GJ, Jackson CB (1996) Check!: a generic and specific industrial inspection tool. VISP'96: proceedings of the conference on vitalizing city centres through integrated spatial, planning 143:241–249
14. Donner R, Reiter M, Langs G, Peloschek P, Bischof H (2006) Fast active appearance model search using canonical correlation analysis. Trans Pattern Anal Mach Intell (TPAMI) 28: 1690–1694
15. Fashing M, Tomasi C (2005) Mean shift as a bound optimization. PAMI 27(3):471–474
16. Felzenszwalb PR, Girshick DM, Ramanan D (2009) Object detection with discriminatively trained part-based models. IEEE Pattern Anal Mach Intell
17. Ginneken BV, Stegmann MB, Loog M (2006) Segmentation of anatomical structures in chest radiographs using supervised methods: a comparative study on a public database. Med Image Anal 10:19–40
18. Gross R, Matthews I, Baker S (2005) Generic vs. person specific active appearance models. IVC 23:1080–1093
19. Gross R, Sweeney L, la Torre Frade FD, Baker S (2006) Model-based face de-identification. In: Proceedings of the 2006 conference on computer vision and pattern recognition, workshop, pp 161–168

20. Gross R, Matthews I, Baker S, Kanade T (2007) The CMU multiple pose, illumination and expression (MultiPIE) database. Technical report, Robotics Institute, Carnegie Mellon University
21. Gu L, Kanade T (2008) A generative shape regularization model for robust face alignment. In: ECCV'08
22. Hansen DW, Hansen JP, Niels M, Stegmann MB (2002) Eye typing using Markov and active appearance models. In: WACV'02: proceedings of the 6th IEEE workshop on applications of computer vision, p 132
23. Horn BKP, Schunck BG (1981) Determining optical flow. Artif Intell 17:185–203
24. Huang G, Ramesh M, Berg T, Learned-Miller E (2007) Labeled faces in the wild: a atabase for studying face recognition in unconstrained environments. Techical report 07–49, University of Massachusetts, Amherst
25. la Torre Frade FD, Romea AC, Cohn J, Kanade T (2007) Filtered component analysis to increase robustness to local minima in appearance models. In: CVPR'07: proceedings of the IEEE computer society conference on computer vision and pattern recognition
26. Liu X (2007) Generic face alignment using boosted appearance model. In: CVPR'07: proceedings of the IEEE computer society conference on computer vision and pattern recognition, pp 1–8
27. Lucas B, Kanade T (1981) An iterative image registration technique with an application to stereo vision. In: Proceedings of imaging understanding workshop
28. Matthews I, Baker S (2004) Active appearance models revisited. IJCV'04 60:135–164
29. Mercer J (1909) Functions of positive and negative type and their connection with the theory of integral equations. Philos Trans R Soc A 209:415–446
30. Mittrapiyanuruk P, DeSouza GN, Kak AC (2005) Accurate 3D tracking of rigid objects with occlusion using active appearance models. In: WACV-MOTION'05: proceedings of the IEEE workshop on motion and video, computing vol 2, pp 90–95
31. Moghaddam B, Pentland A (1997) Probabilistic visual learning for object representation. PAMI 19(7):696–710
32. Nguyen M, De la Torre Frade F (2008) Local minima free parameterized appearance models. In: CVPR'08
33. Ong EJ, Lan Y, Theobald B, Harvey R, Bowden R (2009) Robust facial feature tracking using selected multi-resolution linear predictors. In: computer vision, 2009 IEEE 12th international conference on, pp 1483–1490
34. Peyras J, Bartoli A, Mercier H, Dalle P (2007) Segmented AAMs improve person-independent face fitting. In: BMVC'07: Proceedings of the 18th British machine vision conference
35. Platt JC (1999) Probabilistic outputs for support vector machines and comparisons to regularized likelihood methods. advances in large margin classifiers. MIT Press, Cambridge
36. Roberts M, Cootes T, Adams J (2007) Vertebral morphometry: semi-automatic determination of detailed vertebral shape from DXA images using active appearance models. Invest Radiol 41:849–859
37. Romdhani S, Gong S, Psarrou A (1999) A multi-view nonlinear active shape model using kernel PCA. In: BMVC'99: proceedings of the 10th British machine vision conference, pp 438–492
38. Saragih J (2011) Principal regression analysis. In: CVPR'11: proceedings of the IEEE computer society conference on computer vision and pattern recognition
39. Saragih J, Goecke R (2006) Iterative error bound minimisation for AAM alignment. In: ICPR'06, vol 2, pp 1192–1195
40. Saragih J, Goecke R (2007) A nonlinear discriminative approach to AAM fitting. In: ICCV'07
41. Saragih J, Lucey S, Cohn J (2009) Subspace constrained mean-shift. In: ICCV'09
42. Saragih J, Lucey S, Cohn J (2010) Deformable face fitting using subspace constrained mean shifts. IJCV
43. Saragih JM, Lucey S, Cohn JF (2011) Real-time avatar animation from a single image. In: Automatic face gesture recognition and workshops (FG 2011), 2011 IEEE international conference on, pp 117–124

44. Silverman B (1986) Density estimation for statistics and data analysis. Chapman and Hall/CRC, London
45. Theobald BS, Kruse GC, Bangham JA (2003) Towards a low bandwidth talking head using appearance models. J Image Vision Comput 21:1077–1205
46. Tian Y, Narasimhan SG (2011) Globally optimal estimation of nonrigid image distortion. IJCV
47. Wang S, Wang Y, Li B (2006) Face decorating system based on improved active shape models. In: ACE '06: proceedings of the ACM SIGCHI international conference on advances in computer entertainment technology, p 65
48. Wang Y, Lucey S, Cohn J (2008) Enforcing convexity for improved alignment with constrained local models. In: CVPR'08
49. Wei X, Yin L, Zhu Z, Ji Q (2004) Avatar-mediated face tracking and lip reading for human computer interaction. In: MULTIMEDIA'04: proceedings of the 12th annual international conference on multimedia, pp 500–503
50. Zhang J, Zhou S, McMillan L, Comaniciu D (2007) Joint real-time object detection and pose estimation using probabilistic Boosting Network. In: CVPR'07: proceedings of the IEEE computer society conference on computer vision and, pattern recognition, pp 1–8
51. Zhou X, Comaniciu D, Gupta A (2005) An information fusion framework for robust shape tracking. PAMI 27(1):115–129
52. Zhou SK, Georgescu B, Zhou, XS, Comaniciu D (2005) Image based regression using boosting method. In: ICCV'05: proceedings of the 10th international conference on computer vision, vol 1, pp 541–548
53. Zhou S, Guo F, Park JH, Carneiro G, Jackson J, Brendel M, Simopoulos C, Otsuki J, Comaniciu D (2007) A probabilistic, hierarchical, and discriminant (PHD) framework for rapid and accurate detection of deformable anatomic structure. In: ICCV'07: proceedings of the 11th international conference on computer vision

Learning-Based Detection and Tracking in Medical Imaging: A Probabilistic Approach

Yang Wang, Bogdan Georgescu, Terrence Chen, Wen Wu, Peng Wang, Xiaoguang Lu, Razvan Ionasec, Yefeng Zheng and Dorin Comaniciu

Abstract Medical image processing tools are playing an increasingly important role in assisting the clinicians in diagnosis, therapy planning and image-guided interventions. Accurate, robust and fast tracking of deformable anatomical objects, such as the heart, is a crucial task in medical image analysis. One of the main challenges is to maintain an anatomically consistent representation of target appearance that is robust enough to cope with inherent changes due to target movement, imaging device movement, varying imaging conditions, and is consistent with the domain expert clinical knowledge. To address these challenges, this chapter presents a probabilistic framework that relies on anatomically indexed component-based object models which integrate several sources of information to determine the temporal trajectory of the deformable target. Large annotated imaging databases are exploited to encode the domain knowledge in shape models and motion models and to learn discriminative image classifiers for the target appearance. The framework robustly fuses the prior information with traditional tracking approaches based on template match-

Y. Wang · B. Georgescu (✉) · T. Chen · W. Wu · P. Wang · X. Lu · R. Ionasec · Y. Zheng · D. Comaniciu
Imaging and Computer Vision, Siemens Corporate Research, Princeton, NJ 08540, USA
e-mail: bogdan.georgescu@siemens.com

T. Chen
e-mail: terrence.chen@siemens.com

W. Wu
e-mail: wen.wu@siemens.com

P. Wang
e-mail: peng-wang@siemens.com

X. Lu
e-mail: Xiaoguang.lu@siemens.com

Y. Wang
e-mail: yang-wang@siemens.com

D. Comaniciu
e-mail: dorin.comaniciu@siemens.com

M. González Hidalgo et al. (eds.), *Deformation Models*, Lecture Notes in Computational Vision and Biomechanics 7, DOI: 10.1007/978-94-007-5446-1_9,

ing and registration. We demonstrate various medical image analysis applications with focus on cardiology such as 2D auto left heart, catheter detection and tracking, 3D cardiac chambers surface tracking, and 4D complex cardiac structure tracking, in multiple modalities including Ultrasound (US), cardiac Computed Tomography (CT), Magnetic Resonance Imaging (MRI), and X-ray fluoroscopy.

1 Introduction

Cardiovascular diseases such as cardiomyopathy and heart failure are the leading causes of morbidity and mortality, which account for 1 of every 2.9 deathes and require over 100,000 surgeries in the United States alone every year [22]. To assist diagnosis and evaluation of the progression of diseases, recent advances in medical imaging technologies allow cardiologists to capture morphological and functional information of complex structures, such as heart anatomies, in two, three, and four dimensional dynamic scans. For instance, in real-time echocardiography unstitched volumetric data can be captured in a high volume rate and permit quantification of cardiac strain in a non-invasive manner [10, 12, 40]. Cardiac Magnetic Resonance Imaging (MRI) and Computed Tomography (CT) allow morphological characterization of heart structures with precision [23, 30, 37, 46], and provide a wide topological field of view with visualization of the heart, its internal morphology, and the surrounding mediastinal structures. The X-ray angiography is the primary modality in image-guided interventions, such as percutaneous coronary interventions (PCI) and catheter-based electrical physiology (EP) therapies [38, 41, 42], to precisely visualize and target the surgical site.

As medical imaging becomes more sophisticated and more central to clinical decision-making, there is an evolving need to provide objective, quantitative results for diagnosis, therapy planning, and disease monitoring. However, it remains a time-consuming task for clinicians to extract comprehensive structural and dynamic information from medical imaging. In order to exploit such time-resolved data, fast and precise image processing tools become a crucial part of the analysis workflow.

One of the challenging problems on visual tracking of deformable objects is to maintain a representation of target appearance, which is robust enough to cope with inherent changes due to target movement and/or imaging device movement. Traditional methods based on template matching have to adapt the model template in order to successfully locate and track the target [27, 28]. Without adaptation, tracking is reliable only over short periods of time when the appearance does not change significantly. However, in most applications the target appearance undergoes considerable changes after a long time period and furthermore, accumulated motion error and rapid visual changes make the model to drift away from the tracked target. To improve tracking performance, one can also impose object specific subspace constraints [3, 13] or maintain a statistical representation of the model [20, 29, 31]. This representation, often modeled as a probability distribution function, can be determined a priori or ideally computed online. More sophisticated

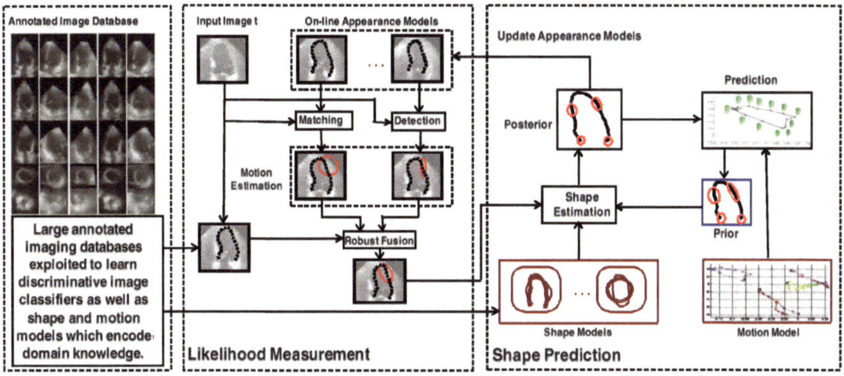

Fig. 1 A block diagram of the probabilistic motion estimation framework including the likelihood measurement and shape prediction processes

approaches, such as adaptive mixture models, have also been proposed to cope with outliers and sudden appearance changes [20].

Recent progress in discriminative learning, along with availability of large medical databases with expert annotation of the structures of interest, make a learning-based approach attractive to achieve robust object detection and tracking in medical imaging. In this chapter, a probabilistic approach is presented to combine learning-based and conventional approaches to obtain the best of both worlds. As illustrated in Fig. 1, a set of component-based models are maintained to determine the next position of the target by combining several sources of information. This approach is a flexible framework to integrate model information across frames through component-based object representations. It can be tailored to perform tracking-by-detection by leveraging domain knowledge encoded in shape models and image based discriminative classifiers, as well as dynamic information encoded in motion models. Alternatively, it can also be tailored towards traditional methods with template based matching/registration, such as optical-flow tracking.

Compared to the existing methods, such as image registration [10, 14, 17] and optical flow [12], this presented framework has the following advantages:

1. Information from multiple cues, such as feature patterns, image gradients, boundary detection, and motion prediction, are fused into a single probabilistic objective function to improve tracking accuracy and robustness.
2. Expert annotations are exploited to learn discriminative image classifiers as well as shape and motion models which encode the domain knowledge.
3. Image quality measurements based on image intensities and feature scores are integrated in a probabilistic framework to handle noise and signal dropouts in the medical imaging data.
4. Efficient optimization is proposed to achieve high speed performance.
5. This system provides a fully automatic solution to track the deformable targets and to provide quantitative analysis of the non-rigid motion.

To demonstrate the performance, we apply this framework in various medical imaging applications with a focus in cardiology, including 2D heart and device (e.g., catheter and guidewire) detection and tracking, 3D cardiac chamber surface tracking in multiple modalities including CT, US, and MRI, and 4D complex cardiac structure tracking, e.g., on heart valves.

2 A Probabilistic Framework for Model-Based Detection and Tracking

In this section a unified framework is introduced for fusing motion estimates from multiple appearance models and fusing a subspace shape model with the system dynamics and measurements with point-dependent noise. The appearance variability is modeled by maintaining several models over time, which can be both learned offline and updated online. This leads to a nonparametric representation of the probability density function that characterizes the object appearance. Inspired by [7], tracking is performed by obtaining independently from each model a motion estimate and its uncertainty through a single probabilistic framework as follows,

$$\arg\max_{\mathbf{X}_t} p(\mathbf{X}_t|\mathbf{Z}_{0:t}) = \arg\max_{\mathbf{X}_t} p(\mathbf{Z}_t|\mathbf{X}_t)p(\mathbf{X}_t|\mathbf{Z}_{0:t-1}) \quad (1)$$

where $\mathbf{Z}_{0:t} = \mathbf{Z}_0, \ldots, \mathbf{Z}_t$ are the image observations from the input image sequence $I_{0:t} = I_0, \ldots, I_t$. In this framework, an anatomy-indexed mesh model is built to represent the object of interest. An example of the underlying anatomy representation is illustrated in Fig. 10. For clarity, we use $\mathbf{X}_t$ to denote a concatenation of the mesh point positions, $\mathbf{X}_t = [X_1, \ldots, X_n]$, which need to be estimated at the current time instance t, and n is the total number of points in the mesh model.

As illustrated in Fig. 1, the probabilistic framework includes the likelihood estimation and shape prediction processes, which leverages the domain knowledge encoded in image based discriminative classifiers and shape and motion models to obtain the final shape estimate. When measurement noise is anisotropic and inhomogeneous, which is often presented in image sequences of deformable objects, joint fusion of all information sources becomes critical for achieving robust and accurate tracking performance.

2.1 Learning-Based Appearance and Shape Models

Given recent advances in medical imaging devices, large databases become available with expert annotation of the structures of interest. Figure 2 shows examples of annotated 2D ultrasound images. This information can be exploited to learn domain knowledge, encoded in the form of shape models and discriminative image classifiers for target appearance.

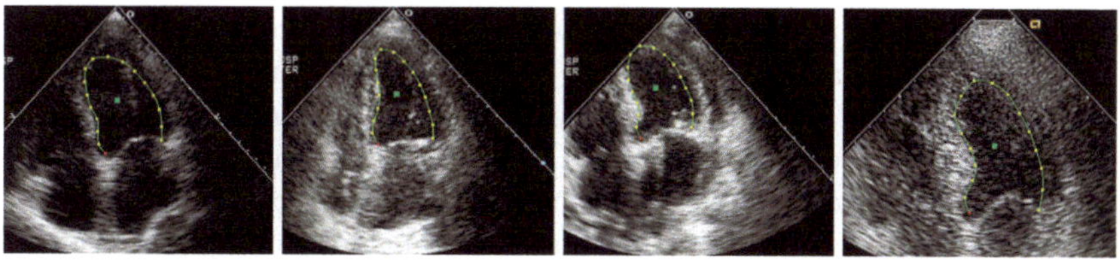

Fig. 2 Examples of 2D ultrasound images with the endocardium boundaries annotated by clinical experts. The images are captured in the apical four chamber view. The annotated endocardium boundaries are highlighted in the *green color*

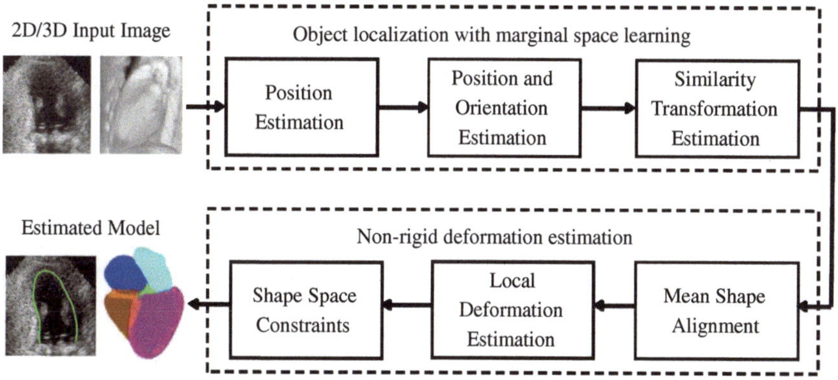

Fig. 3 Diagram for learning-based object detection and non-rigid deformation estimation

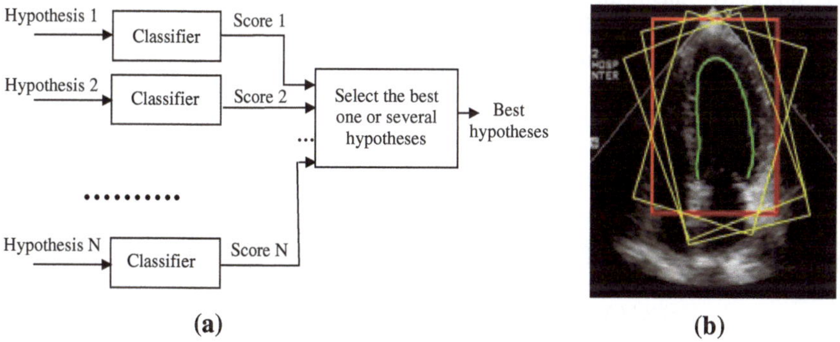

Fig. 4 An example showing the basic idea of a learning-based 3D object detection method: **a** the parameter space is quantized into a large number of discrete hypotheses and the classifier is used to select the best hypotheses in exhaustive search. **b** A few hypotheses of the left ventricle (represented as *boxes*) embedded in an ultrasound image. The *red box* shows the ground truth and the *yellow boxes* show only a few hypotheses

In the presented framework, we apply a learning-based approach for object localization, using marginal space learning (MSL) [46] and the probabilistic boosting-tree (PBT) [33], as illustrated in Fig. 3. Unlike the gradient based search in deformable models or active appearance models (AAM) [9], the full object parameter space is quantized into a large number of hypotheses and the best ones are selected by the image-based classifiers trained in this framework. Figure 4 shows the basic idea of learning-based model estimation in this section.

More specifically, to detect the model pose θ for a target object we need to solve for the similarity transformation, including translation, orientation, and scale, i.e.,

$$\theta = \left\{T^d, R^d, S^d\right\} \tag{2}$$

where T^d, R^d, S^d are the position, orientation and scale parameters in the d dimensional input data, respectively. Therefore, the object localization can be formulated as a classification problem which estimates $\theta(t)$ for each time step t from the corresponding image $I(t)$. The probability $p(\theta(t)|I(t))$ is modeled by a learned detector D, which evaluates and scores a large number of hypotheses for $\theta(t)$. D is trained using the Probabilistic Boosting Tree (PBT) [33] based on positive and negative samples extracted from the ground-truth annotations. For fast computation, efficient 3D Haar wavelet [35] and steerable features [46] can be extracted at each sampling point based on the intensity and gradient from the training data.

The object localization task is then performed by scanning the trained detector D exhaustively over all hypotheses to find the most plausible values for θ in an input data. As the number of hypotheses to be tested increases exponentially with the dimensionality of the search space, a sequential scan in the corresponding transformation parameters might be computationally unfeasible. For example, to find a 3D similarity transform, suppose each dimension in $\theta(t)$ is discretized to n values, the total number of hypotheses is n^9 and even for a small $n = 15$ it becomes extreme $3.98e^{+10}$. To overcome this limitation, we apply a marginal space search (MSL) strategy [46], which groups the original parameter space into subsets of increasing marginal spaces:

$$\Sigma_1 = (T^d),\ \Sigma_2 = (T^d, R^d),\ \Sigma_3 = (T^d, R^d, S^d).$$

Consequently, the target posterior probability can be expressed as:

$$p(\theta_t|I_t) = p(T^d|I_t)p(R^d|T^d, I_t)p(S^d|R^d, T^d, I_t). \tag{3}$$

We train a series of detectors that estimate parameters at a number of sequential stages in the order of complexity, i.e., Σ_1, Σ_2, Σ_3. Different stages utilize different features computed from the input data. Multiple hypotheses are maintained between stages, which quickly removes false hypotheses at the earlier stages while propagates the right hypothesis to the final stage. Only one hypothesis is selected as the final detection result.

With the object pose estimated, we align the mean shape (an average model of all annotations) with data to get an initial estimate of the object shape. To capture the true anatomical morphology of the target object (e.g., LV myocardium), we deform the mean shape by searching the boundary for each vertex of the model. The boundary hypotheses are taken along the normal directions at each vertex of the mean model. Detection is achieved using a boundary detector using PBT with steerable features [33, 46]. The detected boundaries are constrained by projecting the detected model onto a shape subspace obtained by the annotated dataset. As defined in Eq. (1), the shape vectors are formed by concatenating the coordinates of all control points [8, 19]. Thus, the shape space can be constructed using Procrustes analysis and principal component analysis (PCA) [11]. Although more sophisticated representations, such as local affine models [26, 47], can also be applied to constrain shape deformations, we choose the global PCA shape model due to its efficiency during online detection. In particular, the nonrigid deformation has three steps as shown in Fig. 3. First we estimate the deformation of control points which are selected based on image characteristics. The thin-plate-spline (TPS) model [4] is then used to warp the initial mesh toward the refined control points for better alignment. Last, the normal mesh points are deformed to fit the image boundary.

2.2 *Motion Manifold Learning*

Motion characteristics of an anatomical structure encodes morphological and functional properties of the object, which are important in clinical diagnosis and can be used to constrain the deformable tracking process. To obtain these motion characteristics from the pre-annotated databases, we use manifold learning to extract a compact form of the dynamic information [43].

Given a set of training sequences, we first resample a cardiac cycle of each sequence to a fixed number F (typically $F = 16$) of frames through temporal interpolation, and construct motion vectors $M = \{\mathbf{m}_0, \ldots, \mathbf{m}_i, \ldots, \mathbf{m}_n\}$ with $\mathbf{m}_i \in R^m$, where $m = N_f \times d \times F$, N_f is the number of annotation points, and d represents the dimensionality of the input data. Generalized Procrustes analysis (GPA) is then used to align all resampled motion vectors to remove the similarity transformation, including translation, rotation and scaling [11]. Because the actual number of constraints that control the LV motion are much less than its original dimensionality, the aligned 3D shape vectors lie on a low-dimensional manifold, where geodesic distance has to be used to measure the similarities. This property can be exploited by unsupervised manifold learning to discover the nonlinear degrees of freedom that underlie complex natural observations [32]. Figure 5a shows two annotated LV motion sequences. Figure 5b shows several LV motion representations in a low-dimensional manifold. An interesting but expected observation is illustrated in Fig. 5b. The LV motion is almost periodic because one cycle of heart beat starts from ED and returns to ED.

Given the whole set of 3D training shape vectors M, we apply ISOMAP [32] to find a mapping $\mathcal{F}$ which represents $\mathbf{m}_i$ in the low-dimensions as $\mathbf{m}_i = \mathcal{F}(\mathbf{v}_i) + \mathbf{u}_i$,

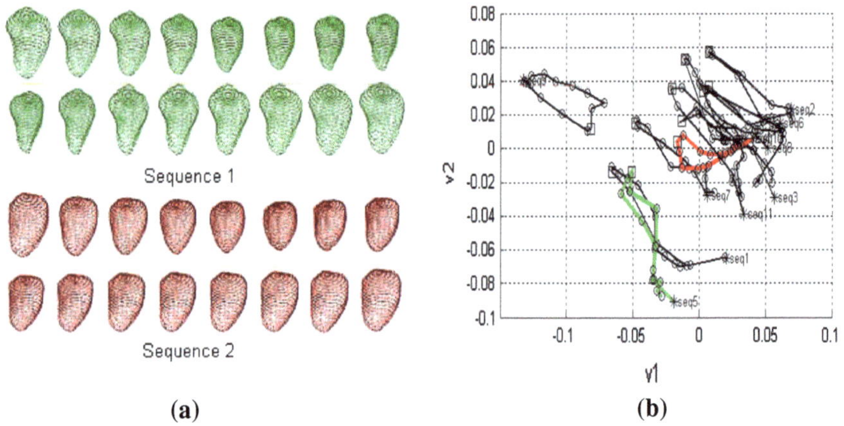

Fig. 5 Examples of manifold embedding for heart motion patterns. **a** Two *left ventricle* surface mesh sequences. **b** 11 sequences embedded in a 2D subspace. *Note* the end diastolic (ED) phase has larger volumes and represented as stars in (**b**), while the end systolic (ES) phase has smaller volumes and represented as *squares* in (**b**)

$i = 1, \ldots, n$, where $\mathbf{u}_i \in R^m$ is the sampling noise and $\mathbf{v}_i \in R^q$ denotes the original ith shape $\mathbf{m}_i$ in the low-dimensional manifold. In the prediction step, the motion prior (state model) $p(\mathbf{X}_t|\mathbf{X}_{t-1})$ is computed using the learned motion modes [43].

3 2D Motion Tracking

Accurate and robust tracking of 2D motion of deformable objects is an important topic in medical imaging. In this section, we apply the probabilistic framework to 2D non-rigid motion estimation in various medical imaging modalities, such as 2D ultrasound in Sect. 3.1 and X-ray fluoroscopy in Sect. 3.2.

3.1 Endocardium Contour Tracking in 2D Echocardiography

Automatic myocardial wall motion tracking in ultrasound images is an important step in analysis of the heart function, such as computing the left ventricle (LV) cavity volume and ejection fraction (EF). This task is difficult due to image noise as well as fast motion of the heart muscle and respiratory interferences. Figure 6 illustrates the difficulties of the tracking task due to signal drop-out, poor signal-to-noise ratio or significant appearance changes. Notice that the endocardium is not always on the strongest edge. Sometimes it manifests itself only by a faint line; sometimes it is

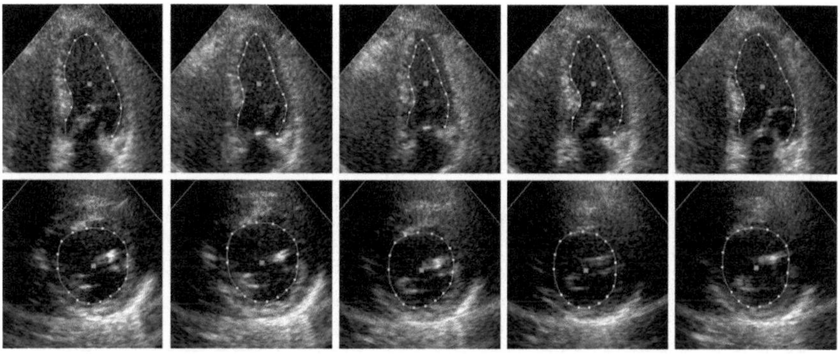

Fig. 6 Two tracking examples in *rows*, with five snapshots per sequence. The *top row* shows the apical four chamber view, which is along the long axis of the left ventricle and passes through the apex tip and the mitral valve. The *bottom row* shows the short axis view, which is perpendicular to the long axis of the left ventricle

completely invisible or buried in heavy noise; sometimes it will cut through the root of the papillary muscles where no edge is present.

To handle occlusions and appearance variations in 2D visual tracking, we apply the learning-based fusion framework presented in Sect. 2, by exploiting expert annotation of the structure of interest in large databases. More specifically, the appearance and shape models are learned by a two-step approach [16]. The first step is to learn a discriminative function between the appearance of the object of interest and the background. The second step is to learn the discriminative features that best associates the shapes to different appearances of the object, and to infer the most likely shape. Consequently, several representatives for the 2D appearance model are maintained to obtain a robust estimate of the target object [15]. When a new image is available, the location $\hat{\boldsymbol{x}}_{ij}$ and its uncertainty $\hat{\mathbf{C}}_{ij}$ are estimated for each model. Thus, the current location $\hat{\boldsymbol{x}}$ can be computed in an iterative manner, e.g., using the Variable-Bandwidth Density-based Fusion (VBDF) method [6]. The optimization process yields a hill-climbing procedure which converges to a stationary point of the underlying density.

To demonstrate the performance of the learning-based fusion method, we apply and evaluate the above framework to track heart contours using very noisy echocardiography data. The tracker was implemented in C++ and is running at about 20 frames per second on a single 2GHz Pentium 4 PC. Our data were selected by a cardiologist to represent normals as well as various types of cardiomyopathies, with sequences varying in length from 18 to 90 frames. Both training and test data were traced by experts, and confirmed by one cardiologist. We used both apical two- or four-chamber views (open contour with 17 control points) and parasternal short axis views (closed contour with 18 control points) for training and testing. PCA is performed and the original dimensionality of 34 and 36 is reduced to 7 and 8, respectively. For the appearance models we maintain 20 templates to capture the appearance variability.

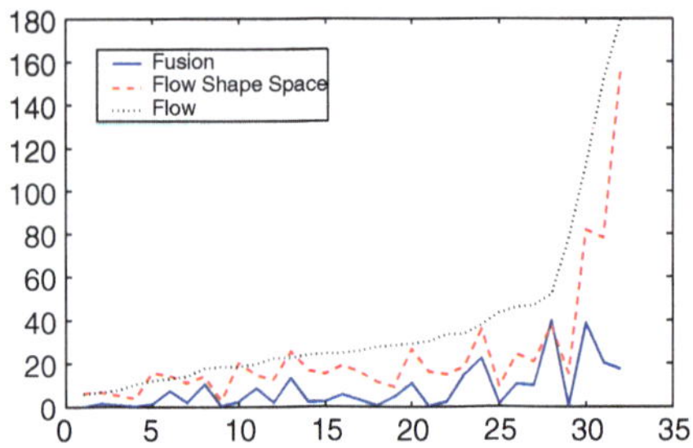

(a)

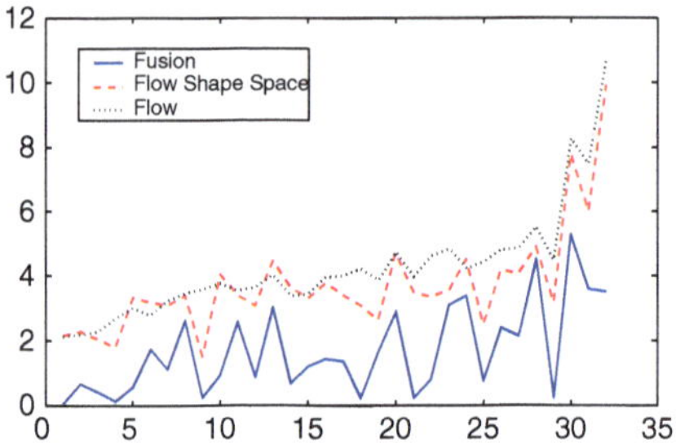

(b)

	All Cases				*Most Difficult Cases*			
Methods	*MSSD*	$\bar{\sigma}_{MSSD}$	*MAD*	$\bar{\sigma}_{MAD}$	*MSSD*	$\bar{\sigma}_{MSSD}$	*MAD*	$\bar{\sigma}_{MAD}$
Flow	38.1	82.9	4.3	3.6	147.9	325.0	8.8	8.2
FlowShapeSpace	24.7	35.5	3.8	2.4	106.0	181.2	7.9	6.3
Fusion	8.3	14.3	1.7	1.6	25.8	34.8	4.1	2.8

(c)

Fig. 7 Comparison experiments: mean distances (**a** Mean sum of squared distance (MSSD) [1], **b** Mean absolute distance (MAD) [24]) between tracked points and the ground truth. **c** Shows the error analysis "All Cases" and "Most Difficult Cases". The learning-based fusion method ("Fusion") significantly outperforms others, with lower average distances and lower standard deviations for such distances

For systematic evaluation, we use a set of 32 echocardiogram sequences outside of the training set for testing, with 18 parasternal short-axis (PS) views and 14 apical two- or four-chamber (AC) views, all with expert-annotated ground-truth contours. Figure 6 shows snapshots from two tracked sequences. Figure 7 reports performance comparison to other existing methods. The learning-based fusion method ("Fusion") is compared with a tracking algorithm without shape constraint ("Flow") or with the same tracker with orthogonal PCA shape space constraints ("FlowShapeSpace").

It should be noted that our results are not indicative for *border localization* accuracies, but rather for *motion tracking* performances given an initial contour. We have set our goal to track control points on the endocardium, with anisotropic confidence estimated at each point at any given time step by using multiple appearance models, and exploit this information when consulting a prior shape model as a constraint. Our framework is general and can be applied to other modalities, including the 2D X-ray fluoroscopy demonstrated in the next section.

3.2 2D Device Tracking in Fluoroscopy

During interventions a medical device might undergo non-rigid deformation due to patients' breathing and cardiac motions, and such 3D motions are complicated

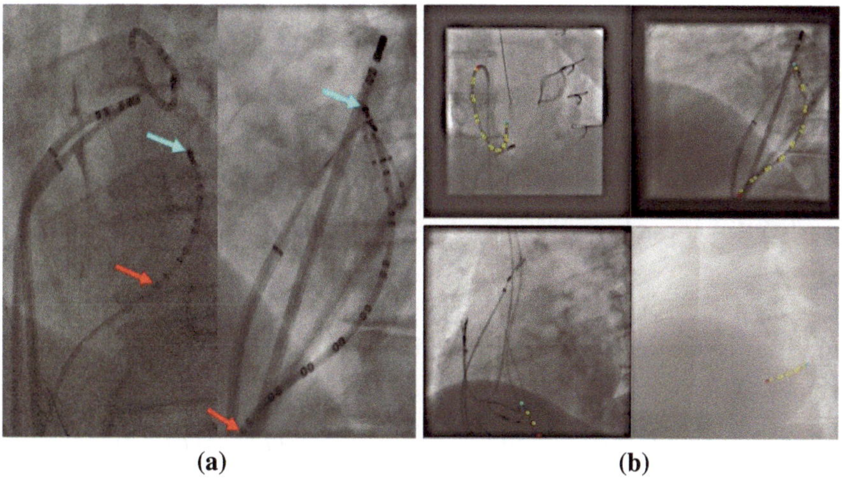

Fig. 8 Examples of coronary sinus (CS) catheters and the tracking results in 2D X-ray fluoroscopy. **a** CS catheters in 2D X-ray fluoroscopic images, which exhibit various appearance and shapes as well as low visibility in different contexts. For clarity the catheter tip and the most proximal electrode (PCS) are highlighted by *cyan* and *red arrows*, respectively. **b** Catheter tracking results in four different sequences. *Cyan*, *yellow*, and *red circles* indicate the catheter tip, intermediate electrodes, and PCSs, respectively

when being projected onto the 2D fluoroscopy. Furthermore, in fluoroscopy there exist severe image artifacts and other wire-like structures. Figure 8a shows several examples of catheters in 2D X-ray fluoroscopy.To tackle the above challenges, the tracking is formalized in the probabilistic inference framework introduced in Sect. 2, to maximize the posterior probability of a tracked target object, i.e.,

$$\hat{\mathbf{X}}_t = \arg\max_{\mathbf{X}_t} p(\mathbf{X}_t|\mathbf{Z}_{0:t}) = \arg\max_{\mathbf{X}_t} p(\mathbf{Z}_t|\mathbf{X}_t)p(\mathbf{X}_t|\mathbf{X}_{t-1})p(\mathbf{X}_{t-1}|\mathbf{Z}_{0:t-1}) \quad (4)$$

The above formula essentially combines two parts: the likelihood term, $P(\mathbf{Z}_t|\mathbf{X}_t)$, which is computed as combination of detection probability and template matching score and the transition term, $P(\mathbf{X}_t|\mathbf{X}_{t-1})$, which captures the motion smoothness. To maximize tracking robustness, the likelihood term $P(\mathbf{Z}_t|\mathbf{X}_t)$ is estimated by learning-based part detectors and appearance-based template matching as follows:

$$P(\mathbf{Z}_t|\mathbf{X}_t) = p^d(\mathbf{Z}_t|\mathbf{X}_t)p_d + p^a(\mathbf{Z}_t|\mathbf{X}_t)p_a \quad (5)$$

where $p^d(\mathbf{Z}_t|\mathbf{X}_t)$ and $p^a(\mathbf{Z}_t|\mathbf{X}_t)$ represents the learning-based and appearance-based measurement models respectively, and p_d and p_a are corresponding priors for the two types of measurement models. In particular, the learning-based measurement model is trained using the probabilistic boosting tree (PBT) [33].

The two measurement models in Eq. (5) can be defined in the following manner as in [38],

$$p^d(\mathbf{Z}_t|\mathbf{X}_t) \propto \frac{e^{f(\mathbf{Z}_t,\mathbf{X}_t)}}{e^{-f(\mathbf{Z}_t,\mathbf{X}_t)} + e^{f(\mathbf{Z}_t,\mathbf{X}_t)}}, \quad where \;\; f(\mathbf{Z}_t,\mathbf{X}_t) = \sum_k \alpha_k H_k(\mathbf{Z}_t,\mathbf{X}_t)$$

$$p^a(\mathbf{Z}_t|\mathbf{X}_t) \propto \exp\left\{ -\frac{\sum_{\mathbf{X}'_t \in S(\mathbf{X}_t)} |\rho(\mathbf{Z}_t(\mathbf{X}'_t) - I^0(\mathbf{X}'_t);\sigma_a)|^2}{2\sigma_a^2} \right\} \tag{6}$$

A good empirical choice for p_d and p_a proposed in [42] is $p_d = 1 - \lambda$ and $p_a = \lambda$, with the weighting parameter λ defined as:

$$\lambda = \frac{1}{1 + e^{-f(T_o^s, D(X_t))}}, \; f(T_o^s, D(X_t)) = \frac{cov(T_o^s, D(X_t))}{\sigma(T_o^s)\cdot\sigma(D(X_t))}, \tag{7}$$

where $cov(T_o^s, D(X_t))$ is the intensity cross-correlation between the catheter model template T_o^s and the image band expanded by X_t. $\sigma(T_o^s)$ and $\sigma(D(X_t))$ are the intensity variance.

Moreover, foreground and background structures in fluoroscopy are constantly changing and moving. In order to cope with it dynamically, the catheter model is updated online by:

$$T_{o,t}^s = (1 - \varphi_w)T_{o,t-1}^s + \varphi_w D(\mathbf{X}_t), if p(\mathbf{Z}_t|\mathbf{X}_t) > \varphi_t \tag{8}$$

where $T_{o,t}^s$ represents the model template in frame t. $D(\mathbf{X}_t)$ is the model obtained at frame t based on the output $\mathbf{X}_t$. φ_w and φ_t are typically set as 0.1 and 0.4 respectively in the experiments.

The tracking algorithm is evaluated on a large database including 1073 sequences collected from Electrophysiology (EP) Afib procedures. The image resolutions vary from 1024 × 1024 to 1440 × 1440 with pixel spacing between 0.154 and 0.183 mm. The test sequences cover a variety of interventional conditions, including low image contrast and contrast injection. Some example frames in the test set are displayed in Fig. 8b.

Quantitative evaluation of the tracking accuracy is reported in Table 1. While the tracking power of the proposed algorithm comes from the robust and efficient measurement models and information fusion, we illustrate and compare the impact of other important components in Table 1 as well. DON is the method by setting $\lambda = 0$ in Eq. (5), which essentially only considers the detection term; ADD is the method using Eq. (5); ARO is ADD with online template update. ARO is the final complete version of our algorithm. During comparison, the number of detected electrode candidates per frame is set as 15 and all other settings are exactly the same. We have tried other options of fusing detection probability and template matching score, such as multiplication of the two terms in Eq. (5). The effectiveness of Eq. (5) is validated through our batch evaluation on 1000+ sequences.

Table 1 CS catheter tracking performance

	Mean	Median	p85	p90	p95	p98
DON	1.16	0.66	0.98	1.12	1.67	4.26
ADD	0.91	0.45	0.72	0.86	1.56	4.45
ADR	0.78	0.48	0.72	0.81	1.10	2.40
ARO	0.76	0.50	0.73	0.82	1.04	2.14

The frame errors are in millimeter (mm) and computed at mean, median, percentile 85th (p85), 90th (p90), 95th (p95) and 98th (p98). Although tracking catheters in real fluoroscopic sequences is a non-trivial task, our algorithm turns out to be very robust against different challenging scenarios and has an error less than 2 mm in 97.8 % of the total evaluated frames. The last row shows the best performance including all essential components

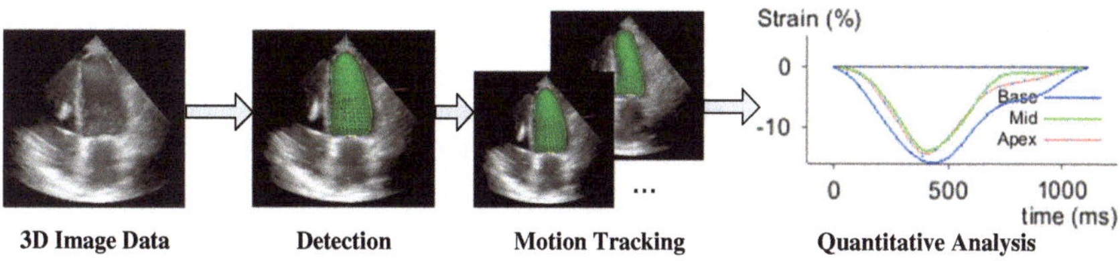

Fig. 9 Diagram of our learning-based 3D detection and tracking framework

4 3D Motion Tracking

To extract dynamic information of anatomical structures from volumetric time-resolved data, such as US, CT, and MRI, a robust tracking system is needed to estimate the 3D non-rigid deformation of the target object. Based on the probabilistic framework introduced in Sect. 2, we present an learning-based detection and tracking approach which includes the following main steps, automatic initialization, dense motion tracking, and 3D measurement computation as illustrated in Fig. 9. We apply and evaluate the presented framework to estimate 3D motion in various modalities, including 3D myocardial mechanics on volume ultrasound in Sect. 4.3, quantification of cardiac flow volume on volume Doppler in Sect. 4.4, joint delineation of left and right ventricles in cardiac MRI in Sect. 4.5, and four chamber tracking in cardiac CT in Sect. 4.6.

4.1 Unified 3D Anatomical Model

To facilitate comprehensive motion estimation and anatomical measurements, an anatomically indexed heart model used in this chapter is illustrated in Fig. 10. The mesh model for the right atrium is shown in Fig. 10b. The left atrium is represented by an open mesh separated by the mitral valve, shown in Fig. 10c. The right ventricle has a more complicated shape and is represented by an open mesh shown in Fig. 10d. Figure 10e shows the left ventricle including both epicardium (magenta) and endocardium (green). The detailed anatomical models can be found in [46].

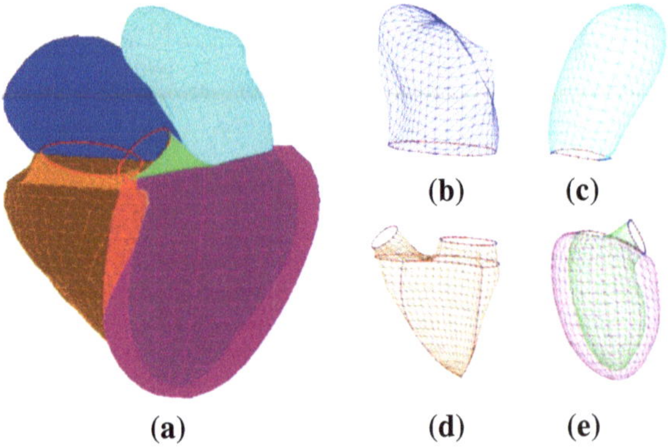

Fig. 10 The anatomically indexed heart model for comprehensive motion estimation and quantitative measurements. **a** The unified heart model with all four chambers. **b** The mesh model for the right atrium (RA). **c** The mesh model for the left atrium (LA). **d** The mesh model for the right ventricle (RV). **e** The mesh model for the left ventricle (LV), with *green* for the LV endocardium and magenta for the LV epicardium

4.2 Learning-Based Detection and Motion Estimation

In order to obtain precise morphological and functional quantification, dense tracking of the cardiac motion is required to establish the inter-frame correspondences for each point on the 3D mesh in Sect. 4.1. To initialize the tracking process, we fit the 3D model automatically in the starting frame (typically the end-systole or end-diastole cardiac phase), using the learning-based detection in Sect. 2.1. Then, we fuse information from multiple cues into the probabilistic framework introduced in Sect. 2, i.e.,

$$\arg\max_{\mathbf{X}_t} p(\mathbf{X}_t|\mathbf{Z}_{0:t}) = \arg\max_{\mathbf{X}_t} \underbrace{p(\mathbf{Z}_t|\mathbf{X}_t)}_{likelihood} \int \underbrace{p(\mathbf{X}_t|\mathbf{X}_{t-1})}_{prediction} p(\mathbf{X}_{t-1}|\mathbf{Z}_{0:t-1}) \qquad (9)$$

where $\mathbf{Z}_{0:t} = \mathbf{Z}_0, \ldots, \mathbf{Z}_t$ are the measurements from the input image sequence $I_{0:t} = I_0, \ldots, I_t$. For clarity, we use $\mathbf{X}_t$ to denote a concatenation of the mesh point positions, $\mathbf{X}_t = [X_1, \ldots, X_n]$, which need to be estimated at the current time instant t and n is the total number of points in the mesh model.

To maximize the accuracy and robustness of the tracking performance, the ***likelihood*** term $p(\mathbf{Z}_t|\mathbf{X}_t)$ is computed from both boundary detection and image template matching as proposed in [39, 40], $p(\mathbf{Z}_t|\mathbf{X}_t) = (1 - \lambda_k)p(y_b|\mathbf{X}_t) + \lambda_k p(T_t|\mathbf{X}_t)$, where T_t is the image pattern template and λ_k is the weighting coefficient of the matching term. The first term $p(y_b|\mathbf{X}_t)$ is the posterior distribution of the endocardial boundary learned in Sect. 2.1, using the steerable features and the probabilistic boosting-tree (PBT) [33]. The second term $p(T_t|\mathbf{X}_t)$ is obtained by a logistic function, $\frac{1}{1+e^{-\|I_t(\mathbf{X}_t)-T_t\|^2}}$, based on image matching: $\|I_t(\mathbf{X}_t) - T_t\|^2 =$

Table 2 In-vitro experiments on both (a) rotation and (b) displacement data

(a) Rotation (degrees)	10	15	20	25	(b) Displacement (mm)	0.82	1.29	2.02
Estimation	9.3	13.5	18.1	21.8	Estimation	0.9	1.54	2.31
Accuracy (%)	93	90	91	87	Accuracy (%)	90	81	91

The ground-truth motion was generated by a rotation device and a water pump controlling the stroke volume. Two crystals were implanted in the apical and middle regions of the left ventricle respectively to measure the myocardial movement. The displacements in (b) were computed based on a 30 mm reference length. Our tracking results are consistent with the ground-truth measurements on both rotation and displacement data

$\sum_{i,j,k}(I_t(\mathbf{X}_t + (i, j, k)) - T_t(i, j, k))^2$, where i, j, and k are the pixel-wise shift in the x, y, and z directions, respectively. λ_k is computed based on the feature measure as follows,

$$\lambda_k = \frac{1}{1 + e^{-fc(I_t(\mathbf{X}_t), T_t)}}, \qquad fc(I_t(\mathbf{X}_t), T_t) = \frac{cov(I_t(\mathbf{X}_t), T_t)}{\sigma(I_t(\mathbf{X}_t))\sigma(T_t)} \tag{10}$$

$cov(I_t(\mathbf{X}_t), T_t)$ is the intensity covariance between the image block $I_t(\mathbf{X}_t)$ centered at $\mathbf{X}_t$ and the image template T_t. $\sigma^2(I_t(\mathbf{X}_t))$ and $\sigma^2(T_t)$ are the intensity variance of the image block $I_t(\mathbf{X}_t)$ and the image template T_t, respectively. In our experiments, the typical image block size is $11 \times 11 \times 11$ voxels, while the typical search range is $7 \times 7 \times 7$ voxels. To handle the temporal image variation, the image template T_t is also updated online using the image intensities $I_t(\mathbf{X}_{t-1})$ from the previous frame $t - 1$.

The ***prediction*** term in Eq. (9), $p(\mathbf{X}_t|\mathbf{X}_{t-1})$, is the transition probability function $\hat{p}(\mathbf{X}_t|\mathbf{X}_{t-1})$ learned directly from the training data set, as explained in Sect. 2.2.

4.3 Myocardial Mechanics on Volume Echocardiography Data

Global and regional cardiac deformation provides important information on myocardial (dys-)function in a variety of clinical settings. Given the recent progress on real-time ultrasound imaging, unstitched volumetric data can be captured at a high volume rate, which allows to quantify cardiac strain in a non-invasive manner. In this section, we demonstrate the performance of the automatic detection and tracking method as well as the myocardial mechanics estimation. In our experiments, high frame-rate 3D+t ultrasound sequences were acquired by a Siemens SC2000 system with the average volume size of $200 \times 200 \times 140$ voxels. The average spatial resolution is 1 mm in the x, y, and z directions, and the average temporal resolution is 44 frames per second.

***In Vitro* Study:** To evaluate the accuracy of the automatic tracking method, we performed an in vitro experiment on animals. The ground-truth motion was generated

Table 3 Comparison of the longitudinal strain estimation between the deformable tracking method and the crystal measurements in the in vitro study

Longitudinal strain (%)	2.63	4.11	6.68
Estimation (%)	3.43	5.19	8.25
Difference (%)	0.8	1.08	1.57

The two crystals were implanted in the apical and middle regions of the left ventricle, such that the longitudinal Lagrangian strain can be computed based on the displacement as the ground-truth measurement in the top row. The estimation results in the middle row are computed from the 3D strain tensor using our method. The low difference values in the bottom row show clearly that the estimation from the deformable tracking method is consistent with the clinical measurements

Table 4 Performance analysis on a large data set including 503 3D+t ultrasound sequences

Measure(mm)	Training (239)	Testing (264)	Training (434)	Testing (69)
Mean/std	2.21/1.57	2.68/2.63	2.26/1.42	2.64/2.23

In the first experiment, the data set was evenly split into a training set with 239 sequences and a testing set with the remaining 264 sequences, while in the second experiment the training set (434) and the testing set (69) were not balanced. The error measurements were computed as the average point distance between the estimated mesh and the ground-truth annotations by experts on both the end-diastolic and end-systolic frames. The consistent evaluation results demonstrate the robustness of the learning-based detection and tracking method

by a rotation device and a water pump controlling the stroke volume. Two crystals were implanted in the apical and middle regions of the left ventricle, respectively, to measure the myocardial movement. Table 2 reports the error analysis on four volumetric ultrasound sequences acquired with 10, 15, 20, and 25 rotation degrees, respectively, and three sequences with different stroke volumes.

Furthermore, to evaluate the results of our myocardial strain estimation, we compare them against the crystal measurements for the same subjects in the in vitro study. The ground-truth longitudinal Lagrangian strain can be computed based on the displacement reported in Table 2b. Table 3 reports the comparison between the estimated strain values and the ones from crystal measurements.

***In Vivo* Study**: To evaluate the robustness of the learning-based detection and tracking method, we tested it on a large data set including 503 volumetric ultrasound sequences from human subjects. The data set was randomly split into a training set and a testing set, where the training set was used to learn the detectors in Sect. 2.1 and the prior distributions in Sect. 2.2, while the testing set reflected the performance for unseen data. The results on both the training and testing sets are reported in Table 4.

Comparison Study: Finally to demonstrate the advantage of the learning-based fusion framework, we compared this method against tracking by 3D optical flow and tracking by detection. The accuracy is measured by the point-to-mesh error [43] reported in Table 5 for all three methods.

Table 5 Comparison between the 3D optical flow, tracking by detection, and learning-based fusion methods

Error (mm)	Mean	Std	Median	Min	Max
3D optical flow	2.68	1.28	2.39	0.94	10.38
Tracking by detection	1.61	1.24	1.31	0.59	9.89
Learning-based fusion	1.28	1.11	1.03	0.38	9.80

The point-to-mesh errors are measured in millimeters. The learning-based fusion method achieved the best accuracy among compared to the other two approaches

4.4 Flow Quantification on 3D Volume Color Doppler Data

The quantification of flow volume is important for evaluation of patients with cardiac dysfunction and cardiovascular disease. However, accurate flow quantification remains a significant challenge for cardiologists [21]. In this section, we apply our automatic tracking framework in cardiac flow volume quantification using instantaneous 3D+t ultrasound data.

To evaluate the performance of the learning-based fusion method, a set of 3D full-volume ultrasound sequences were acquired by a Siemens SC2000 scanner with an average volume rate of 15 vps at the Ohio State University Medical Center. Twenty-two subjects with normal valves were enrolled with the Institutional Review Board (IRB) approval.

Table 6 reports the comparison between the expert measurements using 2D pulsed wave (PW) Doppler and the flow volumes estimated by our method. The LV stroke volume (LVSV) was very close to the volume from LVOT-PW (70.1 ± 20.8 ml, 69.7 ± 16.7 ml) with good correlation ($r = 0.78$). 3D LV inflow and outflow volumes (73.6 ± 16.3 ml, 67.6 ± 14.6 ml) were correlated well with LVSV and LVOT-PW respectively ($r = 0.77, 0.91$).

4.5 Joint Delineation of LV and RV in Cardiac MRI Sequences

Cardiac Magnetic Resonance Imaging (MRI) is now an established, although still rapidly advancing, technique providing information on morphology and function of the cardiovascular system. A typical cardiac MR scan to examine the LV/RV morphology and function contains a short axis stack, which consists of image slices captured at the different positions along the short axis of heart chambers (e.g., the LV). These image slices can be aligned using the physical coordinates (location and orientation) recorded during acquisition. A 3D volume is reconstructed from this stack of aligned image slices. If each image slice is captured in a time sequence and synchronized to each other, a 3D volume sequence is obtained, which is used for 3D chamber segmentation and dynamics extraction in our system. In this section,

Table 6 Flow volume quantification on 22 normal patients

Measure (ml)	Mean	STD	
(a) LVOT-PW	69.7	16.7	
LVSV	70.1	20.8	
3D CD mitral inflow	73.6	16.3	
3D CD LVOT outflow	67.6	14.6	
Measure 1	Measure 2	Correlation	p-value
(b) LVOT-PW	LVSV	0.78	<0.001
3D CD mitral inflow	LVSV	0.77	<0.001
3D CD LVOT outflow	LVOT-PW	0.91	<0.001

(a) Flow measure comparison. The first row shows the LVOT outflow volume measured by a clinical expert using 2D pulsed wave (PW) Doppler. The second row is the estimated LV stroke volume using the delineated LV endocardial boundary on the volumetric b-mode ultrasound data. The last two rows are the de-aliased mitral inflow and LVOT outflow based on the sampled volumetric color Doppler data by our method. (b) Correlation and statistical significance testing of flow measure on 22 normal patients between (1) the LVOT outflow volume measured using 2D pulsed wave (PW) Doppler and the estimated LV stroke volume; (2) the LVOT and the de-aliased Mitral inflow by our method; and (3) the LVOT-PW and the LVOT outflow by our method. The estimated flow volumes are consistent between all four measurements and close to the expert measurements, which demonstrates the accuracy and robustness of the learning-based fusion method

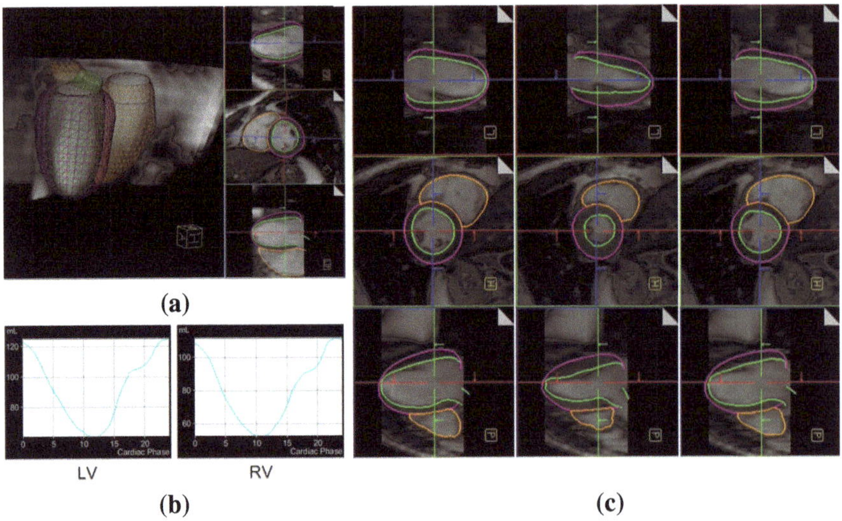

Fig. 11 Models of LV/RV fitted to a 3D reconstructed cardiac MRI volume sequence. **a** Estimated 3D model. **b** Volume measurement across time computed based on the fitted models. **C** 2D views of frame 1, 11, 21 of a single heartbeat cycle (25 frames in total)

we apply the probabilistic framework from Sect. 4.2 to detect the joint LV and RV model and estimate the dynamic motion and quantitative measurements, as illustrated in Fig. 11.

Table 7 Point-to-mesh distance measurements obtained by a 4-fold cross validation

Measure (mm)	Mean	Std	Median
LV endocardium	2.95	4.85	1.84
LV epicardium	3.23	3.94	2.12
RV main	2.99	1.18	2.66

We collected 100 reconstructed volumes from 70 patients with left ventricles annotated, among which 93 reconstructed volumes from 63 patients were also annotated on right ventricles. Volumes were selected to cover a large range of dynamic heart motion, including both end diastole and end systole. The original short-axis stack images have an average in-plane resolution of 1.35 mm, and the distance between slices is around 10 mm.

A 4-fold cross-validation scheme was applied for evaluation. The entire dataset was randomly partitioned into four quarters. For each fold evaluation, three quarters were combined for training and the remaining one was used as unseen data for testing. This procedure was repeated four times so that each volume has been used once for testing. For each segmented mesh, the distance from each vertex to the groundtruth mesh (manual annotation) was computed as point-to-mesh distance. The average distance from all vertices of the segmented mesh was used as the measurement. Three major components, i.e., LV endocardium, LV epicardium, and RV main cavity as illustrated in Fig. 10d, e, were considered in our evaluation as listed in Table 7. Automatic delineation examples are provided in Fig. 11. On average, it took about 3 s to segment both the LV and RV from a single volume (e.g, $256 \times 256 \times 70$ voxels), and about 40 s to fully extract dynamics of the entire sequence (typically 20 frames) on a duo core 2.8 GHz CPU.

4.6 Four Chamber Tracking in Cardiac CT Data

The 3D tracking framework presented in Sect. 4.2 is generic and can be extended to different modalities. In this section we also apply it to tracking all four chambers of the heart, including left ventricle (LV), right ventricle (RV), left atrium (LA), and right atrium (RA), in cardiac Computed Tomography (CT) data, collected from 27 institutes over the world using Siemens Somatom Sensation and Definition scanners. The imaging protocols are heterogeneous with different capture ranges and resolutions. A volume may contain 80 to 350 slices, while the size of each slice is the same with 512×512 pixels. The resolution inside a slice is isotropic and varies from 0.28 to 0.74 mm for different volumes. The ED detector and boundary classifier were trained on 323 static cardiac CT volumes from 137 patients with various cardiovascular diseases. The cardiac motion model was trained on additional 20 sequences (each with ten frames).

During the tracking stage, the learning-based fusion in Sect. 4.2 is applied to calculate the motion displacements. Figure 12 shows the detection and tracking results

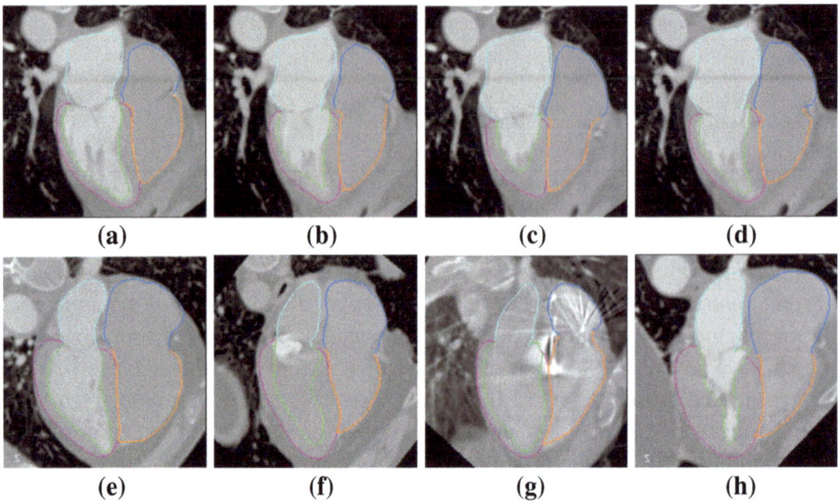

Fig. 12 Examples of heart chamber detection and tracking in 3D CT data. The heart chambers are highlighted in *green* for the LV endocardium, *magenta* for the LV epicardium, *cyan* for the LA, *brown* for the RV, and *blue* for the RA. The *top row* shows example tracking results on a dynamic 3D sequence with 10 frames. Four frames (1, 2, 3, and 6) are shown in **a**,**b**,**c**,**d**, respectively. The *bottom row* includes more results on various CT volumes in our dataset

Table 8 The ejection fraction (EF) estimation accuracy for six dynamic sequences in our dataset

	Patient #1	Patient #2	Patient #3	Patient #4	Patient #5	Patient #6	Mean error	Standard deviation
Ground truth (%)	68.7	49.7	45.8	62.9	47.4	38.9	2.3	1.6
Estimation (%)	66.8	51.8	42.8	64.4	42.3	38.5		

of 3D cardiac CT four chambers (LV-epicardium, LV-endocardium, LA, RV, and RA) in CT volumes. Furthermore, given the tracking result, we can calculate the ejection fraction (EF) as, $\mathrm{EF} = (\mathrm{V}_{ED} - \mathrm{V}_{ES})/\mathrm{V}_{ED}$, where V_{ED} and V_{ES} are the volume measures of the end-diastolic (ED) and end-systolic (ES) phases, respectively. Table 8 reports the EF estimation accuracy for six CT sequences. The estimated EFs are close to the ground truth with a mean error of 2.3 %.

5 4D Trajectory Spectrum Tracking

To extend discriminative learning algorithms for time dependent four-dimensional problems, trajectory-based features have increasingly attracted attention in motion analysis and recognition [36]. It has been shown that the inherent representative power of both shape and trajectory projections of non-rigid motion are equal, but the representation in the trajectory space can significantly reduce the number of parameters to be optimized [2]. This duality has been exploited in motion reconstruction and segmentation [44], structure from motion [2]. In particular, for periodic motion, frequency domain analysis shows promising results in motion estimation and recognition [5, 25]. Although the compact parameterization and duality property are crucial in the context of learning-based object detection and motion estimation, this synergy has not been fully exploited yet.

In this section, we extend the learning-based model estimation in Sect. 2 to the trajectory spectrum learning (TSL) with local-spatio-temporal (LST) features [18]. It includes three main steps: (1) global location and rigid motion estimation which is obtained by the learning-based model fitting technique presented in Sect. 2.1, (2) non-rigid landmark motion estimation using the trajectory spectrum learning (TSL) with local-spatio-temporal (LST) features [18], and (3) non-rigid shape estimation in the same learning-based fusion framework as in Sect. 4.2.

Based on the determined global location and rigid motion from Sect. 2.1, a trajectory spectrum learning algorithm is applied to estimate the non-linear valve movements from volumetric sequences [18]. The objective is to find for each landmark j its trajectory $\mathbf{a^j}$, with the maximum posterior probability from a series of volumes I, given the rigid motion θ. In particular, a trajectory $\mathbf{a^j}$ can be uniquely represented by the concatenation of its discrete Fourier transform (DFT) coefficients, $\mathbf{s^j} = [\mathbf{s^j}(0), \ldots, \mathbf{s^j}(n-1)]$, obtained through the DFT equation, $\mathbf{s^j}(f) = \sum_{t=0}^{n-1} \mathbf{a^j}(t) e^{\frac{-j2\pi tf}{n}}$, where $\mathbf{s^j}(f) \in \mathbb{C}^3$ is the frequency spectrum of the x, y, and z components of the trajectory $\mathbf{a^j}(t)$, and $f = 0, 1, \ldots, n-1$. Therefore, instead of estimating the motion trajectory directly, we apply discriminative learning to detect the spectrum $\mathbf{s^j}$ in the frequency domain by optimizing the following equation:

$$\begin{aligned} &\arg\max_{\mathbf{s^j}}\ p(\mathbf{s^j}|I, \theta) = \arg\max_{\mathbf{s^j}}\ p(\mathbf{s^j}(0), \ldots, \mathbf{s^j}(n-1)| \\ &I(0), \ldots, I(n-1), \theta(0), \ldots, \theta(n-1)) \end{aligned} \tag{11}$$

Inspired by the MSL approach [46], we efficiently perform trajectory spectrum learning and detection in DFT subspaces with gradually increased dimensionality. The intuition is to perform a spectral coarse-to-fine motion estimation, where the detection of coarse level motion (low frequency) is incrementally refined with high frequency components representing fine deformations. More specifically, to obtain object localization and motion estimation in unseen volumetric sequences, the motion parameters are searched in the marginalized spaces $\Sigma_0, \ldots, \Sigma_{r-1}$ using the trained spectrum detectors $D_0, \ldots, D_{r-1}$. Starting from an initial zero-spectrum, we incrementally estimate the magnitude and phase of each frequency component $\mathbf{s}(k)$. At the stage k, the corresponding robust classifier D_k is exhaustively scanned over the potential candidates. The probability of a candidate C_k is computed by the following objective function from the inversed DFT (IDFT):

$$p(C_k) = \prod_{t=0}^{n-1} D_k(IDFT(C_k), I, t) \tag{12}$$

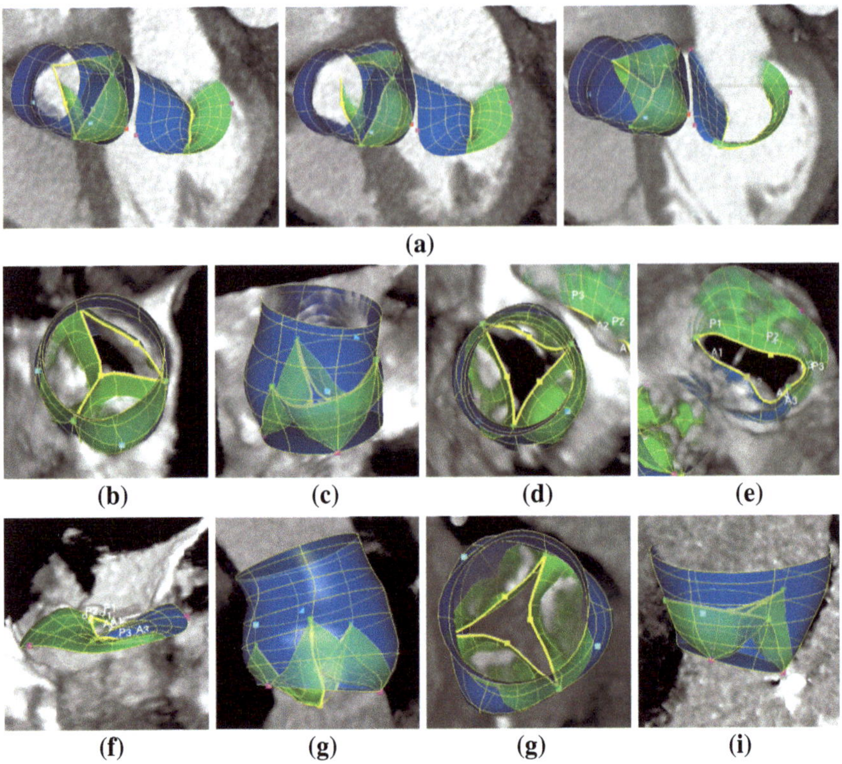

Fig. 13 Examples of estimated patient-specific models from CT and TEE data: **a** healthy valves from three different cardiac phases in the four chamber view. Pathologic valves with **b** bicuspid aortic valve, **c** aortic root dilation and regurgitation, **d** moderate aortic stenosis, **e** mitral stenosis, **f** mitral prolapse, **g** bicuspid aortic valve with prolapsing leaflets, **h** aortic stenosis with severe calcification and **i** dilated aortic root

where $t = 0, \ldots, n-1$ is the time instance (frame index). After each step k, the top 50 trajectory candidates with high probability values are preserved for the next step $k+1$. The procedure is repeated until a final set of trajectory candidates $\mathscr{C}_{r-1}$ are computed. The final trajectory is reported as the average of all elements in $\mathscr{C}_{r-1}$.

Furthermore, to improve learning performance, a Local-Spatial-Temporal (LST) feature is used to incorporate both the spatial and temporal context, by aligning contextual spatial features in time [18]:

$$F^{4D}(\theta(t), T|I, s) = \tau(F^{3D}(I, \theta(t+i*s)), i = -T, \ldots, T) \tag{13}$$

Three-dimensional $F^{3D}()$ features extract simple gradient and intensity information from steerable pattern spatially align with $\theta(t)$ as defined in Eq. (2). The final value of a Local-Spatial-Temporal (LST) feature is the result of time integration using a set of linear kernels τ, which weight spatial features $F^{3D}()$ according to their distance

Table 9 Errors for each estimation stage in TEE and CT

Measure (mm)	TEE Mean	Std.	Median	80 %	CT Mean	Std.	Median	80 %
Global location and rigid motion	6.95	4.12	5.96	8.72	8.09	3.32	7.57	10.4
Non-rigid landmark motion	3.78	1.55	3.43	4.85	2.93	1.36	2.59	3.38
Comprehensive aortic-mitral	1.54	1.17	1.16	1.78	1.36	0.93	1.30	1.53

The "80 %" column represents the 80th percentile of the error values

from the current frame t. A simple example for τ, also used in our implementation, is the uniform kernel over the interval $[-T, T]$, $\tau = 1/(2T+1)\sum_{i=-T}^{T}(F^{3D}(I, \theta(t + i * s))$. For this choice of τ, each F^{3D} contributes equally to the F^{4D}.

To demonstrate the performance of the 4D trajectory spectrum tracking method, we test it on a large and comprehensive data set. More specifically, 690 CT and 1516 TEE volumes were acquired from 134 patients affected by various cardiovascular diseases such as, bicuspid aortic valve, dilated aortic root, stenotic aortic/mitral, regurgitant aortic/mitral as well as prolapsed valves. Example images are shown in Fig. 13. The electrocardiogram (ECG) gated cardiac CT sequences include ten volumes per cardiac cycle, where each volume contains 80–350 slices with 512 × 512 pixels. The in-slice resolution is isotropic and varies between 0.28 to 1.00 mm with a slice thickness from 0.4 to 2.0 mm. TEE data includes an equal amount of rotational (3–5°) and matrix array acquisitions. A complete cardiac cycle is captured in a series of 7–39 volumes, depending on the patient's heart beat rate and scanning protocol. Image resolution and size vary for the TEE data set from 0.6 to 1 mm and 136 × 128 × 112 to 160 × 160 × 120 voxels, respectively.

The performance evaluation was conducted using 3-fold cross-validation in the similar manner as in Sect. 4.5. Table 9 summarizes the model estimation performance averaged over the three evaluation runs. On a standard PC with a quad-core 3.2 GHz processor and 2.0 GB memory, the total computation time for the three estimation stages is 4.8 s per volume (approximately 120 s for an average length volume sequence). Figure 13 shows estimation results on various pathologies for both valves and imaging modalities. Furthermore, we compare the 4D trajectory spectrum tracking method to traditional tracking methods, such as optical flow [12] and tracking-by-detection [45], and report the results in Fig. 14.

Given the tracking results, we can compute quantitative measurements and evaluate them against manual expert measurements. Table 10 shows the accuracy for the Ventriculoarterial Junction, Valsava Sinuses and Sinotubular Junction aortic root diameters as well as for Annular Circumference, Annular-Posterior Diameter and Anterolateral-Posteromedial Diameter of the mitral valve. From a subset of 19 TEE patients, we computed measurements of the aortic-mitral complex and compared those to literature reported values [34]. Distances between the centroids of the aortic and mitral annulae as well as interannular angles were computed. The latter is the angle between the vectors, which point from the highest point of the anterior mitral annulus to the aortic and mitral annular centroids respectively. The mean interannular angle and interannular centroid distance were 137.0±12.2 and 26.5±4.2, respectively compared to 136.2±12.6 and 25.0±3.2 reported in the literature [34].

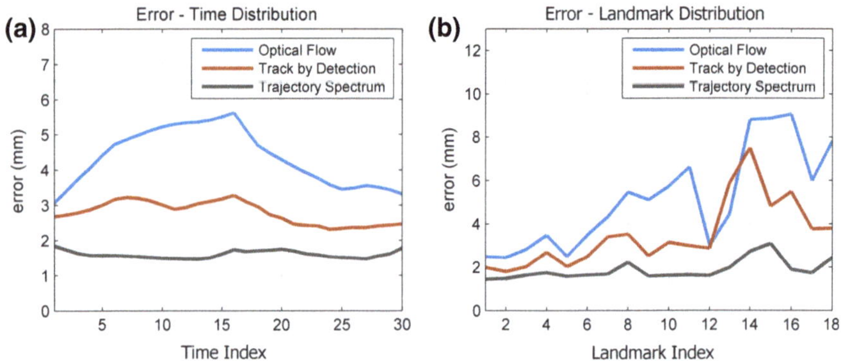

Fig. 14 Error comparison between the optical flow, tracking-by-detection and our trajectory-spectrum approach distributed over (**a**) time and (**b**) detected anatomical landmarks. The *curve* in *black* shows the performance of our approach, which has the lowest error among all three methods

Table 10 System-precision for various measurements of the aortic-mitral apparatus

	Mean	STD
Ventriculoarterial junct. ⊘ (mm)	1.37	0.17
Valsava sinuses ⊘ (mm)	1.66	0.43
Sinotubular junct. ⊘ (mm)	0.98	0.29
Annular ∨ (mm)	8.46	3.0
Annular-posterior ⊘ (mm)	3.25	2.19
Anterolateral–posteromedial ⊘ (mm)	5.09	3.7

⊘ diameter, ∨ circumferential length

6 Conclusions

This chapter presented a probabilistic framework for fast and accurate detection and tracking of deformable objects, with various applications in the medical imaging field. To handle shape and appearance variations in visual tracking, a set of offline and online component-based models are maintained to obtain multiple estimates of the target object, which allows us to combine several sources of information, including domain knowledge encoded in image-based discriminative classifiers, domain knowledge encoded in shape models and motion models, and traditional tracking with template-based matching/registration. The model estimation is automatically performed by applying robust and efficient learning-based algorithms on 2D, 3D and 4D data in various modalities, including US, CT, MRI and X-ray fluoroscopy. Validation experiments on clinical datasets demonstrated the good accuracy and robustness of the presented framework and showed a strong inter-modality and inter-subject correlation for a comprehensive set of model-based measurements. The resulting patient-specific model provides precise morphological and functional quantification of the anatomies to be analyzed, which is a prerequisite during the entire clinical workflow including diagnosis, therapy-planning, surgery or percutaneous intervention as well as patient monitoring and follow-up.

Acknowledgments The authors would like to thank Dr. David Sahn and Dr. Muhammad Ashraf at OHSU for providing the volumetric ultrasound sequences and sonomicrometry data in the in vitro animal study, Dr. Alan Katz from St. Francis Hospital and Dr. Mani Vannan from OSU Medical Center for fruitful interactions and guidance, and SCR colleagues for helpful discussions.

References

1. Akgul Y, Kambhamettu C (2003) A coarse-to-fine deformable contour optimization framework. IEEE Trans Pattern Anal Mach Intell 25(2):174–186
2. Akhter I, Sheikh Y, Khan S, Kanade T (2008) Nonrigid structure from motion in trajectory space. In: Advances in neural information processing systems, pp 41–48
3. Black MJ, Jepson AD (1998) Eigentracking: robust matching and tracking of articulated objects using a view-based representation. Int J Comput Vis 26:63–84
4. Bookstein FL (1989) Principal warps: thin-plate splines and the decomposition of deformation. IEEE Trans Pattern Anal Mach Intell 11(6):567–585
5. Briassouli A, Ahuja N (2007) Extraction and analysis of multiple periodic motions in video sequences. IEEE Trans Pattern Anal Mach Intell 29(7):1244–1261
6. Comaniciu D (2003) Nonparametric information fusion for motion estimation. In: Proceedings of IEEE conference on computer vision and pattern recognition, Madison, Wisconsin, pp 59–66
7. Comaniciu D, Zhou X, Krishnan S (2004) Robust real-time tracking of myocardial border: an information fusion approach. IEEE Trans Med Imaging 23(7):849–860
8. Cootes T, Taylor C (2001) Statistical models of appearance for medical image analysis and computer vision. In Proceedings of SPIE medical imaging, pp 236–248
9. Cootes TF, Edwards GJ, Taylor CJ (2001) Active appearance models. IEEE Trans Pattern Anal Mach Intell 23(6):681–685
10. Craene MD, Camara O, Bijnens BH, Frangi AF (2009) Large diffeomorphic FFD registration for motion and strain quantification from 3D-US sequences. In: Functional imaging and modeling of the heart (2009), vol 5528. Springer, pp 437–446
11. Dryden IL, Mardia KV (1998) Statistical shape analysis. Wiley, New York
12. Duan Q, Parker KM, Lorsakul A, Angelini ED, Hyodo E, Homma S, Holmes JW, Laine AF (2009) Quantitative validation of optical flow based myocardial strain measures using sonomicrometry. In: Proceedings of IEEE international symposium on biomedical imaging, pp 454–457
13. Edwards GJ, Cootes TF, Taylor CJ (1998) Face recognition using active appearance models. In: European conference on computer vision, pp 581–595
14. Elen A, Choi HF, Loeckx D, Gao H, Claus P, Suetens P, Maes F, D'hooge J (2008) Three-dimensional cardiac strain estimation using spatio-temporal elastic registration of ultrasound images: a feasibility study. IEEE Trans Med Imaging 27(11):1580–1591
15. Georgescu B, Zhou XS, Comaniciu D, Rao B (2004) Real-time multi-model tracking of myocardium in echocardiography using robust information fusion. In: Proceedings of international conference on medical image computing and computer assisted intervention
16. Georgescu B, Zhou XS, Comaniciu D, Gupta A (2005) Database-guided segmentation of anatomical structures with complex appearance. In: Proceedings of IEEE conference on computer vision and pattern recognition, pp 429–436
17. Grau V, Becher H, Noble J (2007) Registration of multiview real-time 3-D echocardiographic sequences. IEEE Trans Med Imaging 26(9):11541165

18. Ionasec R, Voigt I, Georgescu B, Wang Y, Houle H, Fernando-Vega H, Navab N, Comaniciu D (2010) Patient-specific modeling and quantification of the aortic and mitral valves from 4-D cardiac CT and TEE. IEEE Trans Med Imaging 29(9):1636–1651
19. Jacob G, Noble J, Behrenbruch C, Kelion A, Banning A (2002) A shape-space-based approach to tracking myocardial borders and quantifying regional left-ventricular function applied in echocardiography. IEEE Trans Med Imaging 21(3):226–238
20. Jepson AD, Fleet DJ, El-Maraghi TF (2003) Robust online appearance models for visual tracking. IEEE Trans Pattern Anal Mach Intell 25:1296–1311
21. Little SH (2010) Quantifying mitral valve regurgitation: new solutions from the 3rd dimension. J Am Soc Echocardiogr 23(1):9–12
22. Lloyd-Jones D, Adams R, Carnethon M, Simone GD, Ferguson TB, Flegal K, Ford E, Furie K, Go A, Greenlund K, Haase N, Hailpern S, Ho M, Howard V, Kissela B, Kittner S, Lackland D, Lisabeth L, Marelli A, McDermott M, Meigs J, Mozaffarian D, Nichol G, ODonnell C, Roger V, Rosamond W, Sacco R, Sorlie P, Stafford R, Steinberger J, Thom T, Wasserthiel-Smoller S, Wong N, Wylie-Rosett J, Hong Y (2009) Heart disease and stroke statistics-2009 update: a report from the american heart association statistics committee and stroke statistics subcommittee. Circulation 119:3
23. Lu X, Wang Y, Georgescu B, Littman A, Comaniciu D (2011) Automatic delineation of left and right ventricles in cardiac MRI sequences using a joint ventricular model. In: Proceedings IEEE international symposium on biomedical, pp 250–258
24. Mikić I, Krucinski S, Thomas JD (1998) Segmentation and tracking in echocardiographic sequences: active contours guided by optical flow estimates. IEEE Trans Med Imaging 17(2):274–284
25. Naftel A, Khalid S (2006) Motion trajectory learning in the DFT-coefficient feature space. In : Proceedings of international conference on computer vision systems, p 47
26. Peters J, Ecabert O, Meyer C, Kneser R, Weese J (2010) Optimizing boundary detection via simulated search with applications to multi-modal heart segmentation. Med Image Anal 14(1):70–84
27. Shi J, Tomasi C (1994) Good features to track. In: IEEE conference on computer vision and pattern recognition, pp 593–600
28. Sidenbladh H, Black MJ, Fleet DJ (2000) Stochastic tracking of 3d human figures using 2D image motion. In: European conference on computer vision, vol 2, pp 702–718
29. Stauffer C, Grimson WEL (1999) Adaptive background mixture models for real-time tracking. In: IEEE conference on computer vision and pattern recognition, vol 2, pp 246–252
30. Sun H, Frangi A, Wang H, Sukno F, Tobon-Gomez C, Yushkevich P (2010) Automatic cardiac mri segmentation using a biventricular deformable medial model. In: Proceedings of international conference on medical image computing and computer assisted intervention
31. Tao H, Sawhney HS, Kumar R (2000) Dynamic layer representation with application to tracking. In: IEEE conference on computer vision and pattern recognition, vol 2, pp 134–141
32. Tenenbaum JB, de Silva V, Langford JC (2000) A global geometric framework for nonlinear dimensionality reduction. Science 290(5500):2319–2323
33. Tu Z (2005) Probabilistic boosting-tree: learning discriminative models for classification, recognition, and clustering. In: Proceedings of international conference on computer vision, part II, pp 1589–1596
34. Veronesi F, Corsi C, Sugeng L, Mor-Avi V, Caiani E, Weinert L, Lamberti C, Lang RM (2009) A study of functional anatomy of aortic-mitral valve coupling using 3D matrix transesophageal echocardiography. Circ Cardiovasc Imaging 2(1):24–31
35. Viola P, Jones M (2001) Rapid object detection using a boosted cascade of simple features. In: IEEE conference on computer vision and pattern recognition, pp 511–518
36. Wang L, Geng X, Leckie C, Kotagiri R (2008) Moving shape dynamics: a signal processing perspective. In: IEEE conference on computer vision and pattern recognition
37. Wang X, Chen T, Zhang S, Metaxas D, Axel L (2008) LV motion and strain computation from tMRI based on meshless deformable models. In: Proceedings of international conference on medical image computing and computer assisted intervention

38. Wang P, Chen T, Zhu Y, Zhang W, Zhou S, Comaniciu D (2009) Robust guidewire tracking in fluoroscopy. In: Proceedings of IEEE conference on computer vision and pattern recognition, pp 691–698
39. Wang Y, Georgescu B, Comaniciu D, Houle H (2010) Learning-based 3D myocardial motion flow estimation using high frame rate volumetric ultrasound data. In: Proceedings of IEEE international symposium on biomedical imaging, pp 1097–1100
40. Wang Y, Georgescu B, Houle H, Comaniciu D (2010) Volumetric myocardial mechanics from 3d+t ultrasound data with multi-model tracking. In: Statistical atlases and computational models of the heart: mapping structure and function (STACOM) + a cardiac electrophysiological simulation, challenge (CESC'10), pp 184–193
41. Wang P, Zheng Y, John M, Comaniciu D (2012) Catheter tracking via online learning for dynamic motion compensation in transcatheter aortic valve implantation. In: Proceedings of international conference on medical image computing and computer assisted intervention
42. Wu W, Chen T, Barbu A, Wang P, Strobel N, Zhou S, Comaniciu D (2011) Learning-based hypothesis fusion for robust catheter tracking in 2D X-ray fluoroscopy. In: Proceedings of IEEE conference on computer vision and pattern recognition, pp 1097–1104
43. Yang L, Georgescu B, Zheng Y, Wang Y, Meer P, Comaniciu D (2011) Prediction based collaborative trackers (PCT): a robust and accurate approach toward 3D medical object tracking. IEEE Trans Med Imaging 30(11):1921–1932
44. Zelnik Manor L, Irani M (2004) Temporal factorization vs. spatial factorization. In: European conference on computer vision, part II, pp 434–445
45. Zhao T, Nevatia R (2002) 3D tracking of human locomotion: a tracking as recognition approach. In: International conference on pattern recognition, part I, pp 546–551
46. Zheng Y, Barbu A, Georgescu B, Scheuering M, Comaniciu D (2008) Four-chamber heart modeling and automatic segmentation for 3-D cardiac CT volumes using marginal space learning and steerable features. IEEE Trans Med Imaging 27(11):1668–1681
47. Zhuang X, Leung K, Rhode K, Razavi R, Hawkes DJ, Ourselin S (2010) A registration-based propagation framework for automatic whole heart segmentation of cardiac MRI. IEEE Trans Med Imaging 29(9):1612–1625

A Supervised Graph-Cut Deformable Model for Brain MRI Segmentation

Laura Igual, Joan Carles Soliva, Antonio Hernández-Vela, Sergio Escalera, Oscar Vilarroya and Petia Radeva

Abstract Accurate automatic segmentation of subcortical brain structures in Magnetic Resonance Images (MRI) is of great interest in the analysis of developmental disorders. Segmentation methods based on a single or multiple atlases have been shown to suitably localize brain structures. However, the atlas prior information may not represent the structure of interest correctly. It may therefore be useful to introduce a more flexible technique for accurate segmentation. A fully-automatic segmentation method for brain MRI is considered, which defines a deformable model combining an atlas-based segmentation strategy with a supervised Graph-cut model. The Graph-cut model is adapted to make it suitable for segmenting small and low-contrast brain structures by defining new data and boundary potentials of the energy function.

L. Igual (✉) · A. Hernández-Vela · S. Escalera · P. Radeva
Universitat de Barcelona, Gran Via de les Corts Catalanes 585,
08007 Barcelona, Spain
e-mail: ligual@ub.edu

A. Hernández-Vela
e-mail: ahernandez@cvc.uab.es

S. Escalera
e-mail: sergio@maia.ub.es

P. Radeva
e-mail: petia@cvc.uab.es

L. Igual · A. Hernández-Vela · S. Escalera · P. Radeva
Computer Vision Center (CVC), Campus UAB, Edifici 0, Bellaterra,
08193 Barcelona, Spain

J. C. Soliva · O. Vilarroya
Unitat de Recerca en Neurociència Cognitiva (URNC) Department of Psychiatry,
Universitat Autònoma de Barcelona (UAB),
IAPS Hospital del Mar. Passeig Marítim, 25-29, 08003 Barcelona, Spain
e-mail: 24744jsv@comb.cat

O. Vilarroya
e-mail: oscar.vilarroya@gmail.com

M. González Hidalgo et al. (eds.), *Deformation Models*, Lecture Notes in Computational Vision and Biomechanics 7, DOI: 10.1007/978-94-007-5446-1_10,

In particular, information concerning the intensity and geometry is exploited, and supervised energies based on contextual brain structures are added. Furthermore, boundary detection is reinforced using a new multi-scale edgeness measure. The method is applied to the segmentation of the brain caudate nucleus in a set of 39 pediatric attention-deficit/hyperactivity disorder (ADHD) patients and 40 control children, as well as in a public database of 18 subjects. We evaluate the quality of the segmentation using several volumetric and voxel by voxel measures, and present a quantitative volumetric analysis of caudate abnormalities in pediatric ADHD. The obtained results show improved performance in terms of segmentation accuracy compared to state-of-the-art approaches.

1 Introduction

Most of the analyses of brain MRI structures, as well as much research in neuroscience, lack an appropriate automated segmentation system, and therefore require physicians to manually segment brain structures (e.g. the caudate) on a slice by slice basis. This process is extremely time consuming, tedious, and prone to inter-rater discrepancies, limiting the statistical power of the analysis. An automated approach would accelerate the analysis and make the procedure feasible for large amounts of data. Automatic segmentation of subcortical structures in the brain is currently an active research area. In contrast to the problem of tissue segmentation (GM, WM, and CSF) in brain MRI, for which acceptable solutions can be found, the issue of subcortical structure segmentation has yet to be satisfactorily addressed. Structures such as the putamen and caudate nucleus are difficult to correctly segment even manually, since they are small and their intensity is non-uniform and non-contrasted. Figure 1 is an example of some brain MRI transversal planes with the caudate nucleus boundary depicted.

Semi-automatic methods for segmenting subcortical structures have been proposed, such as the method developed specifically for neuroanatomical segmentation [46], in which the user specifies two coordinates of the AC-PC line for the segmentation of the caudate. This method is a knowledge-driven two-step algorithm. In the first step, lateral ventricles are extracted to help position a bounding box that contains the caudate nucleus. Region growing of gray matter seed points is performed inside the box to estimate an initial segmentation. A set of anatomical constraints is also defined, based on previous knowledge, and is subsequently imposed on the first result. In the second step, the caudate boundaries are refined outside the bounding box by imposing new anatomical constraints. In [9], the authors use an SPM tool to segment and compute voxel-based morphometry measures. Significant effort has been put into automated segmentation of different structures in brain MRI (see reviews [6, 13]). A good example of these efforts can be found in the Caudate Segmentation Evaluation Challenge (CAUSE07) [40]. In this competition, different algorithms designed to segment the caudate nucleus from brain MRI scans were compared. From among the methods adopted, atlas-based segmentation approaches

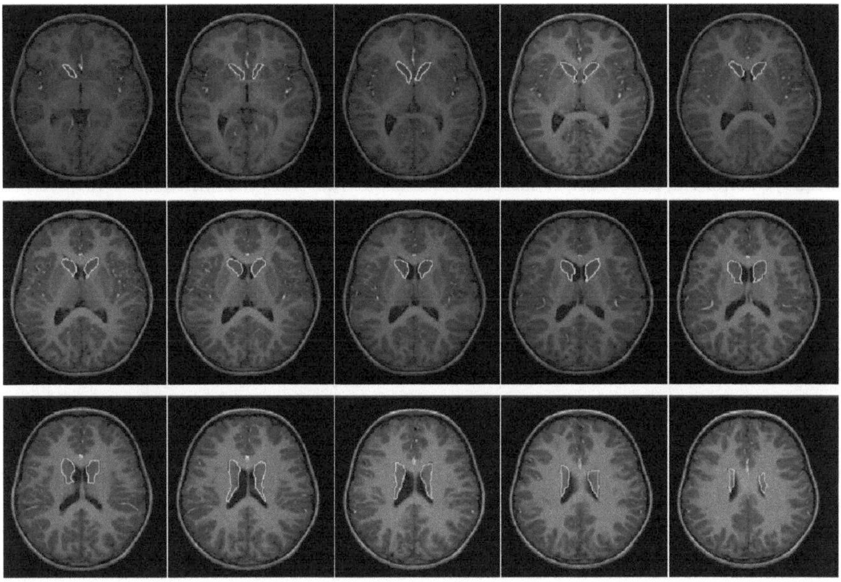

Fig. 1 Example of brain MRI scans. Caudate nucleus are marked in *white*

stand out as a powerful generic technique for automatic delineation of structures in volumetric images. These approaches use data obtained from different subjects to construct an atlas, which acts as a common anatomy for the area (of the brain) and apply it to further segmentations. The results of the CAUSE07 competition show that multi-atlas segmentation methods can outperform schemes based on a single atlas. However, on one hand, running multiple registrations on volumetric data requires a lot of time, and it is difficult to determine the optimum number of atlases to be considered [42]. On the other hand, an important disadvantage of atlas-based methods is that the target object is not necessarily correctly represented by the atlas shapes. In this case, a more flexible and adaptive technique can be useful in order to ensure accurate segmentation results.

The method considered in this work, CaudateCut, combines the power of atlas-based segmentation with a deformable model based on the Graph-cut (GC) framework, to obtain a globally optimal segmentation of the caudate structure in MRI. The GC theory has been used in many computer vision problems [26]. In particular, it has been successfully applied to binary segmentation of images, and has yielded a solution which corresponds to the global minimum of an energy function [7, 8]. The goodness of the solution depends on the suitability of the unary and boundary energy terms and their reliable computation. The original GC definition is limited to image information, and can fail when the caudate structure in MRI is subtle and contrast is low. In order to overcome this problem, supervised contextual information of the caudate nucleus is added and a new multi-scale edgeness measure is used for reinforcing boundary detection.

Segmentation results from two different datasets are presented, and abnormalities in pediatric Attention-Deficit/Hyperactivity Disorder (ADHD) are analyzed. Studies of volumetric brain MRI show neuroanatomical abnormalities in pediatric ADHD [9, 15, 33]. ADHD is a developmental disorder characterized by inattentiveness, motor hyperactivity and impulsiveness, and it represents the most prevalent childhood psychiatric disorder. It is also estimated that half the children with ADHD will display the disorder in adulthood. As stated in several reviews and metanalyses, diminished right caudate volume is one of the most replicated findings among ADHD samples in morphometric MRI studies [39]. As a result of these studies, in [35], the authors proposed a diagnostic test based on the ratio between the volume of parts of the caudate. The first dataset, considered in this work, consists on an MRI dataset of thirty nine children/adolescents with ADHD (ages 6–18) and forty healthy control subjects matched for age, gender, and handedness. The second is a public dataset of 18 healthy controls from the Internet Brain Segmentations Repository provided by the Center for Morphometric Analysis at Massachusetts General Hospital. We show that CaudateCut model, improves segmentation performance with respect to a classical atlas-based approach and a multi-atlas approach proposed recently. Moreover, a quantitative volumetric analysis of pediatric ADHD is provided, and specifications and obtained results are comparable to manual analysis based on caudate nucleus appearance.

The rest of the chapter is organized as follows: Sect. 2 goes through the related work. Section 3 introduces the CaudateCut model and algorithm. Section 4 reports and discusses the results of experiments on caudate nucleus segmentation, as well as the ADHD volumetric quantitative analysis. Finally, Sect. 5 presents the conclusions and describes future lines of research.

2 Related Work

Different strategies can be adopted for fully-automatic segmentation of subcortical structures. Recent techniques can be summarized in four groups: (a) anatomical atlas-based and multi-atlas-based algorithms, (b) supervised learning techniques, (c) statistical model approaches, and (d) energy-based segmentation techniques.

Anatomical atlas-based methods rely on a registration of the query image with a pre-computed anatomical atlas of the brain. After registration, atlas labeling is propagated to give an estimation of the segmentation in the query subject [10, 21, 31]. The main advantage of these methods is that the atlas provides *a priori* information about the spatial distribution of tissue types or structures. Thus, methods directly use knowledge about the structure of the brain. Some techniques for atlas construction can be found in [34]. Automatic methods have been devised to segment any structure of an anatomical image in the native space that had been predefined in an anatomical atlas in a stereotaxic—or normalized—space. These approaches rely on accurate non-linear deformations to find spatial correspondence between either images. In [10], Collins et al. developed a fully automatic procedure, baptised ANIMAL, to segment any structure of an anatomical image in a native space that had been predefined in an anatomical atlas in a stereotaxic—or normalized—space. In that work, it was observed that since the deformation field is bandlimited, irregular structures such as cortical gyri and sulci could not be accurately segmented. It was in their next work ANIMAL+INSECT [11] where the problem was addressed by introducing a post-processing that required tissue classification of the subject, carried out by INSECT, to refine the final segmentation of any labeled structure. Departing from sequential processes like ANIMAL+INSECT, other authors have exploited the benefit of generative models with the aim of reaching optimal solutions. [17] and [3] combined tissue classification, bias correction, and nonlinear warping within the same framework. The latest version of SPM [37] already includes the unified approach of [3] and incorporates scalp extraction into the process.

The main disadvantage of these methods is the computational cost necessary to build an atlas from different subjects, which requires from complex rigid registration. Behind registration with an atlas is time consuming, it might not find a globally optimal solution or even fail completely in the case of strong anatomical anomalies. Moreover, training set selection to build atlas is a difficult issue and most of the methods in Challenge CAUSE07 [40] manually select different training sets to segment the different groups of test data. This converts these methods in semi-automatic. In [31], the influence of the atlas selection is analyzed by comparing different atlases on the segmentation of tissue for brain MRI of young children. In this case, standard expectation-maximization algorithm with registration based segmentation was used [41]. In [16] the authors incorporate structure-specific models using Markov Random Fields and [23] improves the results produced by [16] using diffeomorphic warps. To take advantage of the atlas information in the segmentation process, the image should be registered to the atlas. In this sense, accurate techniques for image registration are crucial for the segmentation procedure [25, 28].

Atlas-based algorithms were first conceived as based on a single mean atlas, as in [21] where subcortical brain structures are segmented using a registration algorithm and a single ad-hoc atlas applied to 14 schizophrenia patients and 14 control patients. Progressively, atlas-based methods evolved to multi-atlases strategies where besides label propagation, decision fusion strategies are involved [2, 19, 42]. When multiple atlases are considered, labels from each atlas are aligned with the query image and treated as classifiers. Classifier fusion, based on the majority vote rule, has been shown to be accurate to segment brain structures. This strategy can be more robust and increasingly accurate with increasing numbers of classifiers. However, it suffers from problems of scale when the number of atlases is large. In [2] the authors compare different classifier selection strategies which are applied to a group of 275 subjects with manually labeled brain MR images. An Adaptive Multi-Atlas Segmentation method (AMAS) was presented in [42]. AMAS includes an automatic decision to select the most appropriate atlases for a target image and stop criterion for registering atlases when no further improvement is expected. This method obtained the best mark in the challenge CAUSE07.

Different ways of exploiting supervised learning in segmentation methods have been incorporated. In [24], the atlas-based segmentation method presented uses segmentation confidence maps, which are learned from a small manually-segmented training set, and incorporated into the cost term. This cost is responsible for weighting the influence of initial segmentations in the multi-structure registration. Moreover, multiple atlases are used both in a supervised atlas-correction step, and multiple atlas propagation. In [29], a two-stage method is presented, which benefits from capabilities of mathematical feature extractors and artificial neural networks. In the first stage, Geometric Moment Invariants (GMIs) are applied at different scales to extract features that represent the shape of the structures. Next, multi-dimensional feature vectors are constructed that contain the GMIs along with image intensity values, Probability Atlas Values (PAVs), and voxel coordinates. These feature vectors are used to estimate Signed Distance Maps (SDMs) of the desired structures. To this end, Multi-Layer Perceptron Neural Networks (MLP-NN) are designed to approximate the SDMs of the target structures. In the second stage, the estimated SDM of each structure is used to design another MLP-NN to classify the image voxels into two classes: inside and outside the structure.

Shape and appearance models involve establishing correspondence across a training set and learning the statistics of shape and intensity variation using PCA models. To segment an image being studied, model parameters which best approximate the structures have to be computed. [4] applies an Active Appearance Model (AAM)-based method to segment the caudate nucleus. A "composite" 3D profile AAM is constructed from the surfaces of several subcortical structures using a training set, and individual AAMs of the left and right caudate are constructed from a different training set. Segmentation starts with affine registration to initialize the composite model within the image. Then, a search is performed using the composite model. This provides a reliable but coarse segmentation, used to initialize a search with the individual caudate models. In [22], the authors use a statistical shape model with elastic deformations to segment the hippocampus, thalamus, putamen, and pallidum. In [5], a comparison of four different strategies of brain subcortical structure segmentation is presented: two of them are atlas-based strategies ([2, 31]) and the other two are based on statistical models of shape and appearance ([4, 32]). The best results are achieved by the multi-atlas classifier fusion and labeling approach [2] which treats atlases as classifiers and combines them using a majority voting rule.

With reference to energy-minimization methods, [38] uses a deformable mesh followed by normalized cuts criterion to segment the caudate and the putamen from PET images. [43] proposes a multiphase level set framework for image segmentation using the Mumford-Shah model, as a generalization of an active contour model. In [27], a method is presented for the segmentation of anatomical structures, which incorporates prior information about the intensity and curvature profile of the structure from a training set of images and boundaries. The intensity distribution is modeled as a function of signed distance from the object boundary instead of modeling only the intensity of the object as a whole. A curvature profile acts as a boundary regularization term specific to the shape being extracted, as opposed to simply penalizing high curvature. Using the prior model, the segmentation process estimates a maximum a posteriori higher dimensional surface whose zero level set converges on the boundary of the object to be segmented. In [36], a graph based method is presented for brain tissue segmentation. In [45], the GC strategy is adapted for segmenting anatomical brain regions of interest in Diffusion Tensor MRI (DT-MRI). An open source application called ITK-SNAP was developed [47] for level set segmentation.

Finally, there exist some libraries, such as Freesurfer [18], Slicer [1], and SPM [37], which have been developed to address the MRI segmentation problem. However, all of them are limited to atlas-based algorithms which lack robustness when dealing with different types of subjects. Hence, constructing an hybrid approach that combines atlas-based and energy-based strategies is a natural extension of state-of-the-art algorithms. The combination presented in this work exploits atlas structure information and an *ad hoc* deformable model. Moreover, the proposed model also takes advantage of supervised learning techniques.

3 CaudateCut

In this section, the GC framework is reviewed and the CaudateCut segmentation presented in [20] is described.

3.1 Graph-Cut Framework

Let us define $\mathscr{X} = (\mathbf{x}_1, \ldots, \mathbf{x}_p, \ldots, \mathbf{x}_{|\mathscr{P}|})$ as the set of pixels for a given grayscale image I; $\mathscr{P} = (1, \ldots, p, \ldots, |\mathscr{P}|)$ as the set of indexes for I; $\mathscr{N}$ as the set of unordered pairs $\{p, q\}$ of neighboring pixels of $\mathscr{P}$ under a 4-(8-) neighborhood system, and $L = (L_1, \ldots, L_p, \ldots, L_{|\mathscr{P}|})$ as a binary vector whose components L_p specify assignments to pixels $p \in \mathscr{P}$. Each L_p can be either "foreground" or "background", or equivalently "cau" or "back" for our problem (abbreviations for caudate and background), indicating whether pixel p belongs to the caudate or background, respectively. Thus, the array L defines a segmentation of image I. The GC model defines the cost function $E(L)$ which describes soft constraints imposed on boundary and region properties of L,

$$E(L) = U(L) + \delta B(L), \tag{1}$$

where $U(L)$ is the unary term (or region properties term),

$$U(L) = \sum_{p \in \mathscr{P}} U_p(L_p), \tag{2}$$

and $B(L)$ is the boundary property term,

$$B(L) = \sum_{\{p,q\} \in \mathscr{N}} B_{\{p,q\}} \, \Omega(L_p, L_q), \text{ where } \Omega(L_p, L_q) = \begin{cases} 0, & \text{if } L_p \neq L_q \\ 1, & \text{otherwise.} \end{cases} \tag{3}$$

The GC method imposes hard constraints on the segmentation results by means of the definition of seed points where labels are predefined and cannot be modified. The subsets $\mathscr{C} \subset \mathscr{P}, \mathscr{B} \subset \mathscr{P}, \mathscr{C} \cap \mathscr{B} = \emptyset$ denote the subsets of caudate and background seeds, respectively. The goal of GC is to compute the global minimum of Eq. (1) from all segmentations L satisfying the hard constraints $\forall p \in \mathscr{C}, L_p =$ "cau", $\forall p \in \mathscr{B}, L_p =$ "back".

3.2 CaudateCut Segmentation

The steps in the CaudateCut algorithm are described in detail in the following subsections. The CaudateCut algorithm is summarized in Algorithm 5.

Algorithm 5 Automatic CaudateCut Segmentation algorithm

1: Initial segmentation using AB method.
2: Set background and caudate seeds from AB mask information.
3: Initialize unsupervised unary potentials UU_p("*cau*") and UU_p("*back*") based on local greylevel intensities.
4: Initialize supervised unary potentials SU_p("*cau*") and SU_p("*back*") based on SVM correlogram classifier.
5: Initialize unary term based on combined unary potentials.
6: Initialize boundary term $B(L)$ based on first and second derivatives of intensities and multi-scale edge map.
7: Estimate caudate segmentation using GC.

3.2.1 Atlas-Based Segmentation

The atlas-based segmentation of the caudate largely follows the strategy proposed by [11], (1) First, a non-uniformity image intensity correction is computed. Then, the corrected image is classified into WM, GM, and CSF. (2) In the next step, the GM image is elastically registered from its original geometrical space to match a template image (which represents the expected distribution of gray matter in the subjects under study) in the so-called normalized space. The deformation field obtained is inverted to map the normalized space onto the original space. (3) This inverted deformation is applied to the caudate segmentation in the normalized space, thus yielding a first segmentation of the caudate nucleus of the subject. (4) Finally, in order to refine this first segmentation, the GM mask of the subject under study is combined with the mask obtained by unwarping the normalized caudate segmentation. They are combined as follows: the GM and caudate probability maps are multiplied and a threshold T_p is imposed over the result: it is considered that a voxel belongs to the caudate only where the product map is larger than T_p.

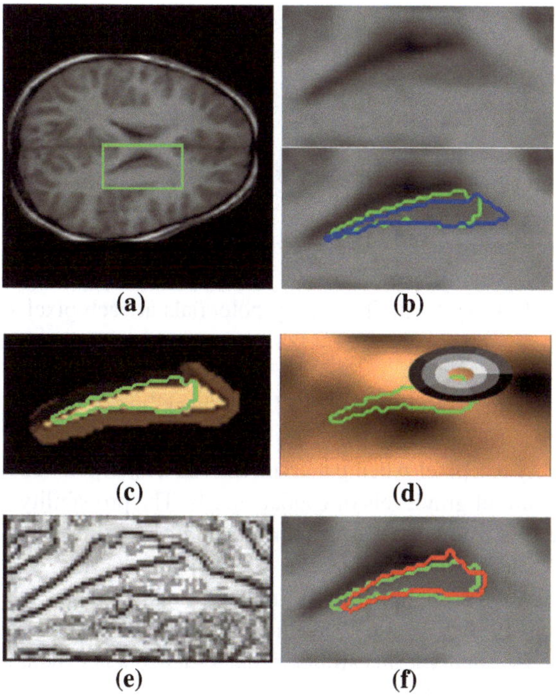

Fig. 2 **a** Original MRI scan, **b** *Up* Crop from the MRI scan, *Down* Atlas-based segmentation (*blue*) and GT (*green*), **c** Unsupervised probability values $P_u(L_p = \text{"cau"})$ and GT (*green*), **d** Supervised probability values $P_s(L_p = \text{"cau"})$ and GT (*green*), **e** Boundary potentials $B(L)$, **f** image crop, GT (*green*) and CaudateCut result (*red*)

This atlas-based segmentation method depends strongly on the atlas definition. In some situations, this can result in a solution that does not fit the target structure well, and a further refinement may be necessary. However, the segmentation obtained may be useful for roughly locating the region of interest, and thus, it can be used to define the seeds for GC application. Figure 2b shows the result of AB segmentation for the input image in Fig. 2a.

3.2.2 Seed Initialization

The segmentation result of the atlas-based method is used to define initial segmentation seeds in order to achieve a fully automatic method. Caudate and background seeds are defined by performing morphological operations on the ROI obtained $\mathscr{R}_0$ in the atlas-based mask. Caudate seeds $\mathscr{C}$, are defined by computing $\mathscr{C} = \mathscr{R}_0 \ominus ST_{k_e}$, where $\ominus$ denotes the erosion operator, and ST_{k_e} is a structural element of k_e pixels. In the case of background seeds, the region $\mathscr{R}_0$ is dilated and the complementary set is kept, $\mathscr{B} = \mathscr{P} \setminus (\mathscr{R}_0 \oplus ST_{k_d})$, where $\oplus$ denotes the dilation operator, and ST_{k_d} is a structural element of k_d pixels. In the example shown in Fig. 2a, the selection of $\mathscr{C}$ and $\mathscr{B}$ seeds is obtained from erosion and dilation of the AB segmentation shown in Fig. 2b.

3.2.3 Unary Energy Term

The unary energy term for the GC energy function is divided into two: an unsupervised part and a supervised part.

Unsupervised unary term. The unary potentials at each pixel p are initialized as,

$$UU_p(\text{“cau”}) = -\ln(\mathrm{P_u}(L_p = \text{“cau”})), \;\; UU_p(\text{“back”}) = -\ln(\mathrm{P_u}(L_p = \text{“back”})).$$

The probability of a pixel p being marked as “cau”, $\mathrm{P_u}(L_p = \text{“cau”})$, is computed using the histogram of graylevels of caudate seeds. The probability of a pixel being marked as “back” is computed using the inverse probability, as $\mathrm{P_u}(L_p = \text{“back”}) = 1 - \mathrm{P_u}(L_p = \text{“cau”})$, since background seeds contain GM, WM and CSF and it is difficult to extract a model directly from them. Figure 2c shows the unsupervised probability values $\mathrm{P_u}(L_p = \text{“cau”})$, for the image in Fig. 2a.

The unsupervised unary term estimates image-dependent caudate pixel probabilities based on caudate seeds. However, given the noisy information of MRI images and the small number of caudate seed pixels, a high generalization based on this term is not always guaranteed. In this context, a combination of the unsupervised and supervised energies is proposed, which is based on learning contextual caudate derivatives from Ground Truth (GT) data.

Supervised unary term. A binary classifier is trained using a set of MRI slices as a training set. In particular, a pixel descriptor using a correlogram structure is extracted. The correlogram structure captures contextual intensity relations from circular bins around the pixel analyzed [14].

Given a pixel p, a correlogram $C_{c \times r}$ is defined, where c and r define the number of circles and radius of the structure. Then, each bin b from the set of n bins, with $n = c \cdot r$, is defined as the area delimited by two consecutive circles of the given radius. Given the pixel p and its correlogram structure $C^p_{c \times r}$, its supervised caudate descriptor is defined as: $\mathbf{d}_p = \{\partial_1, \ldots, \partial_k, \ldots, \partial_{n \cdot (n-1)/2}\}$, where ∂_k is the signed substraction of graylevel information within a pair of bins in $C_{c \times r}$. In this sense, the descriptor contains the $n \cdot (n-1)/2$ graylevel derivatives of all pairs of bins within $C_{c \times r}$, which captures all spatial relations of graylevel intensities in the neighborhood of p. An example of a correlogram structure estimated for a caudate pixel is shown in Fig. 2d.

The descriptors are extracted for a subset of pixels on $\mathscr{C}$ and $\mathscr{B}$ from the training set data. Given the set of descriptors, a linear support vector machines (SVM) classifier is trained in order to predict caudate confidence on image pixels from new test data. In our case, the output confidence of the classifier is used as a measure of the "probability" of a pixel belonging to the caudate. Then, the supervised unary potentials at each pixel p are:

$$SU_p(\text{"cau"}) = -\ln(\mathrm{P_s}(L_p = \text{"cau"})),\; SU_p(\text{"back"}) = -\ln(\mathrm{P_s}(L_p = \text{"back"})).$$

The probability of a pixel being marked as "cau" is computed using the confidence of the SVM classifier over its correlogram descriptor $\mathrm{P_s}(L_p = \text{"cau"}) = \mathrm{SVM}(p)$. The probability of a pixel being marked as "back" is computed as the negative of the output margin of the classifier $\mathrm{P_s}(L_p = \text{"back"}) = -\mathrm{SVM}(p)$. Figure 2d shows the supervised caudate probability values, $\mathrm{P_s}(L_p = \text{"cau"})$, for the image in Fig 2b.

Combined unary term. The final unary term is defined as the addition of the unsupervised and supervised values at pixel p as follows:

$$U_p(\text{"cau"}) = UU_p(\text{"cau"}) + SU_p(\text{"cau"}),\; U_p(\text{"back"}) = UU_p(\text{"back"}) + SU_p(\text{"back"}).$$

3.2.4 Boundary Energy Term

To define boundary potentials, first and second intensity derivatives of the image are used to exploit intensity and geometric information. Moreover, given the high variability in contrast between the caudate and background in different parts of the images, the boundary term is weighted using an image-dependent multi-scale edgeness measure.

Specifically, the boundary potentials are defined as the following convex linear combination:

$$B_{\{p,q\}} = J(\alpha N_{\{p,q\}} + (1-\alpha) O_{\{p,q\}}).$$

First, $N_{\{p,q\}}$ and $O_{\{p,q\}}$ is defined as:

$$N_{\{p,q\}} = \frac{1}{\|\mathbf{x}_p - \mathbf{x}_q\|_2} \exp\left(-\frac{(I_p - I_q)^2}{2\sigma^2}\right),\; O_{\{p,q\}} = \frac{1}{\|\mathbf{x}_p - \mathbf{x}_q\|_2} \exp\left(-\frac{\theta^2_{\{p,q\}}}{2\beta^2}\right). \tag{4}$$

The term $\theta_{\{p,q\}}$ denotes the angle between two unitary vectors codifying the directions of minimum gradient variation in pixel p and q based on the Hessian eigenvectors. In particular, the direction of the eigenvector of the Hessian matrix with the smallest eigenvalue which gives the direction of the smallest variation at each pixel is chosen. The parameter α is empirically set by cross-validation, while σ and β are computed by adapting the image distribution to I_p and $\theta_{\{p,q\}}$, respectively.

Second, the J term is defined as the multi-scale edgeness measure at each pixel: $J = (J_1^*, \ldots, J_p^*, \ldots, J_{|\mathcal{P}|}^*)$. In order to compute J_p^*, the Canny edge detector algorithm is first run on the observed image at different threshold levels. Then, the edge probability is computed at each pixel by linear averaging of the edge thresholds for different scales as follows:

$$J_p^* = \min_j \frac{1}{n} \sum_{k=1}^{n} J_{p,\gamma_k,s_j},$$

where J_{p,γ_k,s_j} is the binary edge map using threshold γ_k and scale s_j for pixel p. If pixel p is labeled as an edge pixel for most of the threshold levels at a significant scale, it has a high probability of being an edge pixel. In order to decrease the smoothness effect at the regions near a boundary, the probability map is convolved with a Gaussian kernel. Figure 2e shows the boundary potential values $B(L)$ for the image in Fig. 2a.

Finally, by applying the min-cut algorithm over the defined energy function and image graph, the final caudate segmentation is obtained. Figure 2f shows the segmentation resulting from applying the CaudateCut algorithm.

4 Experimental Section

Before presenting the results, the material and methods of comparison are firstly described, as well as the validation protocol for the experiments.

4.1 Material

Two different databases were considered, named URNC database and IBSR database, in order to validate the proposed CaudateCut method.

- **URNC database**. This is a new database, which includes 39 children (35 boys and 4 girls) with ADHD, according to DSM-IV, referred from the Unit of Child Psychiatry at the "Vall d Hébron" Hospital in Barcelona, Spain, and coordinated by the Unit of Research in Cognitive Neuroscience (URNC) at the IMIM Foundation, together with 39 control subjects (27 boys and 12 girls) recruited from the community. The mean age of the groups was 10.8 (S.D.: 2.9) and 11.7 (S.D.: 3), respectively. The groups were matched for handedness and IQ. The 1.5-T system was used to acquire brain MRI scans. The resolution of the scan is $256 \times 256 \times 60$ pixels with 2-mm thick slices. Expert segmentation of the 79 individual caudate nuclei was obtained. MRIcro software[1] was used for volume labeling and manipulation.
- **IBSR database**. This dataset is part of a public database released by CAUSE07 Challenge [40]. It is composed by 18 T1-weighted MRI scans from the Internet

[1] www.cabiatl.com/mricro/

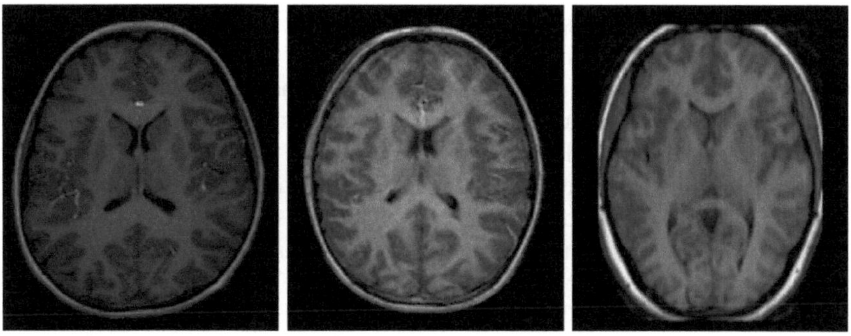

(a) (b) (c)

Fig. 3 Control **a** and ADHD **b** MRI slice example of URNC-database and MRI slice example of IBSR-database **c**

Brain Segmentation Repository (IBSR). It also contains expert segmentations of caudate structure. The MRI scans are of 1.5 mm thickness. Originally, the data size was $256 \times 128 \times 256$ pixels, but in order to prepare data for the later application of the CaudateCut algorithm, they were re-oriented by X-axis rotation and converted it into $256 \times 256 \times 128$ pixels. For more details of the acquisition, visit CAUSE07 Challenge website[2] from where the data were downloaded.

Figure 3 displays a sample control (a) and ADHD (b) MRI from the URNC database and a sample MRI from the IBSR database, (c). As can be appreciated, the quality of the ADHD image is worse than that of the control image, probably due to the movement of the children during image acquisition. Anisotropic filtering [44] was performed on all the slices before CaudateCut was applied.

4.2 Methods

We compared the CaudateCut method to two state-of-the-art methods: a classical atlas-based method, and a multi-atlas segmentation method. We also compared the results with the inter-observer (IO) variability of the expert GT.

AB method: We implemented atlas-based segmentation of the caudate following the strategy presented in [11]. To this end the SPM toolbox implementation of the unified non-linear normalization and tissue segmentation were used. The parameters of the method were set by default as in the SPM8 implementation, except for the threshold T_p, which was estimated using a subset of 5 control subjects from the URNC database and set to $T_p = 0.1$. The method was implemented using Matlab2008.

AMAS method: AMAS, was implemented as presented in [42]. For the atlas selection strategy, the absolute voxelwise difference between the target image and the reg-

[2] www.cause07.org/

istered images from the atlases were computed and they were ordered from smallest to largest. Then, the atlas information was propagated until the stopping criterion was reached. The stopping criterion was defined by the percentage of voxels that change their segmentation label after a new atlas propagation. This threshold was set to 0.05 for all experiments. The rest of the parameters in the AMAS method were set as described in [42]. The method was implemented using Matlab2008, and elastix6 version 3.9[3] was used for volume registration, as suggested in [42].

CaudateCut method: The CaudateCut method was implemented using Matlab2008 and the SPM toolbox. In all the experiments, the parameters were set to $k_e = 4$, $k_d = 10$, $c = 3$, $r = 5$, $\alpha = 0.5$, $\mathscr{S}_p = [1, 1.5, \ldots, 6]$, $\ell = 0$, $\gamma_k \in [0.02, 0.03, \ldots, 0.3]$ and $s_j \in [0.5, 1, \ldots, 5]$. The parameters σ and β were estimated for each image, as explained above. The parameter δ was tuned by cross-validation and was set to 50 for the URNC dataset and 100 for the IBSR database. In order to train the SVM classifiers for computation of the supervised unary term, a subsampling of pixels from each slice was performed. In particular, all the pixels labeled as caudate in the GT were taken, and the same number of background pixels. The background pixels were subsampled in a stratified way, trying to select pixels from all parts of the background.

Manual method: Experts use MRIcro [30] to manually delineate the caudate boundaries slice by slice. See [9] for more details of the procedure.

4.3 Validation

In order to be sufficiently general, several volumetric measures, as well as voxel by voxel comparison measures were evaluated. We focused on the six metrics detailed below, as proposed in [40]. In all of them, R corresponds to the estimated segmentation, G to the GT segmentation and $|\cdot|$ denotes the cardinal of a set.

1. Volumetric similarity index error, in percent $\mathrm{SIE} = \left|1 - 2\frac{\mathrm{R}\cap\mathrm{G}}{R+G}\right| \cdot 100$.
2. Volumetric union overlap error, in percent $\mathrm{VOE} = \left|1 - \frac{\mathrm{R}\cap\mathrm{G}}{\mathrm{R}\cup\mathrm{G}}\right| \cdot 100$.
3. Relative absolute volume difference, in percent $\mathrm{VD} = \left|\frac{\mathrm{VOL_R}-\mathrm{VOL_G}}{\mathrm{VOL_G}}\right| \cdot 100$, where $\mathrm{VOL_R}$ and $\mathrm{VOL_G}$ correspond to the total volume of the R and G segmentations, respectively.
4. Average symmetric surface distance, in millimeters

$$\mathrm{AD} = \frac{\left(\sum_{i=1}^{N} \mathrm{d}(\mathrm{B_{S}}_i, \mathrm{B_R})^2 + \sum_{i=1}^{M} \mathrm{d}(\mathrm{B_S}, \mathrm{B_R}_i)^2\right)}{|B_S| \cdot |B_R|},$$

---

[3] http://elastix.isi.uu.nl

where B_S and B_R correspond to the set of border voxels in R and G, respectively, and $d(\cdot,\cdot)$ returns the minimum Euclidean distance between two sets of voxels.

5. Root Mean Square (RMS) symmetric surface distance, in millimeters RMSD = $\sqrt{AD}$.
6. Maximum symmetric surface distance, in millimeters

$$\mathrm{MD} = \max_{i,j}\left(\mathrm{d}(\mathrm{B}_{\mathrm{S}i}, \mathrm{B}_{\mathrm{R}}), \mathrm{d}(\mathrm{B}_{\mathrm{S}}, \mathrm{B}_{\mathrm{R}j})\right).$$

Note that for all the volumetric and voxel measures, the perfect value is 0. SIE and VOE have 0 as a perfect segmentation and 100 as the lowest possible value, when there is no overlap at all between the estimated segmentation and GT. In the case of VD, the perfect value is 0, which can also be obtained for a non-perfect segmentation, as long as the volume of that segmentation is equal to the volume of the reference.

In order to validate the AMAS and CaudateCut methods (SVM classifiers for supervised unary term computation), a leave-one-out strategy was followed. Finally, Student's paired t-test [12] was used to evaluate the statistical significance between pairs of segmentation algorithms with a particular dataset (threshold of $p < 0.05$). The null hypothesis corresponds to the hypothesis that the two groups belong to the same distribution and is called H_0. Matlab2008 was used to perform this test.

4.4 Results and Discussion

The results are divided into two sections corresponding to two related experiments: segmentation evaluation and ADHD volumetric quantitative analysis.

4.4.1 Segmentation Evaluation

(*A*) *Quantitative segmentation results*. The performance of the CaudateCut, AMAS and AB methods is compared. Figure 4 shows the results obtained in the experiments on both URNC and IBSR datasets for the six validation measures. SIE, VOE and VD are measured in %, AD, RMSD amd MD are measured in mm, and all the measures are displayed in base 10 logarithm. For all validation measures, CaudateCut produced better results than both AB and AMAS for both databases. With regard to the volumetric measures, CaudateCut achieved good mean rates of 19.25 % for SIE (equivalently, 80.75 % SI), 31.98 % for VOE (equivalently, 68.02 % VO), and 16.22 for VD. Voxel by voxel mean measures are also acceptable, with 0.0024 mm for AD, 0.0733 mm for RMSD, and 35.70 mm for MD. The large MD values are due to the recurrent errors present in the internal boundaries of the caudate defined between caudate head and body, as is clarified in the visual results below. For the IBSR database, the AMAS method obtained larger VO and SI values than the AB method, whereas, in the URNC database, the AB method improved on the result of

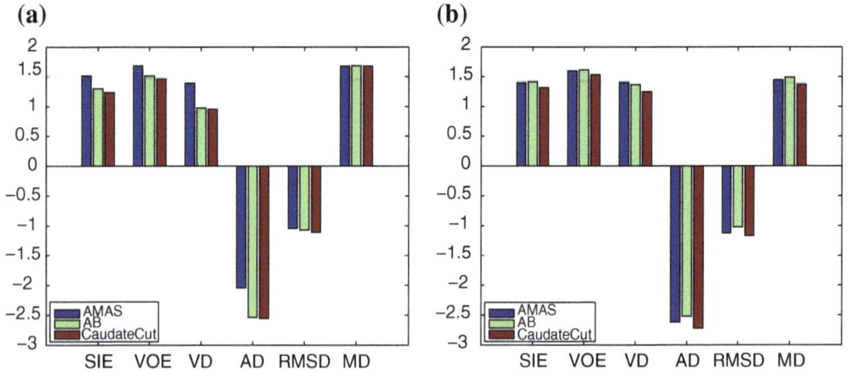

Fig. 4 Quantitative results of AB, AMAS and CaudateCut methods applied to URNC **a** and IBSR **b** databases. Validation measures are, *SIE* volumetric similarity index error (in %); *VOE* volumetric overlap error (in %); *VD* relative absolute volume difference (in %); *AD* average symmetric surface distance (in mm); *RMSD* root mean square symmetric surface distance (in mm); *MD* maximum symmetric surface distance (in mm)

the AMAS method. This could be due to the fact that the AB parameters were tuned in the URNC database. In this sense, CaudateCut was able to properly overcome this inconvenience and improve on the AMAS results in the IBSR database. It is important to note that CaudateCut showed robustness to AB

(*B*) *Qualitative segmentation results.* Figure 5 shows qualitative CaudateCut results for the MRI slices of a control subject. In most of the slices, the CaudateCut segmentation result (red line) is highly comparable to the GT (green line). However, segmentation differences occur in the first and last caudate frames, where some voxels are classified as caudate by CaudateCut, but not by the GT (false positives). The inherent ambiguity of the caudate boundaries makes the expert's task of manually defining the caudate start and end slices arduous. This introduces variability and produces errors in MRI atlas information corresponding to the end slices. It is difficult for CaudateCut to rectify this kind of error. The AB method introduces fake seeds in these positions and CaudateCut propagates these errors, since it can not remove the seeds. In the second column of the second row, some voxels are not classified as caudate, while they should be, according to GT (false negatives). This particular sample slice corresponds to the transition between caudate head and body, where the caudate shape changes abruptly from the rounded head to the elongated body [35].

Figure 6 compares qualitative results of left caudate segmentation of URNC database MRI slices using the AMAS (second column), AB (third column) and CaudateCut (fourth column) methods. Note that the best segmentation results were obtained by the novel CaudateCut segmentation method, followed by AB, and finally, by the AMAS strategy. In general, CaudateCut improves AB segmentation and obtains a better fit to the caudate boundaries. Only in a few cases (examples in rows 2 and 3), CaudateCut agrees with the AB segmentation, and the GC strategy did

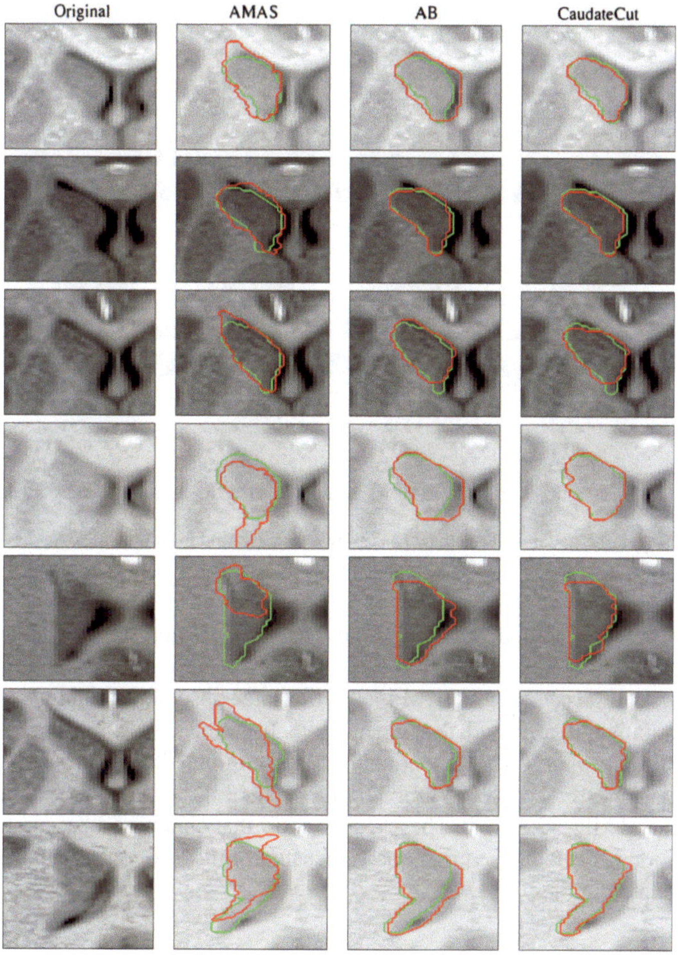

Fig. 5 Some *left* caudate segmentation results of URNC-database. *First column* Original image crop. *Second column* AMAS result. *Third column* AB result. *Forth column* CaudateCut result. GT is shown in *green* and automatic segmentation result in *red*

not apply changes to the final segmentation. It can be seen that the registration strategy applied for the AMAS method was unable to correctly fit the caudate boundaries.

(*C*) *Computational differences*. Concerning the computational time, AMAS was the most costly method in terms of testing time, since multiple registration had to be performed for each subject segmentation. On average, 2–3 registrations were performed for each volume segmentation and each registration took 7 min on a standard high-end PC, thus making 17.5 min for the whole volume segmentation. The AB method was the fastest, taking around 5 min on average for the whole volume segmentation. CaudateCut involves applying the AB method and later the GC min-

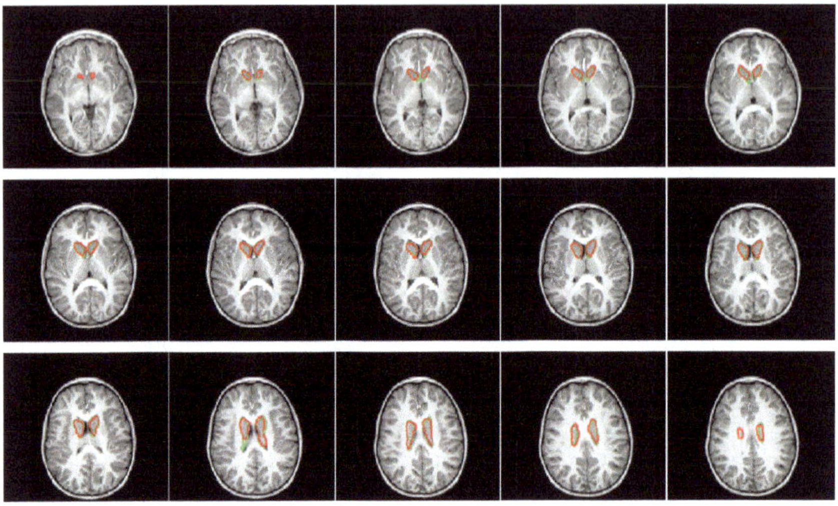

Fig. 6 Example of CaudateCut results. GT is shown in *green* and CaudateCut segmentation in *red*

imization process method. The total time was around 6 min for the whole volume segmentation.

(*D*) *Inter-observer variability*. Finally, the inter-observer variability was computed using manual left caudate segmentations from the URNC database by means of two different experts. The validation measures computed using these two GT segmentations are, SIE: 19.44 %, VOE: 32.20 % VD: 22.84 %, AD: 0.003 mm, RMSD: 0.092 mm and MD: 92.54 mm. These results show that obtaining an accurate manual segmentation is difficult even for experts, because of the low contrast and resolution of the caudate regions.

4.4.2 ADHD Volumetric Quantitative Analysis

The *a priori* hypothesis that developmental anomalies exist in the caudate nucleus of people with ADHD is generally accepted. Previous imaging studies have analyzed this hypothesis [15, 33, 39]. In this work, right and left caudate volumetric differences between ADHD and control subjects were analyzed in the URNC database. To this end, we performed a comparison of mean volume values applying Student's t-test for independent samples (with a threshold of $p < 0.05$). The aim of this experiment was to show that the analysis performed using automatic CaudateCut segmentation was coherent with the results of manual analysis. To carry out the manual and automated statistical analysis GT and CaudateCut segmentations were considered, respectively. ROI measures in voxels were transformed into cubic millimeters, mm^3 (ROI total number of voxels multiplied by voxel dimensions). Tables 1 and 2 shows the results of the manual and automatic analyses, respectively. Both tables contain mean volume

Table 1 Manual analysis of control and ADHD volume statistical differences for right and left caudate volume

Manual analysis	Group	N	M	Std	d	t-test	t	p	CI (95 %)
R caudate	Control	39	5031.44	660.18	312.29	**False**	1.9983	0.0493	1.03 to 623.56
	ADHD	39	4719.15	718.81					
L caudate	Control	39	4882.45	643.81	195.11	**True**	1.1946	0.2360	−130.19 to 520.42
	ADHD	39	4687.34	791.17					

N, sample size; *M*, mean volume (in mm^3); *std*, standard deviation, *d* mean difference; *t-test*, t-test result; t, Student's t statistic;p, p-value; *CI*, confidence interval of mean difference

Table 2 Automatic analysis of control and ADHD volume statistical differences for right and left caudate volume

Automatic analysis	Group	N	M	Std	d	t-test	t	p	CI (95 %)
R caudate	Control	39	4636.72	596.66	430.19	**False**	2.74	0.0075	118.05 to 742.35
	ADHD	39	4206.52	775.86					
L caudate	Control	39	4426.24	615.69	288.14	**True**	1.93	0.0571	−8.90 to 585.19
	ADHD	39	4138.10	698.89					

N, sample size; *M*, mean volume (in mm^3); *std*, standard deviation, *d* mean difference; t, Student's t statistic; p, p-value; *CI*, confidence interval of mean difference

measures, and standard deviation of control and ADHD groups for the right and left caudate, separately. Moreover, the results of Student's t-tests are presented: the t-test corresponds to a true (accept H_0) or false (reject H_0) result, t is Student's t statistic, p represents the p-value, and CI means the confidence interval of differences. As can be observed, the ADHD group has lower right and left mean caudate volume than the control group in both the manual and automatic analysis. Moreover, the results of the statistical test were the same in the manual and automatic analysis: the volume measure was found to be statistically different between the groups for the right caudate but not for the left. Comparing volume values, it can be seen that the automatic CaudateCut segmentation method under-segments the caudate nucleus compared with the manual delineation. However, these discrepancies in the segmentations do not prevent coherent results between the two methods in the statistical analysis of the groups considered.

Finally, the manual and CaudateCut automatic analysis were qualitatively compared. Figure 7 shows both control and ADHD caudate volume distributions using GT segmentation (a, b) and CaudateCut segmentation (c, d). First column plots (a, c) correspond to right caudate volume measures and second column plots (b, d) to left caudate volume measures. The histogram of caudate volume for the ADHD and control groups are depicted in dashed black and solid red lines, respectively. Two Gaussian functions were fitted to the histograms. It can be appreciated that the differences between the ADHD and control distributions were larger for the right caudate in both the manual and the automatic analysis. The immediate conclusion is that CaudateCut generates results that are comparable to gold-standard analyses

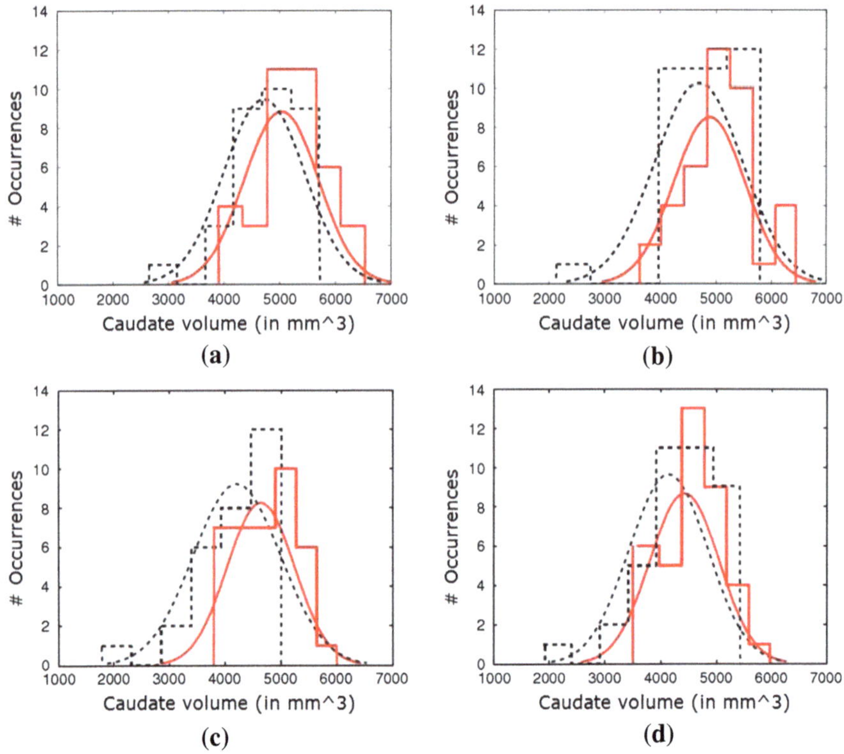

Fig. 7 Control and ADHD caudate volume distributions for *right* (**a, c**) and *left* (**b, d**) caudate volume measures. First row corresponds to the manual analysis and second row to automatic analysis. Control (*red solid line*) and ADHD (*black dashed line*) histograms are shown together with two Gaussian functions fitted to the histograms

in differentiating neuroanatomical abnormalities between healthy controls and the group of individuals with ADHD.

5 Conclusion

In this work, a new model for structure segmentation in brain MRI was presented. The method combines the power of atlas-based strategy and Graph Cut energy-minimization framework to adapt the final segmentation to the small and low-contrast brain structures. A new energy function was defined with data potentials exploiting intensity and geometry information, as well as using supervised learned local brain structures. Boundary potentials were also redefined using a new multi-scale edgeness measure. The method has different advantages for different neuroimaging researchers. First of all, it is fully automatic, and secondly, the algorithm is reliable.

The results are 100 % reproducible in subsequent runs with the same data, avoiding the inaccuracies of intra-rate and inter-rater drift.

The method was tested on two different datasets. Although the method was tuned on the novel URNC dataset, it provided outstanding results on IBSR, showing the inherent robustness of the approach. Moreover, comparable results to manual volumetric analysis of ADHD children were obtained based on automatic caudate nucleus volume measures.

Future lines of research include the use of multiple-hypotheses for seed initialization in order to increase robustness to possible errors of atlas application and the incorporation of 3D information in the caudate segmentation. From the clinical point of view, new features based on the caudate appearance can be added to analyze ADHD abnormalities in an automatic way.

References

1. 3d slicer. http://www.slicer.org/
2. Aljabar P, Heckemann R, Hammers A, Hajnal J, Rueckert D (2007) Classifier selection strategies for label fusion using large atlas databases. In: Ayache N, Ourselin S, Maeder A (eds): MICCAI 2007, Part I, Lecture notes in computer science, Springer, Heidelberg, pp 523–531
3. Ashburner J, Friston K (2005) Unified segmentation. NeuroImage 26:839–851
4. Babalola KO, Petrovic V, Cootes TF, Taylor CJ, Twining CJ, Mills A (2007) Automatic segmentation of the caudate nuclei using active appearance models. Proceedings of the 10th International Conference on Medical Image Computing and Computer-Assisted Intervention: 3D Segmentation in the Clinic: A Grand Challenge, pp 57–64
5. Babalola KO, Patenaude B, Aljabar P, Schnabel J, Kennedy D, Crum W, Smith S, Cootes T, Jenkinson M, Rueckert D (2009) An evaluation of four automatic methods of segmenting the subcortical structures in the brain. Neuroimage 47:1435–1447
6. Balafar M, Ramli A, Saripan M, Mashohor S (2010) Review of brain mri image segmentation methods. Artif Intell Rev 33:261–274
7. Boykov Y, Funka-Lea G (2006) Graph cuts and efficient n-d image segmentation. Int J Comput Vis 70(2):109–131
8. Boykov Y, Kolmogorov V (2001) An experimental comparison of min-cut/max-flow algorithms for energy minimization in vision. IEEE Trans Pattern Anal Mach Intell 26:359–374
9. Carmona S, Vilarroya O, Bielsa A, Trèmols V, Soliva JC, Rovira M, Tomàs J, Raheb C, Gispert J, Batlle S, Bulbena A (2005) Global and regional gray matter reductions in ADHD: a voxel-based morphometric study. Neurosci Lett 389(2):88–93
10. Collins DL, Holmes CJ, Peters TM, Evans AC (1995) Automatic 3d model-based neuroanatomical segmentation. Human Brain Mapping 3(3):190–208
11. Collins DL, Zijdenbos AP, Baaré WFC, Evans AC (1999) Animal+insect: improved cortical structure segmentation. In: IPMI, Springer, pp 210–223
12. Dietterich TG (1998) Approximate statistical tests for comparing supervised classification learning algorithms. Neural Comput 10:1895–1923
13. Duncan JS, Member S, Ayache N (2000) Medical image analysis: progress over two decades and the challenges ahead. IEEE Trans Pattern Anal Mach Intell 22:85–106
14. Escalera S, Fornés A, Pujol O, Lladós J, Radeva P (2010) Circular blurred shape model for multiclass symbol recognition. IEEE Trans Syst Man Cybern 41(2):497–506
15. Filipek PA, Semrud-Clikeman M, Steingard RJ, Renshaw PF, Kennedy DN, Biederman J (1997) Volumetric mri analysis comparing subjects having attention-deficit hyperactivity disorder with normal controls. Neurology 48(3):589–601

16. Fischl B, Salat DH, Busa E, Albert M, Dieterich M, Haselgrove C, van der Kouwe A, Killiany R, Kennedy D, Klaveness S, Montillo A, Makris N, Rosen B, Dale AM (2002) Whole brain segmentation: automated labeling of neuroanatomical structures in the human brain. Neuron 33(3):341–355
17. Fischl B, Salat DH, van der Kouwe AJW, Makris N, Ségonne F, Quinn BT, Dalea AM (2004) Sequence-independent segmentation of magnetic resonance images. Neuroimage 23(Suppl 1):S69–S84
18. Freesurfer. http://surfer.nmr.mgh.harvard.edu/
19. Heckemann RA, Hajnal JV, Aljabar P, Rueckert D, Hammersc A (2006) Automatic anatomical brain mri segmentation combining label propagation and decision fusion. Neuroimage 33: 115–126
20. Igual L, Soliva JC, Hernandez-Vela A, Escalera S, Jimenez X, Vilarroya O, Radeva P (2012) A fully-automatic caudate nucleus segmentation of brain mri: application in pediatric attention-deficit/hyperactivity disorder volumetric analysis. BioMed Eng Online 10(105)
21. Iosifescu DV, Shenton ME, Warfield SK, Kikinis R, Dengler J, Jolesz FA, Mccarley RW (1997) An automated registration algorithm for measuring MRI subcortical brain structures. Neuroimage 6(1):13–25
22. Kelemen A, Székely G, Gerig G (1999) Elastic model-based segmentation of 3-d neuroradiological data sets. IEEE Trans Med Imaging 18(10):828–839
23. Khan AR, Wang L, Beg MF (2008) FreeSurfer-initiated fully-automated subcortical brain segmentation in MRI using large deformation diffeomorphic metric mapping. NeuroImage 41(3):735–746
24. Khan AR, Chung MK, Beg MF (2009) Robust atlas-based brain segmentation using multi-structure confidence-weighted registration. In: Proceedings of the 12th international conference on medical image computing and computer-assisted intervention: part II, MICCAI '09. Springer, pp 549–557
25. Klein A, Andersson J, Ardekani BA, Ashburner J, Avants B, Chiang MC, Christensen GE, Collins DL, Gee J, Hellier P, Song JH, Jenkinson M, Lepage C, Rueckert D, Thompson P, Vercauteren T, Woods RP, Mann JJ, Parsey RV (2009) Evaluation of 14 nonlinear deformation algorithms applied to human brain mri registration. NeuroImage 46(3):786–802
26. Kolmogorov V, Zabih R (2004) What energy functions can be minimized via graph cuts. IEEE Trans Pattern Anal Mach Intell 26:65–81
27. Leventon ME, Grimson WEL, Faugeras O, III WMW (2000) Level set based segmentation with intensity and curvature priors. In: Proceedings of the IEEE workshop on mathematical methods in biomedical image analysis (MMBIA '00), pp 4–12
28. Mäkelä T, Clarysse P, Sipilä O, Pauna N, Pham QC, Katila T, Magnin IE (2002) A review of cardiac image registration methods. IEEE Trans Med Imaging 7(3):1011–1021
29. Moghaddam JM, Zadeh SH (2009) Automatic segmentation of brain structures using geometric moment invariants and artificial neural networks. In: Prince J, Pham D, Myers K (eds) Information processing in medical imaging, Lecture notes in computer science, vol 5636. Springer, Berlin, pp 326–337
30. MRIcro and MRIcron medical image viewer softwares. http://www.cabiatl.com/mricro/mricro/, http://www.cabiatl.com/mricro/mricron/
31. Murgasova M, Dyet L, Edwards D, Rutherford M, Hajnal J, Rueckert D (2007) Segmentation of brain MRI in young children. Acad Radiol 14(11):1350–1366
32. Patenaude B, Smith SM, Kennedy DN, Jenkinson M (2011) A bayesian model of shape and appearance for subcortical brain segmentation. NeuroImage 56(3):907–922
33. Reiss A, Abrams M, Singer H, Ross J, Denckla M (1996) Brain development, gender and iq in children: a volumetric imaging study. Brain 119:1763–1774
34. Rohlfing T, Brandt R, Menzel R, Maurer CR (2004) Evaluation of atlas selection strategies for atlas-based image segmentation with application to confocal microscopy images of bee brains. NeuroImage 21(4):1428–1442
35. Soliva JC, Fauquet J, Bielsa A, Rovira M, Carmona S, Ramos-Quiroga JA, Hilferty J, Bulbena A, Casas M, Vilarroya O (2010) Quantitative mr analysis of caudate abnormalities in pediatric

adhd: proposal for a diagnostic test. Psychiatry Research: Neuroimaging 182(3):238–243. doi:10.1016/j.pscychresns.2010.01.013
36. Song Z, Tustison NJ, Avants BB, Gee JC (2006) Integrated graph cuts for brain mri segmentation. In: MICCAI (2), Lecture notes in computer science, vol 4191. Springer, pp 831–838
37. Statistical parametric mapping (spm). http://www.fil.ion.ucl.ac.uk/spm/
38. Tohka J, Wallius E, Hirvonen J, Hietala J, Ruotsalainen U (2006) Automatic extraction of caudate and putamen in [^{11} C] raclopride pet using deformable surface models and normalized cuts. IEEE Trans Nucl Sci 53(1):220–227
39. Tremols V, Bielsa A, Soliva JC, Raheb C, Carmona S, Tomas J, Gispert JD, Rovira M, Fauquet J, Tobeña A, Bulbena A, Vilarroya O (2008) Differential abnormalities of the head and body of the caudate nucleus in attention deficit-hyperactivity disorder. Psychiatry Res 163(3):270–278
40. van Ginneken B, Heimann T, Styner M (2007) 3d segmentation in the clinic: a grand challenge. In: MICCAI workshop on 3D segmentation in the clinic: a grand, challenge In: Heimann T, Styner M, van Ginneken B (eds)
41. Van Leemput K, Maes F, Vandermeulen D, Suetens P (1999) Automated model-based tissue classification of mr images of the brain. IEEE Trans Med Imaging, 18(10):897–908
42. van Rikxoort E, Isgum I, Arzhaeva Y, Staring M, Klein S, Viergever M, Pluim J, van Ginneken B (2010) Adaptive local multi-atlas segmentation: application to the heart and the caudate nucleus. Med Image Anal 14(1):39–49
43. Vese LA, Chan TF, Tony, Chan F (2002) A multiphase level set framework for image segmentation using the mumford and shah model. Int J Comput Vision 50:271–293
44. Weickert J (1998) Anisotropic diffusion in image processing. ECMI Series, Teubner, stuttgart
45. Weldeselassie YT, Hamarneh G (2007) Dt-mri segmentation using graph cuts. In: Proceedings of the SPIE, San Diego
46. Xia Y, Bettinger K, Shen L, Reiss AL (2007) Automatic segmentation of the caudate nucleus from human brain mr images. IEEE Trans Med Imaging 26:509–517
47. Yushkevich PA, Piven J, Cody Hazlett H, Gimpel Smith R, Ho S, Gee JC, Gerig G (2006) User-guided 3D active contour segmentation of anatomical structures: significantly improved efficiency and reliability. Neuroimage 31(3):1116–1128

Elastic Registration of Edges Using Diffuse Surfaces

Stefan Fürtinger, Stephen Keeling, Gernot Plank and Anton J Prassl

Abstract In this work, edge sets are mapped one to the other by representing these zero area sets as *diffuse images* which have positive measure supports that can be registered elastically. The driving application for this work is to map a Purkinje fiber network in the endocardium of one heart to the endocardium of another heart. The approach is to register sufficiently accurate diffuse surface representations of two endocardia and then to apply the resulting transformation to the points of the Purkinje fiber network. To create a diffuse image from a given edge set, a region growing method is used to approximate diffusion of brightness from an edge set to a given point. To be minimized is the sum of squared differences of the transformed diffuse images along with a linear elastic penalty for the registration transformation. A Newton iteration is employed to solve the optimality system, and the degree of diffusion is larger in initial iterations while smaller in later iterations so that a desired local minimum is selected by means of vanishing diffusion. Favorable results are shown for registering highly detailed cardiac geometries.

S. Fürtinger (✉) · S. Keeling
Institute for Mathematics and Scientific Computing, Karl-Franzens University, Graz, Austria
e-mail: stefan.fuertinger@uni-graz.at

S. Keeling
e-mail: stephen.keeling@uni-graz.at

G. Plank · A. J. Prassl
Institute for Biophysics, Medical University Graz, Graz, Austria
e-mail: gernot.plank@medunigraz.at

A. J. Prassl
e-mail: anton.prassl@medunigraz.at

M. González Hidalgo et al. (eds.), *Deformation Models*, Lecture Notes in Computational Vision and Biomechanics 7, DOI: 10.1007/978-94-007-5446-1_11,

1 Introduction

The heart is an electrically controlled mechanical organ whose main function is to pump blood around the circulatory system. Under healthy conditions the heart fulfills this vital objective with remarkable efficiency thanks to a highly organized sequence of events referred to as electro-mechanical coupling. The fast and regular distribution of the electrical impulse which leads to electrical activation of the cardiac tissue is of critical importance in this context. The Purkinje system (PS), that is the specialized conduction system of the main pumping chambers of the heart referred to as ventricles, plays an essential role in this process by conducting electrical excitation wavefronts from the atrio-ventricular node to the endocardium, i.e. the inner surfaces of the ventricular cavities. Due to the high conduction velocities and the network-like topology of the PS the entire endocardium is electrically activated almost simultaneously, thus triggering a highly synchronized mechanical action of the heart.

Histological investigations revealed that the PS consists of a highly ramified network of thin cable-like structures beneath the epithelial layer of the endocardium. The PS is electrically isolated from the ventricular muscle, except at discrete endpoints referred to as Purkinje-Ventricular junctions (PVJs) [27]. Transmission of the electrical signals at these discrete junctional sites is essential to excite the ventricular mass [15], since a loss of electrical synchronicity may entail a severe impairment of cardiac function, which, ultimately, may even lead to sudden cardiac death. The electro-anatomical characteristics of the PS varies significantly between species, but may also vary inter-individually. For instance, the location of PVJs within the ventricular wall is quite distinct between species. In sheep [2] and pig [14] the penetration depth of the PS appears almost fully transmural, whereas in the ventricles of dogs, humans and rabbits [27] PVJs tend to be located rather in sub-endocardial layers.

While recent advances in experimental methodology have allowed more detailed characterizations of cardiac function, quite often further progress is hampered by the inability of current experimental techniques to resolve, with sufficient accuracy, electrical behavior confined to the depth of the ventricles or within the PS. Computer models quite naturally suggest themselves as a surrogate technique to bridge the gap between experimental observations, typically recorded from the surfaces of the heart, and electrical events occurring within the PS or in the depth of ventricular walls. Owing to the physiological importance of the PS it is therefore crucial to account for its role in cardiac function in health and disease, however, despite major recent advancements in modeling technology [24, 25], the integration of the complex PS topology with anatomically realistic and biophysically detailed models of the ventricles remains a challenge.

The main goal of the presented work is the development of a mathematical framework suitable for mapping the endocardial PS between different ventricular surface geometries. Since detailed experimental characterizations of the network topology of the PS were not available, a reduced model of the PS based on data from literature [29], was constructed and integrated with the San Diego rabbit heart model

[28] which served as a template model. For the sake of developing and testing the mapping technique, a recent anatomically realistic model of the rabbit ventricles [4] served as a reference geometry (see models in Fig. 2). We first seek a geometric transformation to match the template heart model to the given reference model. In the context of mathematical image processing this can be seen as 3D registration problem. Once the transformation is found it can be applied to map structures (like the PS) within the template model onto the reference model. This approach guarantees that not only topological features of the PS but also its position relative to the ventricle are preserved and projected onto the reference heart.

Given the sheer size of the models considered here would necessitate a massive computational effort to calculate only simple transformations. Hence we developed a method that is capable of computing even highly non-linear transformations within reasonable time frames while requiring only moderate computational resources. We reduced the dimensionality of the problem by treating the 3D models as sequence of 2D edges. This strategy reduces memory consumption considerably while simultaneously allowing us to use very efficient techniques to solve the occurring 2D registration problems.

2 Edges as Binary Images

We slice up 3D models and thus obtain two-dimensional cuts. From a mathematical point of view these cuts can be seen as curves. Hence let Γ_0 and Γ_1 denote two curves in $\mathbb{R}^2$. Depending on the chosen cutting direction these curves are potentially very non-smooth. However, we may safely assume that both curves are not infinitely long which in mathematical terms means that their Hausdorff-measure is finite, i.e., $\mathcal{H}^1(\Gamma_i) < \infty$ for $i = 0, 1$. Since we want to approach the problem using image processing techniques we interprete Γ_0 and Γ_1 not as curves but as *edges*. In doing so let $\Omega := (0, 1)^2 \subset \mathbb{R}^2$ denote our image domain and define I_0 and I_1 to be the characteristic functions of Γ_0 and Γ_1 respectively. In this manner I_0 and I_1 can be seen as *edge maps* since $\mathrm{Rg}(I_i) = \{0, 1\}$. In other words I_0 and I_1 are *binary images* on Ω. As stated above the objective is to find a transformation to match one heart model to another. Since we do no longer consider 3D models but rather 2D cuts of those models the reformulated objective is now to find a displacement $\boldsymbol{w} : \mathbb{R}^2 \to \mathbb{R}^2$ such that $I_0(\boldsymbol{x} + \boldsymbol{w}(\boldsymbol{x})) \approx I_1(\boldsymbol{x})$ for all $\boldsymbol{x} \in \Omega$.

One approach to the computation of the desired displacement is to treat points on Γ_0 as if connected to one another by elastic springs which are perturbed minimally in order to meet the target set Γ_1. However, the computational complexity of such a formulation is very high in relation to the conceptually comparable strategy pursued here. The present strategy is to embed the edges into images which are then registered elastically. Minimizing a linear elastic potential in the registration scheme involving Γ_0 and Γ_1 can be seen as an approximation to the afore mentioned approach. Furthermore, elastic potential energy being used by many authors to regularize image registration is well established and investigated; see, e.g., [16, 23] and particularly

for a fixed $\varepsilon > 0$. The cost J is well defined for $\boldsymbol{w} \in H^1(\Omega)$. Thus we want to solve the minimization problem

$$\min_{\boldsymbol{w} \in H^1(\Omega)} J^\varepsilon(\boldsymbol{w}). \tag{6}$$

Here several important observations should be made. The fuzzy edges I_i^ε vary with the value of ε. Hence for each ε, (6) forms a stand-alone minimization problem. This immediately gives rise to the following questions. For fixed $\varepsilon > 0$, does (6) have a solution, i.e., is the use of min instead of inf justified? Assuming that the first question can be answered positively, what are the asymptotic properties of these solutions as $\varepsilon \to 0$? We devote ourselves to both questions in the following subsection.

2.1 Properties of J^ε

Due to the lack of space the first question concerning the existence of solutions to (6) for a given $\varepsilon > 0$ cannot be answered in full detail. However, in the following we will outline the basic idea of the proof.

Below, we use the notation $\boldsymbol{A} : \boldsymbol{B} = \sum_{ij} A_{ij} B_{ij}$ for matrices $\boldsymbol{A} = \{A_{ij}\}$ and $\boldsymbol{B} = \{B_{ij}\}$. We introduce the space of infinitesimal rigid motions

$$\mathrm{RM} = \left\{ \boldsymbol{w} = \boldsymbol{c} + \boldsymbol{W}\boldsymbol{x} : \boldsymbol{c} \in \mathbb{R}^d, \boldsymbol{W} \in \mathbb{S}^d \right\},$$

where

$$\mathbb{S}^d = \left\{ \boldsymbol{W} \in \mathbb{R}^{d \times d} : \boldsymbol{W} + \boldsymbol{W}^\mathrm{T} = 0 \right\}.$$

Furthermore, we define the following closed linear subspace of $H^1(\Omega)$:

$$\mathcal{H} = \left\{ \boldsymbol{w} \in H^1(\Omega) : \int_\Omega \boldsymbol{w}(\boldsymbol{x}) d\boldsymbol{x} = 0, \int_\Omega \left[\boldsymbol{w}(\boldsymbol{x})\boldsymbol{x}^\mathrm{T} - \boldsymbol{x}\boldsymbol{w}(\boldsymbol{x})^\mathrm{T} \right] d\boldsymbol{x} = 0 \right\}.$$

Straight forward calculations show that J^ε is bounded on $H^1(\Omega)$. However, it is easy to see that the whole of RM is in the kernel of the elastic penalty E in (5), i.e., $\mathrm{RM} \subset \ker(E)$. Thus

$$c\|\boldsymbol{u}\|^2_{H^1(\Omega)} \not\leq J^\varepsilon(\boldsymbol{u}) = S^\varepsilon(\boldsymbol{u}) \leq 2|\Omega|, \quad \forall \boldsymbol{u} \in \mathrm{RM},$$

which means that J^ε is not coercive on $H^1(\Omega)$. However, it can be shown that the cost J^ε *is* coercive and bounded on $\mathcal{H}$. Therefore we established the following direct sum decomposition of $H^1(\Omega)$.

Theorem 0.1 $H^1(\Omega) = \mathrm{RM} \oplus \mathcal{H}$.

Proof A rigorous proof is given in [11].

This decomposition is key to proving existence of a minimizing element of (6) in $H^1(\Omega)$. Under further assumptions existence of (in a certain sense) minimizing rigid motions can be shown. Conversely, given $\boldsymbol{u} \in \mathrm{RM}$ there exists a $\boldsymbol{v}^* \in \mathcal{H}$ such that

$$J^\varepsilon(\boldsymbol{u} + \boldsymbol{v}^*) = \inf_{\boldsymbol{v} \in \mathcal{H}} J^\varepsilon(\boldsymbol{u} + \boldsymbol{v}).$$

For the proof we also refer to [11]. Following Theorem 0.1 and using these two results existence of a minimizer of J^ε for $\varepsilon > 0$ fixed is guaranteed by the following

Theorem 0.2 *Let $\varepsilon > 0$ be fixed, $\mathcal{I}_0^\varepsilon \in W^{1,\infty}(\Omega)$, $\mathcal{I}_1^\varepsilon \in L^\infty(\Omega)$ and $\lambda, \mu > 0$. Then there exists a $\boldsymbol{w}^\star \in H^1(\Omega)$ such that*

$$J^\varepsilon(\boldsymbol{w}^\star) = \min_{\boldsymbol{w} \in H^1(\Omega)} J^\varepsilon(\boldsymbol{w}). \tag{7}$$

Proof A rigorous proof is given in [11].

Note that by definition (3) both fuzzy edges $\mathcal{I}_i^\varepsilon$ satisfy the regularity assumptions stated in Theorem 0.2, provided the edge sets Γ_i are sufficiently regular. Under such assumptions, (6) is a well posed minimization problem for each $\varepsilon > 0$.

Having guaranteed existence of minimizers for each fixed ε we may now address the question of asymptotic behavior of solutions to (6) as $\varepsilon \to 0$. Looking at definition (3) it is easy to see that the fuzzy edges $\mathcal{I}_i^\varepsilon$ converge pointwise to the binary images I_i as $\varepsilon \to 0$. However, as pointed out above for $\varepsilon = 0$ the trivial deformation $\boldsymbol{w} = 0$ minimizes the similarity measure (4). Indeed, we could show the following convergence of minimizers as $\varepsilon \to 0$.

Theorem 0.3 *Assume the conditions of Theorem 0.2. For every $\varepsilon > 0$, let $\boldsymbol{w}_\varepsilon \in H^1(\Omega)$ denote the minimizer of $J^\varepsilon(\boldsymbol{w})$. Let $\boldsymbol{u}_\varepsilon$ and $\boldsymbol{v}_\varepsilon$ denote the projections of $\boldsymbol{w}_\varepsilon$ onto RM and $\mathcal{H}$, respectively. Then*

$$\lim_{\varepsilon \to 0} \boldsymbol{v}_\varepsilon = 0. \tag{8}$$

Also, there exists a $\boldsymbol{u}_0 \in RM$ such that

$$\lim_{\varepsilon \to 0} S^\varepsilon(\boldsymbol{u}_0 + \boldsymbol{v}_\varepsilon) = \lim_{\varepsilon \to 0} S^\varepsilon(\boldsymbol{u}_\varepsilon + \boldsymbol{v}_\varepsilon) = 0. \tag{9}$$

Proof A rigorous proof is given in [11].

For a better understanding of the behavior of the minimization problem (6) as $\varepsilon \to 0$ we present a simplified 1D-example.

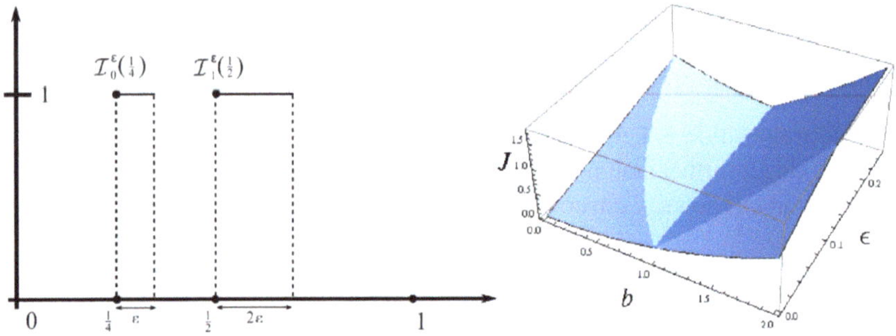

(a) (b)

Fig. 1 **a** Sketch of the simplified 1D-example discussed in Remark 0.1. **b** Graph of $G(\varepsilon, b)$ on $[0, \frac{1}{4}] \times [0, 2]$ for $\mu = \frac{1}{4}$

Remark 0.1 Let $\Omega := (0, 1) \subset \mathbb{R}$ and consider the "binary images" I_0 and I_1 defined by

$$I_0(x) := \begin{cases} 1, & x = \frac{1}{4}, \\ 0, & \text{otherwise}, \end{cases} \qquad I_1(x) := \begin{cases} 1, & x = \frac{1}{2}, \\ 0, & \text{otherwise}. \end{cases}$$

The objective is to find a deformation w such that $I_0(x + w(x)) \approx I_1(x)$. In analogy to the approach presented above we start with the cost functional

$$J(w) := \int_0^1 |I_0(x + w(x)) - I_1(x)|^2 \, dx + \mu \int_0^1 |w'(x)|^2 \, dx$$
$$= \mathcal{S}(I_0, I_1; w) + \mathcal{P}(w),$$

with a regularization parameter $\mu > 0$. Similar to the 2D case described above $\mathcal{S}(I_0, I_1; 0) = 0$ since $\mathrm{supp}(I_i)$ consists only of discrete points. Employing the same strategy as in 2D we proceed to fuzzy "edges" by defining

$$\mathcal{I}_0^\varepsilon(x) := \begin{cases} 1, & \frac{1}{4} \le x \le \frac{1}{4} + \varepsilon, \\ 0, & \text{otherwise}, \end{cases} \quad \mathcal{I}_1^\varepsilon(x) := \begin{cases} 1, & \frac{1}{2} \le x \le \frac{1}{2} + 2\varepsilon, \\ 0, & \text{otherwise}, \end{cases} \quad 0 < \varepsilon \le \frac{1}{4},$$

and thus modify the cost accordingly:

$$J^\varepsilon(w) := \mathcal{S}(\mathcal{I}_0^\varepsilon, \mathcal{I}_1^\varepsilon; w) + \mathcal{P}(w).$$

Since $\mathcal{I}_i^\varepsilon$ are step functions (see Fig. 1a) we obviously look for deformations of the form $w_\varepsilon(x) := bx$ with some non-negative scalar b and a fixed $0 < \varepsilon \le \frac{1}{4}$. In this very simple case it is possible to explicitly write down what the desired deformation should be in order to get a perfect registration: $w_\varepsilon^\star(x) := x$, thus $b^\star = 1$. In the following we investigate the properties of the modified cost J^ε given the desired solution $w_\varepsilon^\star(x) = x$ versus the trivial deformation $w = 0$.

First note that the penalty $\mathscr{P}$ for deformations of the form $w_\varepsilon = bx$ is given by

$$\mathscr{P}(w_\varepsilon) = \mu \int_0^1 |b|^2 \, dx = \mu b^2,$$

thus

$$\mathscr{P}(w_\varepsilon^\star) = \mu, \quad \text{and} \quad \mathscr{P}(0) = 0. \tag{10}$$

Furthermore, the desired solution minimizes the (adapted) similarity measure

$$\mathscr{S}(\mathscr{I}_0^\varepsilon, \mathscr{I}_1^\varepsilon; w_\varepsilon^\star) = 0. \tag{11}$$

On the other hand for the trivial deformation we compute

$$\mathscr{S}(\mathscr{I}_0^\varepsilon, \mathscr{I}_1^\varepsilon; 0) = \left(\left(\frac{1}{4} + \varepsilon\right) - \frac{1}{4}\right) + \left(\frac{1}{2} + 2\varepsilon - \frac{1}{2}\right) = 3\varepsilon. \tag{12}$$

Now choose $0 < \varepsilon < \min(\frac{1}{4}, \frac{\mu}{3})$. Then relations (10–12) imply

$$\begin{aligned} J^\varepsilon(0) = \mathscr{S}(\mathscr{I}_0^\varepsilon, \mathscr{I}_1^\varepsilon; 0) + \mathscr{P}(0) &= 3\varepsilon + 0 \\ &< 0 + 3 \cdot \frac{\mu}{3} \\ &= \mathscr{S}(\mathscr{I}_0^\varepsilon, \mathscr{I}_1^\varepsilon; w_\varepsilon^\star) + \mathscr{P}(w_\varepsilon^\star) = J^\varepsilon(w_\varepsilon^\star). \end{aligned}$$

The crucial observation to make here is that although $\varepsilon > 0$ the trivial deformation minimizes the cost J^ε. A closer look at the calculations shows that there is an important relation between the constant μ and ε: in the above example ε was chosen "too small" compared to the regularization parameter μ so that the similarity measure was minimized by $w = 0$ instead of the desired deformation. To illustrate this phenomenon Fig. 1b shows the graph of $G(\varepsilon, b) := J^\varepsilon(bx)$ for $\varepsilon \in [0, \frac{1}{4}]$, $b \in [0, 2]$ and $\mu = \frac{1}{4}$. It can be seen that for small values of ε the global minimizer of J^ε is $b = 0$ (and hence $w = 0$) whereas for even very large values of ε the cost J^ε is (globally) minimized by $b = 1$, i.e., the desired deformation.

Note that the behavior manifested in the 1D example can be seen in 2D as well. For vanishingly small values of ε (small compared to the Navier–Lamé constants) J^ε is minimized by $\boldsymbol{w} = 0$. On the other hand, local minima provide a better registration of the edge sets. This knowledge has led us to the development of an iterative solution strategy that allows to compute first global minima for ε large, which become desired local minima as ε gets smaller. Starting with sufficiently diffuse images, we use the computed minimizer for larger values of ε as an initial guess to solve the registration problem for smaller values of ε. By doing so the computed transformations are refined in the course of the iteration but simultaneously do not get close to zero as $\varepsilon \to 0$ (details are given in Sect. 3.1). This strategy proved to be remarkably robust in practice and produced very promising results.

2.2 Optimality Conditions

We start by deducing the necessary optimality conditions for the minimization problem (6). The Gâteaux derivative of J^ε in an arbitrary direction $\boldsymbol{v} \in C^\infty(\bar{\Omega})$ is defined by

$$\frac{\delta J^\varepsilon}{\delta \boldsymbol{w}}(\boldsymbol{w}; \boldsymbol{v}) := \frac{d}{ds} J^\varepsilon(\boldsymbol{w} + s\boldsymbol{v})\Big|_{s=0}.$$

For the sake of clarity we compute the derivatives of S^ε and E separately. Starting with S^ε we get

$$\frac{\delta S^\varepsilon}{\delta \boldsymbol{w}}(\boldsymbol{w}; \boldsymbol{v}) = \int_\Omega (\mathscr{I}_0^\varepsilon \circ (\mathrm{Id} + \boldsymbol{w}) - \mathscr{I}_1^\varepsilon)\boldsymbol{v} \cdot \nabla \mathscr{I}_0^\varepsilon \circ (\mathrm{Id} + \boldsymbol{w}), \tag{13}$$

and for $E(\boldsymbol{w})$ we obtain

$$\begin{aligned} \frac{\delta E}{\delta \boldsymbol{w}}(\boldsymbol{w}; \boldsymbol{v}) = &\int_\Omega \mu \left(\nabla \boldsymbol{w}^\top + \nabla \boldsymbol{w}\right) : \left(\nabla \boldsymbol{v}^\top + \nabla \boldsymbol{v}\right) dx \\ &+ \int_\Omega \lambda\, (\nabla \cdot \boldsymbol{w})\, (\nabla \cdot \boldsymbol{v})\,. \end{aligned} \tag{14}$$

Assuming sufficient regularity of $\boldsymbol{w}$ we will deduce a strong optimality formulation for (6). Therefore we first use partial integration in (14)

$$\begin{aligned} \frac{\delta E}{\delta \boldsymbol{w}}(\boldsymbol{w}; \boldsymbol{v}) =& \mu \int_{\partial\Omega} \sum_{\ell=1}^{2} v_\ell \left(\nabla w_\ell + \partial_{x_\ell} \boldsymbol{w}\right) \cdot \boldsymbol{n} - \mu \int_\Omega \left(\Delta \boldsymbol{w} + \nabla\, (\nabla \cdot \boldsymbol{w})\right) \cdot \boldsymbol{v} \\ &+ \lambda \int_{\partial\Omega} (\nabla \cdot \boldsymbol{w})\, (\boldsymbol{v} \cdot \boldsymbol{n}) - \lambda \int_\Omega \boldsymbol{v} \cdot \nabla\, (\nabla \cdot \boldsymbol{w})\,, \end{aligned} \tag{15}$$

where $\boldsymbol{n}$ denotes the outer unit normal vector on $\partial\Omega$. The (weak) necessary optimality condition associated to (6) is given by

$$\frac{\delta J}{\delta \boldsymbol{w}}(\boldsymbol{w}; \boldsymbol{v}) = \frac{\delta S^\varepsilon}{\delta \boldsymbol{w}}(\boldsymbol{w}; \boldsymbol{v}) + \frac{\delta E}{\delta \boldsymbol{w}}(\boldsymbol{w}; \boldsymbol{v}) = 0, \quad \forall \boldsymbol{v} \in C^\infty(\bar{\Omega}). \tag{16}$$

Since (16) holds for any variation $v \in C^\infty(\bar{\Omega})$ we may apply the fundamental Lemma of calculus of variations. Thus using (13) and (15) we obtain the Euler–Lagrange equations, i.e., the strong optimality formulation, associated to the minimization problem (6)

$$\begin{cases} \mathscr{E}w - f(x, w; \mathscr{I}_0^\varepsilon, \mathscr{I}_1^\varepsilon) = 0, & \text{in } \Omega, \\ \lambda n_\ell \nabla \cdot w + \mu \left(\nabla w_\ell + \partial_{x_\ell} w\right) \cdot n = 0, & \text{on } \partial\Omega, \end{cases} \tag{17}$$

where $\mathscr{E}$ is the elasticity operator defined by

$$\mathscr{E}w := \mu \Delta w + (\mu + \lambda)\nabla (\nabla \cdot w), \tag{18}$$

and

$$f(x, w; \mathscr{I}_0^\varepsilon, \mathscr{I}_1^\varepsilon) := (\mathscr{I}_0^\varepsilon \circ (\mathrm{Id} + w) - \mathscr{I}_1^\varepsilon)\nabla \mathscr{I}_0^\varepsilon \circ (\mathrm{Id} + w), \tag{19}$$

is the driving force of the registration. Here several important observations should be made. First note that the strength of the driving force f determines the magnitude of a deformation field w that satisfies (17). Furthermore, f is the Gâteaux derivative of the adapted SSID distance measure (4). Thus if we use a very small ε (or the raw binary edge maps I_i instead of the fuzzy edges $\mathscr{I}_i^\varepsilon$ in the SSID-distance) the resulting driving force of the registration is zero, a.e., and hence the trivial deformation solves the Euler–Lagrange equations (17) (compare the 1D-example given in Remark 0.1).

On the other hand large values of ε indeed generate a sufficiently big driving force in order to prevent the trivial deformation from solving (17). However, choosing ε too large leads to blurred fuzzy edges $\mathscr{I}_i^\varepsilon$ and thus a loss of potentially important features of the original edges Γ_i.

The vanishing diffusion strategy proposed here (see Algorithm 6) is designed to address both problems. Starting with ε large the driving force in (17) provokes a non trivial initial solution w. By iteratively refining the initial deformation field w we are able to preserve characteristic features of the original edges Γ_i. Thus our solution strategy is robust against the initial choice of (a possibly large) $\varepsilon > 0$ while simultaneously allowing us to compute only the desired local minima of J^ε as ε gets smaller.

2.3 Solution Strategy

The strong optimality conditions (17) are a system of nonlinear partial differential equations (PDEs) in w. Thus we linearize (17) by employing Newton's method on the functional J which takes the form

$$\begin{cases} \dfrac{\delta^2 J}{\delta w^2}(w_k; v, \delta w_k) = -\dfrac{\delta J}{\delta w}(w_k; v), \quad \forall v \in C^\infty(\bar{\Omega}), \\ w_{k+1} = w_k + \tau \delta w_k, \end{cases} \tag{20}$$

where $\tau > 0$ denotes a given step-size and $k = 1, 2, \ldots$ is the iteration index. The Lax–Milgram Lemma [5] may be used to show that (20) admits an unique solution for each fixed k.

Theorem 0.4 *Let* $v \in C^\infty(\bar{\Omega})$, $\mathscr{I}_0^\varepsilon \in W^{1,\infty}(\Omega)$, $\mathscr{I}_1^\varepsilon \in L^\infty(\Omega)$ *and* $\lambda, \mu > 0$. *If for given* $\boldsymbol{w} \in H^1(\Omega)$

$$\int_\Omega \left|(\boldsymbol{a} + M\boldsymbol{x}) \cdot \nabla \mathscr{I}_0^\varepsilon \circ (Id + \boldsymbol{w})\right|^2 \, d\boldsymbol{x} = 0 \;\; implies \;\; \boldsymbol{a} + M\boldsymbol{x} = 0, \tag{21}$$

for every skew-symmetric matrix $\boldsymbol{M} \in \mathbb{R}^{N\times N}$ *and every vector* $\boldsymbol{a} \in \mathbb{R}^N$ *then there exists a unique* $\boldsymbol{u} \in H^1(\Omega)$ *satisfying* (20).

Proof A rigorous proof is given in [11].

Assumption (21) essentially says that the image $\mathscr{I}_0^\varepsilon \circ (\mathrm{Id} + \boldsymbol{w}_k)$ manifests sufficiently few symmetries. Convergence of Newton's method (20) as $k \to \infty$ for a fixed $\varepsilon > 0$ and a suitable initial guess can be shown using the Newton–Kantorovich Theorem (see e.g. [21]).

Similar to the strategy presented in the previous section we will now deduce a strong formulation of (20). For the sake of clarity we compute again the derivatives of S^ε and E separately. Starting with S^ε we get

$$\frac{\delta^2 S^\varepsilon}{\delta \boldsymbol{w}^2}(\boldsymbol{w}_k; \boldsymbol{v}, \delta\boldsymbol{w}_k) = \int_\Omega \boldsymbol{v}[\nabla \mathscr{I}_0^\varepsilon \circ (\mathrm{Id} + \boldsymbol{w}_k)][\nabla \mathscr{I}_0^\varepsilon \circ (\mathrm{Id} + \boldsymbol{w}_k)]^\top \delta\boldsymbol{w}_k.$$

By using the first variational derivative (14) of the linear elastic potential we obtain further

$$\begin{aligned} \frac{\delta^2 E}{\delta \boldsymbol{w}^2}(\boldsymbol{w}_k; \boldsymbol{v}, \delta\boldsymbol{w}_k) = &\int_\Omega \mu(\nabla \boldsymbol{v}^\top + \nabla \boldsymbol{v}) : (\nabla \delta\boldsymbol{w}_k^\top + \nabla \delta\boldsymbol{w}_k) \\ &+ \int_\Omega \lambda(\nabla \cdot \boldsymbol{v})(\nabla \cdot \delta\boldsymbol{w}_k) \, d\boldsymbol{x}, \end{aligned}$$

Under stronger regularity assumptions on $\boldsymbol{w}$ we may again use partial integration and apply the fundamental Lemma of calculus of variations to obtain the strong formulation of (20)

$$\begin{cases} \left(-\mathscr{E} + [\nabla \mathscr{I}_0^\varepsilon \circ (\mathrm{Id} + \boldsymbol{w}_k)][\nabla \mathscr{I}_0^\varepsilon \circ (\mathrm{Id} + \boldsymbol{w}_k)]^\top\right) \delta\boldsymbol{w}_k = \mathscr{E}\boldsymbol{w}_k - \boldsymbol{f}(\boldsymbol{x}, \boldsymbol{w}_k; \mathscr{I}_0^\varepsilon, \mathscr{I}_1^\varepsilon), & \Omega, \\ \lambda n_\ell \nabla \cdot \boldsymbol{w}_k + \mu \left(\nabla w_{k_\ell} + \partial_{x_\ell} \boldsymbol{w}_k\right) \cdot \boldsymbol{n} = 0, & \partial\Omega. \end{cases} \tag{22}$$

which is the desired linear PDE-system for the unknown function $\delta\boldsymbol{w}_k$.

3 Numerical Approximation

We will first introduce a discretization scheme for the strong formulation (22) of the Newton step and then explain the discrete realization of Newton's method (20). Let from now on $\boldsymbol{w}(\boldsymbol{x}) := (u(\boldsymbol{x}), v(\boldsymbol{x}))^\top$, and $\boldsymbol{x} = (x, y) \in \Omega$. Then we may rewrite the elasticity operator $\mathscr{E}$

$$\begin{aligned}\mathscr{E}\boldsymbol{w} =& \mu\left(\frac{\partial^2 \boldsymbol{w}}{\partial x^2} + \frac{\partial^2 \boldsymbol{w}}{\partial y^2}\right) + (\mu+\lambda)\nabla\left(\frac{\partial u}{\partial x} + \frac{\partial v}{\partial y}\right)\\ =& \begin{pmatrix} (\lambda+2\mu)\frac{\partial^2}{\partial x^2} + \mu\frac{\partial^2}{\partial y^2} & (\lambda+\mu)\frac{\partial^2}{\partial x \partial y} \\ (\lambda+\mu)\frac{\partial^2}{\partial x \partial y} & \mu\frac{\partial^2}{\partial x^2} + (\lambda+2\mu)\frac{\partial^2}{\partial y^2} \end{pmatrix}\begin{pmatrix} u \\ v\end{pmatrix} =: \begin{pmatrix} \mathscr{E}_{11} & \mathscr{E}_{12} \\ \mathscr{E}_{21} & \mathscr{E}_{22}\end{pmatrix}\begin{pmatrix} u \\ v\end{pmatrix}.\end{aligned}$$

For the sake of simplicity we drop the iteration index k here. Since we are working with digital images we define a grid $\Omega_h := \{1, \ldots, N\}^2$, where N denotes the resolution of the images. We use a unit step size $h := 1$, i.e., the width of a cell is one, and employ standard central finite differences to discretize the Newton step (22). In particular let $\boldsymbol{j} := (j_1, \ldots, j_N) \in \mathbb{R}^N$ be an integer component multi index, $\mathbf{1} := (1, \ldots, 1)^\top \in \mathbb{R}^N$ and the cell centroids be given by $\boldsymbol{x}_{\boldsymbol{j}} := \boldsymbol{j}$, $\mathbf{1} \leq \boldsymbol{j} \leq N \cdot \mathbf{1}$. We denote the array arising from evaluating e.g. u at each grid point by $u(\Omega_h) \in \mathbb{R}^{N\times N}$. Hence $U_{\boldsymbol{j}} \approx u(\boldsymbol{x}_{\boldsymbol{j}})$ and $\mathbf{u} \in \mathbb{R}^{N^2}$ denotes the vector of values $\{U_{\boldsymbol{j}}\}$ corresponding to the lexicographic ordering in which j_1 increments first from 1 to N, then j_2 and so on. Further, let $\boldsymbol{D}(\mathbf{u}) \in \mathbb{R}^{N^2 \times N^2}$ be the diagonal matrix arising from situating the values $\{U_{\boldsymbol{j}}\}$ along the diagonal according to lexicographic ordering.

Due to the lack of space we cannot give the discretization of the elasticity operator $\mathscr{E}$ in full detail. Thus we show, for instance, the discretization of $\mathscr{E}_{11}$ near the lower left corner of Ω_h

$$\begin{matrix} \vdots & & \vdots & \\ \begin{pmatrix} 0 & 0 & -2\mu \\ 0 & 8\mu+4\lambda & -4\mu-4\lambda \\ 0 & 0 & -2\mu \end{pmatrix} & & \begin{pmatrix} -2\mu & 0 & -2\mu \\ -4\mu-4\lambda & 16\mu+8\lambda & -4\mu-4\lambda \\ -2\mu & 0 & -2\mu \end{pmatrix} & \cdots \\ \begin{pmatrix} 0 & 0 & -2\mu \\ 0 & 2\lambda+4\mu & -2\mu-2\lambda \\ 0 & 0 & 0 \end{pmatrix} & & \begin{pmatrix} -2\mu & 0 & -2\mu \\ -2\mu-2\lambda & 8\mu+4\lambda & -2\mu-2\lambda \\ 0 & 0 & 0 \end{pmatrix} & \cdots \end{matrix} \qquad (23)$$

The upper right block represents the stencil weights for neighbors of a field cell. Similarly the other blocks show for boundary cells the stencil weights for their neighbors. With the same format we represent the stencils of $\mathscr{E}_{12}$:

$$
\begin{array}{cc}
\vdots & \vdots \\
\begin{pmatrix} 0 & \mu-\lambda & -\mu-\lambda \\ 0 & 0 & 0 \\ 0 & -\mu-\lambda & \mu+\lambda \end{pmatrix} & \begin{pmatrix} \mu+\lambda & 0 & -\mu-\lambda \\ 0 & 0 & 0 \\ -\mu-\lambda & 0 & \mu+\lambda \end{pmatrix} \cdots \\
\begin{pmatrix} 0 & \mu-\lambda & -\mu-\lambda \\ 0 & \mu+\lambda & -\mu+\lambda \\ 0 & 0 & 0 \end{pmatrix} & \begin{pmatrix} \mu+\lambda & 0 & -\mu-\lambda \\ \mu-\lambda & 0 & -\mu-\lambda \\ 0 & 0 & 0 \end{pmatrix} \cdots
\end{array}
\tag{24}
$$

The stencils for $\mathscr{E}_{22}$ and $\mathscr{E}_{21}$ are constructed by adequate copying and mirroring of (23) and (24) respectively. This gives rise to matrices $\boldsymbol{E}_{k.\ell} \in \mathbb{R}^{N^2 \times N^2}$ with $1 \le k, \ell \le 2$ which form the discrete version of the operator $\mathscr{E}$ under lexicographic ordering, i.e.,

$$
\mathscr{E}(u(\Omega_h), v(\Omega_h)) \approx \begin{pmatrix} E_{11}\mathbf{u} + E_{12}\mathbf{v} \\ E_{21}\mathbf{u} + E_{22}\mathbf{v} \end{pmatrix},
$$

so that we may write

$$
\boldsymbol{E}\mathbf{w} := \begin{pmatrix} \boldsymbol{E}_{11} & \boldsymbol{E}_{12} \\ \boldsymbol{E}_{21} & \boldsymbol{E}_{22} \end{pmatrix} \begin{pmatrix} \mathbf{u} \\ \mathbf{v} \end{pmatrix}.
$$

The discrete version $\tilde{\mathscr{I}}_1 \in \mathbb{R}^{N \times N}$ of $\mathscr{I}_1^\varepsilon$ is readily established by setting $\tilde{\mathscr{I}}_{1,j} := \mathscr{I}_1^\varepsilon(\boldsymbol{x}_j)$. For approximating $\mathscr{I}_0^\varepsilon(\boldsymbol{x} + \boldsymbol{w})$ and $\nabla \mathscr{I}_0^\varepsilon \circ (\mathrm{Id} + \boldsymbol{w})$ we use a bilinear interpolation scheme whenever $\mathscr{I}_0^\varepsilon$ is evaluated at non-grid points and assign the extrapolation value zero if $\boldsymbol{x} + \boldsymbol{w}(\boldsymbol{x})$ lies outside of Ω. Thus we obtain a matrix $\tilde{\mathscr{I}}_0 \in \mathbb{R}^{N \times N}$. Let $\boldsymbol{G}(\tilde{\mathscr{I}}_0) \in \mathbb{R}^{2N^2 \times 2N^2}$ denote the approximation to $[\nabla \mathscr{I}_0^\varepsilon \circ (\mathrm{Id} + \boldsymbol{w})][\nabla \mathscr{I}_0^\varepsilon \circ (\mathrm{Id} + \boldsymbol{w})]^\top$ and define $\mathbf{f} \in \mathbb{R}^{2N^2}$ to be the discrete version of the force field $\boldsymbol{f}(\boldsymbol{x}, \mathbf{w}_k; \mathscr{I}_0^\varepsilon, \mathscr{I}_1^\varepsilon)$ (further details are given in [11] and [12]). Then the Newton step (22) is discretized by

$$
\left(-\boldsymbol{E} + \boldsymbol{G}(\tilde{\mathscr{I}}_0)\right) \delta\mathbf{w} = -\boldsymbol{E}\mathbf{w} - \mathbf{f}, \tag{25}
$$

which is a linear equation system in the unknown $\delta\mathbf{w} \in \mathbb{R}^{2N^2}$.

3.1 The Discrete Newton Iteration

We use $\mathbf{w}_1 := 0 \in \mathbb{R}^{2N^2}$ as initial guess and employ the discrete Newton step (25) to obtain the following iteration

$$
\begin{cases} \left(-\boldsymbol{E} + \boldsymbol{G}(\tilde{\mathscr{I}}_{0,k})\right) \delta\mathbf{w}_k = -\boldsymbol{E}\mathbf{w}_k - \mathbf{f}_k, \\ \qquad\qquad \mathbf{w}_{k+1} = \mathbf{w}_k + \tau_k \delta\mathbf{w}_k, \end{cases} \quad k = 1, 2, \ldots \tag{26}
$$

We use a backtracking-like line search [7] to determine the step size τ_k. Let J_h denote a discrete approximation of the cost J then τ_k is computed as follows

$$\begin{cases} \tau_k = & \min_{\tau \in \mathscr{T}} J_h(\mathbf{w}_k + \tau \delta \mathbf{w}_k), \\ \mathscr{T} := \left\{ \tau = \frac{2\ell}{L} | \ell = 1, \dots, L \right\}, \end{cases} \tag{27}$$

where $\mathbf{w}_k$ denotes the current iterate and $\delta \mathbf{w}_k$ the currently computed Newton direction. This method has proven to provide good performance and less total computational cost than standard Armijo–Goldstein or Wolfe–Powell techniques [20] (especially since no expensive evaluations of $\boldsymbol{G}(\tilde{\mathscr{I}}_{0,k})$ are needed).

A combination of residual error and smallest change of iterates is used as stopping criterion in (26). The right hand side of (25) corresponds to the discretized Euler–Lagrange equations (17) of the original minimization problem (6). This motivates stopping the iteration (26) if the (relative) residual error $r_b := |-E\mathbf{w}_k - \mathbf{f}_k| / |-E\mathbf{w}_1 - \mathbf{f}_1|$ is smaller than some tolerance. Additionally we compute the relative change of iterates $r_e := |\mathbf{w}_k - \mathbf{w}_{k-1}|/|\mathbf{w}_k|$ (where $r_e := 0$ if $|\mathbf{w}_k| = 0$) and combine these two notions in a stopping criterion.

As mentioned above the choice of ε may have negative effects on the outcome of the registration. Too small values of ε may produce a trivial solution. Very large values of ε can lead to a loss of potentially important information on the original edges Γ_i. Thus we developed a solution strategy of the registration problem (6) which is robust against the choice of ε. To compensate for the approximation error introduced by replacing the binary edges I_i with the fuzzy edge maps $\mathscr{I}_i$ we augmented Newton's method (26) with an outer loop that reduces blurring in $\mathscr{I}_i$ and thus iteratively refines the computed deformations. We start by computing a "rough" initial guess $\mathbf{w}^\star$ by using a low number of maximal iterations $k_{\max}$ in (26). Then we compute the element-wise-squared of $\tilde{\mathscr{I}}_i^\varepsilon$. According to definition (3) the fuzzy edge maps have intensity one at all points where the raw edges I_i are nonzero, i.e., on Γ_i, and have an intensity less than one everywhere else. Thus by computing the element-wise squared of $\tilde{\mathscr{I}}_i^\varepsilon$ edge-set-pixels still have intensity one whereas their blurred surroundings end up having a lower intensity. Hence the original edge sets Γ_i are accentuated. Then we restart Newton's method using the images $\tilde{\mathscr{I}}_i^\varepsilon(\boldsymbol{x}_j)^2$, the initial guess $\mathbf{w}_1 = \mathbf{w}^\star$ and increase $k_{\max}$. Repeating this procedure sufficiently many times yields fuzzy edges that are very close approximations to the raw edges. Details are given in Algorithm 6.

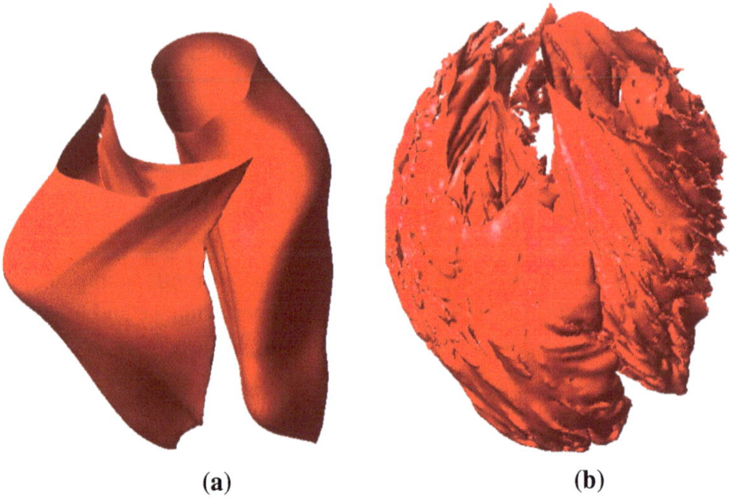

Fig. 2 3D surface representations of the ventricular cavities: **a** The San Diego rabbit ventricles [28] with a spatial discretization Δx of ~250 µm serves as template geometry. **b** An anatomically highly realistic model of the ventricular cavities [4] with a Δx of ~100 µm provides the target geometry during the mapping process

---

Algorithm 6 Iterative method to solve the elastic registration problem (6).

1: **Choose:** $\varepsilon > 0$, $k_{\text{inc}} \in \mathbb{N} : k_{\text{inc}} \geq 2$, tol > 0 and $\mathbf{w} \in \mathbb{R}^{2N^2}$.
2:
3: Given the edge-sets Γ_0 and Γ_1 embed them in the center of images I_0 and I_1.
4: Compute diffuse versions $\mathscr{I}_i^\varepsilon$ of I_i.
5: **for** $\kappa = 1, \ldots K$ **do**
6: **Set** $k_{\max} = k_{\text{inc}} \cdot \kappa$
7: Compute $\mathbf{w}^\star$ using (26) with the line search (27) until either $\min(r_b, r_e) <$ tol or $k > k_{\max}$.
8: **Set** $\tilde{\mathscr{I}}_i^\varepsilon(\boldsymbol{x}_j) := \mathscr{I}_i^\varepsilon(\boldsymbol{x}_j)^2$ and $\mathbf{w}_1 = \mathbf{w}^\star$.
9: **end for**
10: **Set** $\mathbf{w} = \mathbf{w}^\star$.

---

4 Computational Results

All computations were carried out on a 64bit Linux workstation in MATLAB™ 2009b running on a Dell Optiplex 745 equipped with 8 GB of RAM.

The objective of this work was to map a literature-based PS available for the San Diego rabbit ventricles [28, 29] onto the endocardial surfaces of an anatomically highly realistic heart model [4] (referred to as "Oxford-heart"). Both models are shown in Fig. 2. Following an elastic registration approach, we were able to determine the transformation field between the complex endocardial surfaces of both hearts. Since the PS was embedded in the surfaces of the template heart model, it was

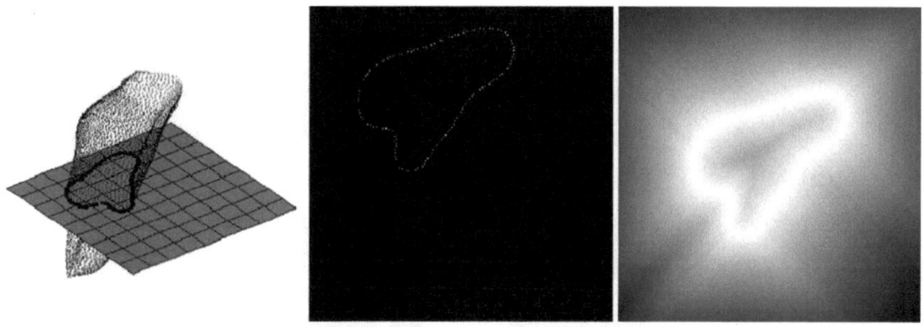

(a) (b) (c)

Fig. 3 Illustration of the dissection of the 3D heart models. **a** shows the surface point cloud (*left cavity*) of the San Diego rabbit heart with a cutting plane present. First the model is cut in horizontal direction (**a**) to generate a 2D edge set (**b**). We center the image and apply our region-growing algorithm (**c**)

self-evident to apply this transformation to the given PS to map the network-like structure onto the endocardial surfaces of the Oxford-heart. In this way topological features as well as the relative position of single network nodes on the ventricular surfaces could be preserved.

Following steps were pursued to apply the proposed solution strategy presented above: the 3D models were sliced to obtain sets of edges in 2D. Each slice was then registered consecutively employing Algorithm 6. The deformation fields computed for each slice were used to map the literature-based Purkinje fiber network onto the endocardium of the Oxford-heart.

Prior to slicing the hearts we applied a (linear) translation to the models to get a maximal spatial overlap. Furthermore, to ensure a clear discrimination between left and right cavity we separated the 3D heart models accordingly. Once left and right cavities were separated, the z-axis was the obvious choice for the normal of the cutting plane.

In this fashion we generated a stack of *binary images* onto which Algorithm 6 was applied. After centering the edges within the binary images we employed our "region-growing" method to obtain blurred approximations $\mathscr{I}_i^\varepsilon$. Note that we approximated the distance function d_{Γ_i} appearing in the definition (3) of $\mathscr{I}_i^\varepsilon$ by a marching procedure using successive discrete convolutions with a 3×3 kernel of ones. A representative result of this region-growing is depicted in Fig. 3c. Thus we generated two image stacks: the blurred San Diego rabbit ventricle slices $\{\mathscr{I}_0^{\varepsilon,m}\}_{m=1}^M$ (the *template image stack*) and the blurred Oxford-heart-cuts $\{\mathscr{I}_1^{\varepsilon,m}\}_{m=1}^M$ (the *reference image stack*). The next step was to compute elastic deformations $\boldsymbol{w}^m$ such that $\mathscr{I}_0^{\varepsilon,m}(\boldsymbol{x} + \boldsymbol{w}^m) \approx \mathscr{I}_1^{\varepsilon,m}(\boldsymbol{x})$ for $m = 1, \ldots, M$ which was done by using a MATLAB-implementation of the augmented Newton method depicted in Algorithm 6.

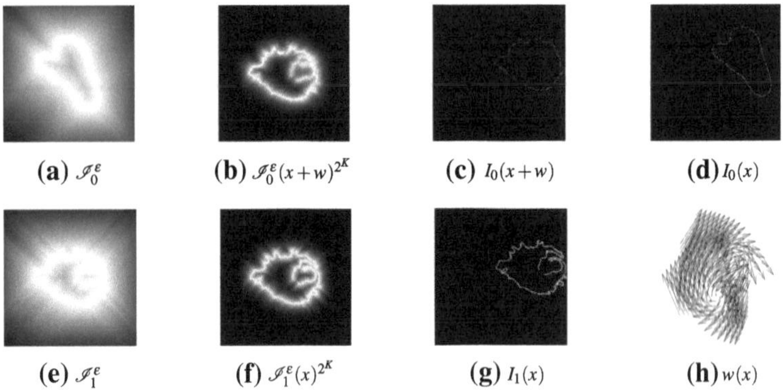

Fig. 4 The different stages of Algorithm 6 for the template image I_0 (*top row*) and the reference image I_1 (*bottom row*). Panels **a** and **e** show the centered and region grown versions of the images. Panel **b** presents the resulting image $\mathscr{I}_0^\varepsilon(\boldsymbol{x} + \boldsymbol{w})^{2^K}$ after completion of Algorithm 6 versus the final reference $\mathscr{I}_1^{\varepsilon 2^K}(\boldsymbol{x})$ (**f**). Panel **c** depicts the original binary image I_0 after application of the computed transformation $\boldsymbol{w}$ versus the original binary target I_1 (**g**). The original binary template is given in panel **d**. Panel **h** shows the computed deformation field $\boldsymbol{w}(\boldsymbol{x})$

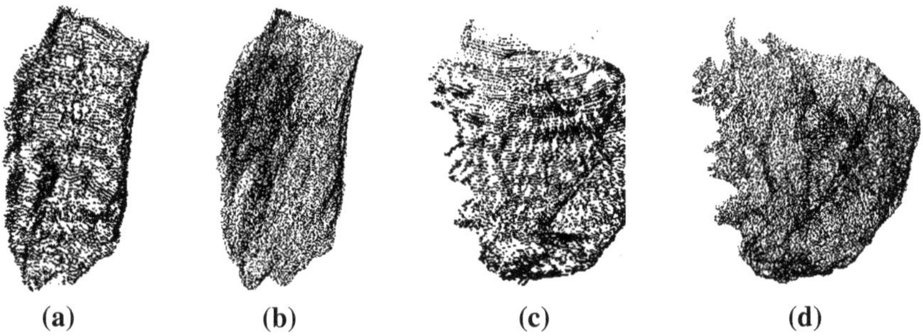

Fig. 5 3D Reconstruction of the registered San Diego rabbit ventricle slices for the *left* (**a**) and *right* (**c**) cavity and the *left* (**b**) and *right* (**d**) cavity of the Oxford heart reference geometry, respectively

The computed deformations $\boldsymbol{w}^m$ were then applied to the raw binary images I_0^m to register the edges Γ_0^m to Γ_1^m. Figure 4 sketches the procedure. Figure 5 shows the 3D reconstruction of the registered San Diego rabbit ventricle slices in comparison to the 3D reference, i.e. the Oxford heart, for both left and right cavities. Finally we used the computed deformation fields to map the literature-based Purkinje fiber network (given as Cartesian coordinates of points representing the spatial locations of the nodes of a 3D graph) from the San Diego heart model onto the Oxford heart. The result is depicted in Fig. 6. A list of all parameters used and their values is given in Table 1.

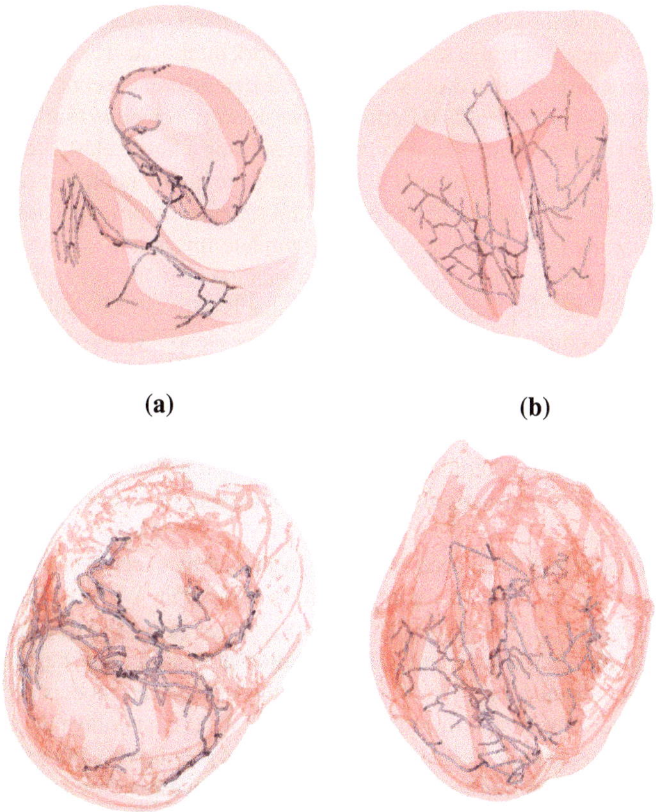

(a) (b) (c) (d)

Fig. 6 The artificial Purkinje fiber network in the San Diego rabbit ventricle model (**a, b**) and the registered network in the Oxford heart (**c, d**)

Table 1 The used parameter values

Parameter	Value	Meaning
λ	1e-2	see (5)
μ	1e-2	see (5)
L	10	see (27)
tol	1e-3	see Algorithm 6
K	4	see Algorithm 6
k_{inc}	5	see Algorithm 6

5 Discussion and Conclusion

The application addressed here required the registered PS to be a subset of the Oxford heart, i.e. the computed deformations had to be sufficiently accurate to avoid network nodes being projected off the endocardial walls. This requirement together with the very complex geometries of the considered heart models made the construction of affine linear mappings to project the network infeasible. The computed elastic deformations guarantee that despite even large differences in the endocardial geometries of both models the artificial Purkinje fiber network is mapped sufficiently close to the Oxford endocardium. The disadvantage of using a *linear* elastic regularization in this context is that the computed deformations are highly nonlinear (compare for instance Fig. 4h). In a linear elastic registration scheme the principal assumption is, however, that the reference image and the template image are two observations of the same elastic body before and after a deformation. The linear elastic potential is thus a model for the displacement of the body that is only valid for sufficiently *small* deformations (compare [18]). This assumption is clearly violated in our context. However, here the elastic registration framework is embedded into our vanishing diffusion strategy; the most pronounced non linearities in the deformation fields are due to the first (outer) loop of Algorithm 6. But in the first loop we only use a very low number of maximal iterations $k_{\max}$ in (25). This together with a suitable choice of the Navier–Lamé constants allowed us to circumvent the restrictions of the linear elastic penalizer.

Though our results have been positively evaluated by experts, the lack of experimental observations makes a rigorous validation difficult. While our 2D approach depends strongly upon the pre-registrations performed in 3D and in 2D, this dependence might be relaxed by a full 3D registration at considerably higher cost. Our simulations confirmed that separating the cavities of the heart models is crucial. By cutting the hearts in vertical direction one obtains slices containing edges from both the left and right cavities. This fact seriously impairs the outcome of the registration: the computed deformation fields may "pull" edges corresponding to the left cavity to an edge arising from a cut through the right cavity and vice versa. Hence we separated the 3D hearts into left and right cavities.

Despite the application presented here our method proved to be a highly efficient and reliable technique to register 2D edges. Forthcoming work even shows that our vanishing diffusion strategy is a really promising method in the very general context of edge set registration problems. In contrast to methods employing the Hausdorff–distance our approach allows us to use the much simpler SSID-distance which is computationally very cheap. Due to the plain structure of the associated cost functional the derivation of necessary optimality conditions by means of variational calculus is straight forward. The use of variational derivatives further enables us to employ fast and theoretically well-founded optimization routines such as Newton's method. Furthermore, the driving force of the registration can be quickly evaluated and is easy to interpret. In relation to other works employing blurring strategies in registration problems we want to emphasize that our novel vanishing diffusion strategy described in Algorithm 6 still proves to be very robust in practical use.

Acknowledgments All authors are supported by the Austrian Science Fund *Fond zur Förderung der Wissenschaftlichen Forschung* (FWF) under grant SFB F032 ("Mathematical Optimization and Applications in Biomedical Sciences" http://math.uni-graz.at/mobis).

References

1. Ambrosio L, Tortorelli V (1990) Approximation of functionals depending on jumps by elliptic functionals via γ-convergence. Commun Pure Appl Math 43:999–1036
2. Ansari A, Ho SY, Anderson RH (1999) Distribution of the purkinje fibres in the sheep heart. Anat Rec 254(1):92–97
3. Aubert G, Kornprobst P (2006) Mathematical problems in image processing. In: Applied mathematical sciences, vol 147. 2nd edn. Springer, Berlin
4. Bishop MJ, Plank G, Burton RA, Schneider JE, Gavaghan DJ, Grau V, Kohl P (2010) Development of an anatomically detailed MRI-derived rabbit ventricular model and assessment of its impact on simulations of electrophysiological function. Am J Physiol Heart Circ Physiol 298(2):H699–718
5. Ciarlet P (1978) The finite element method for elliptic problems. North-Holland, Amsterdam
6. De Giorgi E (1979) Convergence problems for functionals and operators. In: Proceedings of the international meeting on recent methods in non-linear analysis, pp 131–188, Rome, 1978
7. Dennis J, Schnabel R (1983) Numerical methods for unconstrained optimization and nonlinear equations. Prentice-Hall, Englewood Cliffs
8. Droske M, Ring W (2006) A mumford-shah level-set approach for geometric image registration. SIAM J Appl Math 66(6):2127–2148
9. Fitzpatrick J, Hill D, Maurer C (2000) Image registration, medical image processing. In: Chapter 8: Handbook of Medical Imaging, vol 2. SPIE Press, Bellingham
10. Fuchs M, Jüttler B, Scherzer O, Yang H (2009) Shape metrics based on elastic deformations. J Math Imaging Vis 35(1):86–102
11. Fürtinger S, Keeling SL, Plank G, Prassl AJ (2010) Registration of edge sets for mapping a purkinje fiber network onto an endocardium. Technical report, Institute for Mathematics and Scientific Computing, Karl-Franzens University, Graz, Austria
12. Fürtinger S, Keeling SL, Plank G, Prassl AJ (2011) Elastic registration of edge sets by means of diffuse surfaces—with an application to embedding purkinje fiber networks. In: Mestetskiy L, Braz J (eds) VISAPP, SciTePress, pp 12–21
13. Hill W, Baldock RA (2006) The constrained distance transform: interactive atlas registration with large deformations through constrained distances. In: Workshop on image registration in deformable environments (DEFORM'06), Edinburgh
14. Holland RP, Brooks H (1976) The qrs complex during myocardial ischemia. an experimental analysis in the porcine heart. J Clin Invest 57(3):541–550 doi:10.1172/JCI108309
15. Huelsing DJ, Spitzer KW, Cordeiro JM, Pollard AE (1998) Conduction between isolated rabbit purkinje and ventricular myocytes coupled by a variable resistance. Am J Physiol 274(4 Pt 2):H1163–H1173
16. Keeling S, Ring W (2005) Medical image registration and interpolation by optical flow with maximal rigidity. J Math Imaging Vis 23(1):47–65
17. Knauer C, Kriegel K, Stehn F (2009) Minimizing the weighted directed hausdorff distance between colored point sets under translations and rigid motions. In: Proceedings of the 3d international workshop on frontiers in algorithmics (FAW '09), Springer, Berlin, pp 108–119
18. Modersitzki J (2004) Numerical methods for image registration. Oxford Science Publications, New York
19. Mumford D, Shah J (1989) Optimal approximations by piecewise smooth functions and associated variational problems. Commun Pure Appl Math 42(5):577–685
20. Nocedal J, Wright S (2000) Numerical optimization. Springer, New York
21. Ortega JM (1968) The Newton–Kantorovich theorem. The Am Math Monthly 7(6):658–660
22. Paragios N, Ramesh MRV (2002) Matching distance functions: a shape-to-area variational approach for global-to-local registration. In: Proceedings of the 7th European conference on computer vision-part II
23. Peckar W, Schnörr C, Rohr K, Stiehl HS (1999) Parameter-free elastic deformation approach for 2d and 3d registration using prescribed displacements. J Math Imaging Vis 10(2):143–162

24. Plank G, Burton RA, Hales P, Bishop M, Mansoori T, Bernabeu MO, Garny A, Prassl AJ, Bollensdorff C, Mason F, Mahmood F, Rodriguez B, Grau V, Schneider JE, Gavaghan D, Kohl P (2009) Generation of histo-anatomically representative models of the individual heart: tools and application. Philos Trans A Math Phys Eng Sci 367(1896):2257–2292
25. Prassl AJ, Kickinger F, Ahammer H, Grau V, Schneider JE, Hofer E, Vigmond EJ, Trayanova NA, Plank G (2009) Automatically generated, anatomically accurate meshes for cardiac electrophysiology problems. IEEE Trans Biomed Eng 56(5):1318–1330
26. Toriwaki J, Yoshida H (2009) Fundamentals of three-dimensional digital image processing. Springer, New York
27. Tranum-Jensen J, Wilde AA, Vermeulen JT, Janse MJ (1991) Morphology of electrophysiologically identified junctions between purkinje fibers and ventricular muscle in rabbit and pig hearts. Circ Res 69(2):429–437
28. Vetter F, McCulloch A (1998) Three-dimensional analysis of regional cardiac function: a model of rabbit ventricular anatomy. Prog Biophys Mol Biol 69(2–3):157–183
29. Vigmond EJ, Clements C (2007) Construction of a computer model to investigate sawtooth effects in the purkinje system. IEEE Trans Biomed Eng 54(3):389–399
30. Yang S, Kohler D, Teller K, Cremer T, Le Baccon P, Heard E, Eils R, Rohr K (2008) Nonrigid registration of 3-d multichannel microscopy images of cell nuclei. IEEE Trans Image Process 17:493–499

Errors in Estimates of Motion and Strain-Tensor in Ultrasound Elastography

Mehmet Bilgen

Abstract Error analysis was performed for quantifying noise in ultrasound elastography images of biologically soft tissue. The interaction of ultrasonic bean with tissue was modeled in 3D. Static tissue deformation applied by an external mechanical source was represented by a second order strain tensor. Complex motion induced in response to deformation was tracked using standard cross correlation based estimator. Pre- and postcompression echo signals were windowed by the same kernel and cross correlated. The amount of shift where the cross correlation function peaked was considered as the estimate of the local tissue motion. Covariance matrix of the errors made in estimating motions within two windows with certain amount of overlap was derived analytically. The components of the covariance matrix were related to the variances of the displacement errors and the errors made in estimating the elements of the strain tensor. The results were combined to investigate the dependencies of these errors on the experimental and signal-processing parameters as well as to determine the effects of one strain component on the estimation of the other. The expressions were evaluated for special cases of axial strain estimation in the presence of axial, axial-shear and lateral-shear type deformations. The signals were shown to decorrelate with any of these deformations; with strengths depending on the reorganization and interaction of the tissue scatterers with the ultrasonic point spread function following the deformation. The loss of signal coherence resulted in more degradation in the estimation performance. The precision of the estimates was sensitive to the direction of the motion. Motion parallel to the transducer array axis produced more error than the motion perpendicular to this axis, and axial shear type deformation introduced more error than that of the lateral shear depending on the features of ultrasonic point spread function as it was defined by its width, length and wavelength. Conditions that favor the improvements in the motion estimation performance were discussed, and advantages gained by signal companding and pulse compression were also illustrated.

M. Bilgen (✉)
Radiology Department, Faculty of Medicine, Erciyes University, 38039 Kayseri, Turkey
e-mail: mehmet.bilgen@yahoo.com

M. Gonzälez Hidalgo et al. (eds.), *Deformation Models*, Lecture Notes in Computational Vision and Biomechanics 7, DOI: 10.1007/978-94-007-5446-1_12,

Ratio 0.5)] and the applied boundary conditions on the tissue surfaces. In any case, the tensor **A** can be decomposed into its symmetric and asymmetric parts. Elasticity theory in classical mechanics calls the symmetric part as strain tensor defining the true deformation while the asymmetric part as pure rotation or translation. Mathematically, the noise-free echo signals received before and after the compression are the signals f_I and f_{II} filtered by the pulse-echo impulse response function of the ultrasonic imaging system h (PSF). In general, h is time dependent and spatially variant as the shape of the PSF can change with depth. With focusing, the curvature of the phase front and width of the PSF also change due to the near-field and farfield effects. By focusing on a specific region of interest located at depth $\mathbf{x}$, the time dependence of the PSF can be absorbed into the spatial description of the function h. The signals received from this region before and after the compression can then be written as

$$\begin{aligned} r_I(\mathbf{x}) &= p_I(\mathbf{x}) + n_I(\mathbf{x}), \\ r_{II}(\mathbf{x}) &= p_{II}(\mathbf{x}) + n_{II}(\mathbf{x}). \end{aligned} \tag{1}$$

where $p_I(\mathbf{x}) = h_I(\mathbf{x}) * f_I(\mathbf{x})$ and $p_{II}(\mathbf{x}) = h_{II}(\mathbf{x}) * f_{II}(\mathbf{x})$ are the noise free signals, $*$ denotes convolution, and $n_I(\mathbf{x})$ and $n_{II}(\mathbf{x})$ are the signal independent additive noise. Here, we allowed the properties of PSF to change during the acquisition of the postcompression echo signal and thus indicated the PSF before and after the compression by different subscripts. Equation (1) represents continuous signals acquired with an ultrasonic linear array. In elevational and lateral directions, the sampled echo field has lower resolution due to coarser spacing between the array elements and higher temporal resolution is obtained along the axial direction due to finer digital sampling. Consequences of sampling on motion estimation performance can be minimized by employing subsample interpolation at the crosscorrelation peak. The analysis in the following was performed with the continuous signals.

3 Correlation Coefficient and Covariance Matrix of Motion Estimates

The strain tensor components, a_{ij}s, are calculated from the differential motion estimated from the pre- and postcompression signals in Eq. (1). For this purpose, window functions $w_1(\mathbf{x})$ and $w_2(\mathbf{x})$ were used for segmenting the echo signal field over volume in 3D and areas in 2D. The functions were separated by a vector $\Delta\mathbf{Z} = (\Delta Z_x, \Delta Z_y, \Delta Z_z)$ between their centers and were made to overlap by choosing the length of $\Delta\mathbf{Z}$ small compared to the window size. Since the windows were separated, the displacement vectors affecting the signals under each window were different and denoted respectively by $\mathbf{D}_1$ and $\mathbf{D}_2$. The motion in tissue was obtained by crosscorrelating the windowed waveforms via

$$\phi_i(\tau_i) = \int d\mathbf{x} w_i(\mathbf{x}+\tau_i) r_I(\mathbf{x}+\tau_i) w_i(\mathbf{x}) r_{II}(\mathbf{x}) \tag{2}$$

where the subscript $i = 1$ and 2 was used to identify the windows from which the displacement vector was estimated. The shift vector τ_i maximizing the cross-correlation function was considered as the estimate of the displacement vector D_i, i.e.,

$$\phi_i(\hat{\mathbf{D}}_i) = \max_{\tau_i} \phi_i(\tau_i). \tag{3}$$

Signal decorrelation was measured from the attenuation of the correlation coefficient at its peak

$$\rho = \frac{E\left[\phi_1\left(\hat{\mathbf{D}}_1\right)\right]}{\sqrt{\phi_{I-I}\phi_{II-II}}}, \tag{4}$$

where the terms in the denominator were given by

$$\phi_{j-j} = E\left[\int d\mathbf{x} w_1(\mathbf{x}) r_j^2(\mathbf{x})\right], \quad \text{for} \quad j = I \text{ and } II. \tag{5}$$

Taylor series expansion of the cross correlation function at its peak in Eq. (3) allows writing the error between the true displacement $\mathbf{D}$ and its estimate $\hat{\mathbf{D}}$ in matrix-vector representation as

$$\hat{\mathbf{D}}_i - \mathbf{D}_i \approx -\mathbf{H}_i^{-1} \nabla \phi_N(\tau_i)|_{\tau_i = \hat{\mathbf{D}}_i}. \tag{6}$$

Here $\nabla = (\partial/\partial x, \partial/\partial y, \partial/\partial z)$ denotes the gradient operation and

$$\mathbf{H}_i = \nabla \nabla^\top E\left[\phi_{ip}(\tau_i)\right]_{\tau_i = \mathbf{D}_i}, \tag{7}$$

(7) is the expected value of the Hessian matrix of the cross-correlation function of the noise free signals

$$\phi_{ip}(\tau_i) = \int d\mathbf{x} w_1(\mathbf{x}+\tau_i) p_I(\mathbf{x}+\tau_i) w_2(\mathbf{x}) p_{II}(\mathbf{x}). \tag{8}$$

The function

$$\begin{aligned}\phi_N(\tau_i) = \int & d\mathbf{x} w_1(\mathbf{x}+\tau_i) w_2(\mathbf{x}) \left(p_I(\mathbf{x}+\tau_i) n_{II}(\mathbf{x})\right. \\ & \left. + n_I(\mathbf{x}+\tau_i) p_{II}(\mathbf{x}) + n_I(\mathbf{x}+\tau_i) n_{II}(\mathbf{x})\right),\end{aligned} \tag{9}$$

is the contribution from the cross terms associated with the signals and the noise. Equation (6) was reached by making the following assumptions:

1. Variations in $\phi_{ip}(\tau_i)$ about the ensemble mean $E\left[\phi_{ip}(\tau_i)\right]$ are small, so that its gradient and Hessian matrix can be approximated by $\nabla\phi_{ip}(\tau_i) \approx 0$ and $\nabla\nabla^\top\phi_{ip}(\tau_i) \approx E\left[\nabla\nabla^\top\phi_{ip}(\tau_i)\right]$.
2. Contributions from the third and higher terms of the Taylor series expansion of the correlation function are small.
3. Components of the strain tensor $\mathbf{A}$ are small so that the displacement estimates are unbiased and the PSF is spatially invariant under each window.
4. $n_I(\mathbf{x})$ and $n_{II}(\mathbf{x})$ in Eq. (1) are zero mean random noise processes that are signal-independent and uncorrelated.

With these assumptions, the expected deviation of the displacement estimates becomes zero, i.e.,

$$E\left[\hat{\mathbf{D}}_i - \mathbf{D}\right] \approx 0, \tag{10}$$

indicating that $\hat{\mathbf{D}}_i$ is an unbiased estimate of $\mathbf{D}_i$. $E[\cdot]$ in the above equation denotes the expectation operation. The covariance matrix of the displacement estimates can be derived from the above expressions as

$$\begin{aligned} cov(\hat{\mathbf{D}}_1, \hat{\mathbf{D}}_2) &= E\left[(\hat{\mathbf{D}}_1 - \mathbf{D}_1)(\hat{\mathbf{D}}_2 - \mathbf{D}_2)^\top\right] \\ &\approx \mathbf{H}_1^{-1}\nabla\nabla^\top E\left[\phi_{1N}(\tau_1)\phi_{2N}(\tau_2)\right]|_{\tau_1=\hat{\mathbf{D}}_1, \tau_2=\hat{\mathbf{D}}_2}\mathbf{H}_2^{-\top}. \end{aligned} \tag{11}$$

The correlation coefficient in Eq. (4) and the covariance matrix in Eq. (11) were calculated explicitly. The window functions $w_1(\mathbf{x})$ and $w_2(\mathbf{x})$ were chosen as Gaussian

$$w_1(\mathbf{x}) = w(\mathbf{x}) = \exp(-4\mathbf{x}^\top\mathbf{Z}^{-1}\mathbf{x}), \quad \text{and} \tag{12}$$

$$w_2(\mathbf{x}) = w(\mathbf{x} - \Delta\mathbf{Z}) = \exp(-4(\mathbf{x} - \Delta\mathbf{Z})^\top\mathbf{Z}^{-1}(x - \Delta\mathbf{Z})). \tag{13}$$

Here the matrix $\mathbf{Z}$ determines the size of the window functions and $\Delta\mathbf{Z}$ is the window shift vector defined in the previous section. To make the mathematical analysis tractable, we further assume that $w(\mathbf{x} + \tau)w(\mathbf{x}) \approx w^2(\mathbf{x})$ and $w(\mathbf{x}_1 + \tau_1)w(\mathbf{x}_1)w(\mathbf{x}_2 + \tau_2 - \Delta\mathbf{Z})w(\mathbf{x}_2 - \Delta\mathbf{Z}) \approx w^2(\mathbf{x}_1)w^2(\mathbf{x}_2 - \Delta\mathbf{Z})$. These are valid approximations when $|\tau|$, $|\tau_1|$ and $|\tau_2| \ll |\mathbf{Z}|$.

The PSF h_I in Eq. (1) was modeled by a Gaussian modulated sinusoid

$$H_I(\mathbf{k}, \mathbf{L}, \mathbf{k}_0) = \exp\left(-(\mathbf{k} + \mathbf{k}_0)^\top\mathbf{L}(\mathbf{k} + \mathbf{k}_0)/2\right) + \exp\left(-(\mathbf{k} - \mathbf{k}_0)^\top\mathbf{L}(\mathbf{k} - \mathbf{k}_0)/2\right), \tag{14}$$

which represents a band-pass spectrum in the 3D domain. The matrix $\mathbf{L}$ determines the size of the PSF while $\mathbf{k}_0$ is the center frequency and direction of ultrasound propagation. The PSF h_{II} assumes the same functional form as h_I but defined by different parameters $\mathbf{L}'$ and $\mathbf{k}_0'$. This primed notation was conveniently used to distinguish the PSF shaping effect during the acquisition of the postcompression signals. The reflectivity function f_I and the additive noise n

components in echo signals were considered to be Gaussian white noise processes with variances σ_f^2 and σ_n^2, respectively. The signal-to-noise ratio was defined as $SNR = \sigma_f^2/\sigma_n^2 \gg 1$. The above considerations lead to the following derivations. Based on Eq. (8),

$$E\left[\phi_{ip}(\tau_i)\right] = \frac{\sigma_f^2|\mathbf{Z}||\mathbf{L}|^{1/2}|\mathbf{L}'|^{1/2}}{8\sqrt{2}|\Sigma|^{1/2}} \times \mathrm{Re}\Big[\exp\Big(-(\mathbf{k}_0^\top\mathbf{L}\mathbf{k} + \mathbf{k}_0'^\top\mathbf{L}'\mathbf{k}_0')/2 + (j(\tau_i - \mathbf{D}_i) + \mathbf{b})^\top \Sigma^{-1}(j(\tau_i - \mathbf{D}_i) + \mathbf{b})/4\Big)\Big], \tag{15}$$

where Re[·] denotes the real part of its argument and

$$\Sigma = \left[\mathbf{L}\mathbf{k}_0 + (\mathbf{I}+\mathbf{A})\mathbf{L}'(\mathbf{I}+\mathbf{A})\right]/2 + \mathbf{A}\mathbf{Z}\mathbf{A}^\top/32, \quad \text{and,} \tag{16}$$

$$\mathbf{b} = \mathbf{L}\mathbf{k}_0 + (\mathbf{I}+\mathbf{A})\mathbf{L}\mathbf{k}_0'. \tag{17}$$

And based on Eq. (11) and $SNR \gg 1$,

$$E[\phi_{1N}(\tau_1)\phi_{2N}(\tau_2)] = \frac{\sigma_f^2\sigma_n^2|\mathbf{Z}|^{1/2}}{SNR}\exp(-4\Delta\mathbf{Z}^\top\mathbf{Z}^{-1}\Delta\mathbf{Z}) \times \mathrm{Re}\left[\frac{\exp(-\mathbf{c}_1^\top\mathbf{L}^{-1}\mathbf{c}_1 + j\mathbf{c}_1\mathbf{k}_0)}{|\mathbf{L}|} + \frac{\exp(-\mathbf{c}_2^\top\mathbf{L}'^{-1}\mathbf{c}_2 + j\mathbf{c}_2\mathbf{k}_0')}{|\mathbf{I}+\mathbf{A}||\mathbf{L}|}\right] \tag{18}$$

where

$$\mathbf{c}_1 = \tau_1 - \tau_2, \tag{19}$$

$$\mathbf{c}_2 = \tau_1 - \tau_2 - (\mathbf{I}+\mathbf{A})^{-1}(\mathbf{D}_1 - \mathbf{D}_2). \tag{20}$$

The calculation of the covariance matrix in Eq. (11) further requires the Hessian matrices $\mathbf{H}_1$ and $\mathbf{H}_2$ through Eqs. (7) and (15). Notice that Eq. (18) is a function of unknown displacements $\mathbf{D}_1$ and $\mathbf{D}_2$ through Eq. (20). The algebraic complexity in evaluating Eq. (11) can further be reduced by the approximation $\mathbf{c}_2 \approx \tau_1 - \tau_2$. The error committed with this substitution into the calculation of the covariance matrix was negligible since the displacements introduce shift effect via Eq. (20), but the partial derivatives in the gradient operations are taken with respect to the variables τ_1 and τ_2 in Eq. (11).

Based on the observation that, the shift dependent term in the first line of Eq. (18) is not affected by the gradient operations and therefore it can be factored out to render the displacement covariance matrix

$$cov(\hat{\mathbf{D}}_1, \hat{\mathbf{D}}_2) = \exp(-\Delta\mathbf{Z}^\top\mathbf{Z}^{-1}\Delta\mathbf{Z})\mathrm{var}(\hat{\mathbf{D}}_1), \tag{21}$$

where $\mathrm{var}(\hat{\mathbf{D}}_1) = cov(\hat{\mathbf{D}}_1, \hat{\mathbf{D}}_1)$ is the displacement variance matrix. Equation (21) relates the displacement covariance to the displacement variance in a simple fashion through multiplication by an exponential with window shift and size dependency. The correlation coefficient in Eq. (4) can be expressed in closed form as

$$\rho = \frac{|\mathbf{L}|^{1/4}|\mathbf{L}'|^{1/4}|\mathbf{I}+\mathbf{A}|^{1/2}}{|\Sigma|^{1/2}} \exp\left(-(\mathbf{k}_0^\top \mathbf{L}\mathbf{k} + \mathbf{k}_0'^\top \mathbf{L}'\mathbf{k}_0')/2 + \mathbf{b}^\top \Sigma^{-1}\mathbf{b}/4\right). \quad (22)$$

4 Errors in Strain Tensor Estimates

The diagonal term $E\left[(\hat{D}_{1z} - D_{2z})^2\right]$ of the covariance matrix in Eq. (11) is essential for predicting the variance of the axial strain estimates, i.e., a_{zz} component of the tensor $\mathbf{A}$,

$$\hat{a}_z z = \frac{(\hat{D}_{2z} - \hat{D}_{1z})}{\Delta Z_z}. \quad (23)$$

The other diagonal terms can similarly be used to predict the variances of the lateral and elevational strain estimates $\hat{a}_{ii} = (\hat{D}_{2i} - \hat{D}_{1i})/\Delta Z_i$ for $i = x, y$. The off-diagonal elements are related to the variances for the estimates of the off-diagonal components of $\mathbf{A}$. For example, the estimate of the shear strain a_{xz} are calculated from the displacement estimates according to

$$\hat{a}_{xz} = \frac{1}{2}\left(\frac{\partial D_z}{\partial x} + \frac{\partial D_x}{\partial z}\right) = \frac{(\hat{D}_{2z} - \hat{D}_{1z})}{2\Delta Z_x} + \frac{(\hat{D}_{2x} - \hat{D}_{1x})}{2\Delta Z_z} = \frac{\Delta Z_z}{2\Delta Z_x}\hat{a}_{zz} + \frac{\Delta Z_x}{2\Delta Z_z}\hat{a}_{xx}. \quad (24)$$

The variance of this estimate can explicitly be written as

$$\mathrm{var}(\hat{a}_{xz}) = \frac{\Delta Z_x^2}{4\Delta Z_z^2}\mathrm{var}(\hat{a}_{xx}) + \frac{\Delta Z_z^2}{4\Delta Z_x^2}\mathrm{var}(\hat{a}_{zz}) + \frac{2cov(\hat{D}_{1x}, \hat{D}_{1z}) - cov(\hat{D}_{1x}, \hat{D}_{2z}) - cov(\hat{D}_{2x}, \hat{D}_{1z})}{4\Delta Z_x \Delta Z_z}, \quad (25)$$

where

$$\mathrm{var}(\hat{a}_{ii}) = 2\left(cov(\hat{D}_{1i}, \hat{D}_{1i}) - cov(\hat{D}_{1i}, \hat{D}_{2i})\right)/\Delta Z_i^2 \quad \text{for } i = x, z, \quad \text{and} \quad (26)$$

equalities $cov(\hat{D}_{1i}, \hat{D}_{1i}) = cov(\hat{D}_{2i}, \hat{D}_{2i})$, $cov(\hat{D}_{1x}, \hat{D}_{1z}) = cov(\hat{D}_{2x}, \hat{D}_{2z})$ hold. The $cov(\cdot, \cdot)$ terms in the right hand side of Eq. (25) are determined from the off diagonal $x - z$ and $z - x$ elements of the covariance matrix via Eq. (11).

5 Calculation Results

In the following, the utilities of the formulas derived in the previous section are illustrated with examples. The geometry was restricted to 2D to reduce the complexity of computations while, at the same time, providing basic understanding of the interactions between the experimental and signal processing parameters. The extension to 3D can be carried out trivially by invoking the elevational parameters into the calculations. Three 2D strain tensors were considered to represent the basic deformations in the forms of axial compression, pure axial shear, and pure lateral shear that are respectively defined by the following matrices

$$\mathbf{A} = \begin{pmatrix} a_{xx} & 0 \\ 0 & a_{zz} \end{pmatrix}, \begin{pmatrix} a_{xx} & 0 \\ a_{az} & a_{zz} \end{pmatrix} \text{ and } \begin{pmatrix} a_{xx} & a_{xz} \\ 0 & a_{zz} \end{pmatrix} \tag{27}$$

The last two can be combined to obtain a rotation tensor. The condition $|\mathbf{I}+\mathbf{A}| = 1$ simulates an incompressible tissue, which leads to the substitution $a_{xx} = -a_{zz}$ since $a_{zz}^2 \ll 1$ for small axial strains. When dealing with other types of tissue, e.g. poroelastic tissue, the values for the individual tensor components are determined from the Poisson ratio and boundary conditions. Unlike the convention used in classical elastodynamics, the positive strain values in this paper represented compression along the axial z-direction and negative strains were expansion along the lateral x-direction. The PSF was described by a diagonal matrix

$$\mathbf{L} = \begin{pmatrix} L_x^2 & 0 \\ 0 & L_z^2 \end{pmatrix}, \tag{28}$$

where the terms L_x and L_z, respectively, define the width and length of PSF. The vector $\mathbf{k}_0$ was considered as $(0, k_0)$ to represent ultrasound beam propagating along the z-direction. The corresponding parameters were indicated by primed variables, i.e., $\mathbf{L}$ and $\mathbf{k}_0'$, for the PSF that is shaped during the postcompression signal acquisition.

The parameters in the calculations were set to simulate 5 MHz transducer, no PSF shaping and large signal-to-noise ratio ($SNR = 100$). The PSF parameters were $L_z = 300\,\mu\text{m}$, $k_0 L_z = 2\pi$, $L_z' = L_z$ and $k_0 = k0$.

Figures 1, 2 and 3 illustrate the changes in the axial displacement variance $var(\hat{D}_{1z})$ and the behaviour of the correlation coefficient ρ as the experimental conditions are varied for a range of parameter values described in the figure captions. Figure 1 depicts the dependence of the variance on the ratio Z_z/L_z (equivalent to time-bandwidth product) when the axial- and lateral-shear strains were absent in the tissue volume (i.e., $a_{xz} = a_{zx} = 0$). For the condition of zero axial strain, a unit volume translates with its dimensions unchanged as shown in Fig. 1a. According to Fig. 1b, the variance of estimating the axial component of the motion decreases linearly with the ratio Z_z/L_z in the log scale as in the case of the Cramer-Rao lower bound. This decreasing trend can be represented by the formula

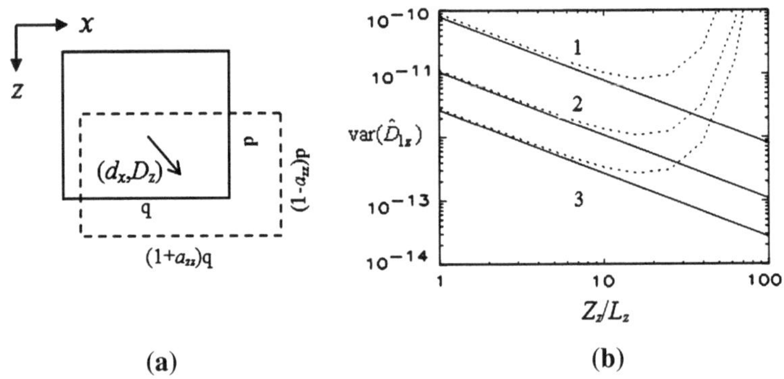

Fig. 1 Shows **a** the deformation of a unit volume with axial compression and **b** the variations of axial displacement estimation variance for $a_{zz} = 0$ (*solid lines*) and $a_{zz} = 0.025$ (*dotted lines*) when (*1*) $L_x = L_z, d_x = 2L_x$, (2) $L_x = L_z, d_x = 0$ and (*3*) $L_x = L_z/4, dx = 0$. The other parameter values used in the calculations are $a_{xz} = a_{zx} = 0$, $Lz = 300\,\mu\mathrm{m}$, $k_0 L_z = 2\pi$, $Lz = Lz$, $k_0' = k_0$ and $SNR = 100$

$$\mathrm{var}(\hat{D}_{1z}) = \left\{ \frac{8L_z^3}{SNR Z_z (1 + 2k_0^2 L_z^2)} \right\} L_x \exp(d_x^2 / 2L_x^2), \quad \text{for} \quad a_{zz} = 0. \tag{29}$$

The parameters L_x (PSF width) and d_x (lateral motion induced in the tissue after the compression) scale the variance, as illustrated in the figure.

For nonzero axial strains $a_{zz} \neq 0$, the unit volume in Fig. 1a is subjected to a deformation and consequently its dimension changes depending on the amount of axial strain. The displacement variance curves attain a minimum at

$$\frac{Z_z}{L_z} = \frac{\sqrt{10}}{a_{zz}\sqrt{1 + k_0^2 L_z^2}}. \tag{30}$$

Conditions simulating Z_z/L_z larger than this optimal value produce less precise displacement estimates.

The effects of shear strain components (both axial and lateral) and the width of PSF on the axial displacement error and the correlation coefficient are illustrated effectively and compactly in Figs. 2 and 3. Figure 2a shows the associated deformations and axial motions of a unit volume under the mentioned shear strain conditions. Figure 2b and c are plotted as a function of axial strain with data points corresponding to two different L_x values under three combinations of shear strains. The results in Fig. 2b indicates that wide PSF favors larger errors and the presence of shear strain may alter the estimation performance. For example, for $L_x = 4L_z$ (dotted lines), the variances represented by the symbols + and Δ overlap and, at the same time, attain lower values compared to those indicated by the symbol □. For the case of $L_x = L_z/4$ (solid lines), the curves with the symbols + and □ overlap while crossed by the curve with symbol Δ at 5 %. Displacement errors are larger when $a_{zz} < 5\,\%$ and lower when $a_{zz} > 5\,\%$. All the overlapping data points in Fig. 2b produce

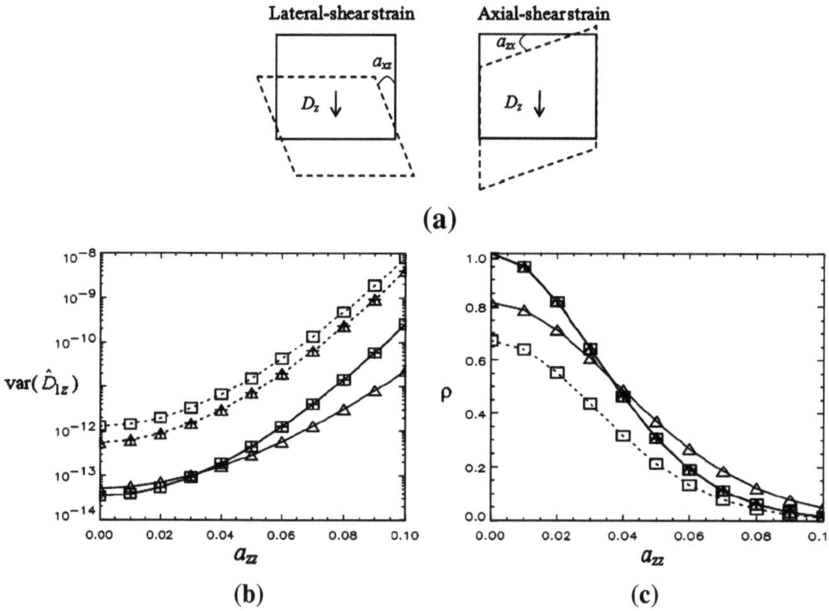

Fig. 2 Shows **a** deformations of a unit volume under lateral- and axial-shear strains, and variations of **b** axial displacement estimation variance and **c** signal decorrelation with axial strain for $L_x = L_z/4$ (*solid lines*) and $L_x = 4L_z$ (*dotted lines*). The symbols represent; $+ : a_{xz} = a_{zx} = 0$, Δ : $a_{xz} = 0.05$, $a_{zx} = 0$, and $a_{xz} = 0$, $a_{zx} = 0.05$. The other parameter values used in the calculations are $L_z = 300\,\mu\text{m}$, $k_0 L_z = 2\pi$, $L_z' = L_z$, $k_0' = k_0$, $Z_z = 20L_z$ and $SNR = 100$. Notice that in **b**, *solid lines* with + and □ overlap, indicating minimal a_{zx} effect for $L_x = L_z/4$, and *dotted lines* with + and Δ overlap, indicating minimal a_{xz} effect for $L_x = 4L_z$. All the lines overlapping in **b** also overlap in **c**

identical correlation coefficient as seen in Fig. 2c. From these results, it is clear that the accuracy of displacement estimates are determined by the complex interactions of underlying strain parameters. To further investigate these interactions, similar calculations were repeated in Fig. 3a and b, but this time, the horizontal axis represented either a_{xz} or a_{zx}, and the data points were for the two different values of L_x and a_{zz}. According to this data, the errors exhibited varying degrees of sensitivity to the strain components a_{xz} and a_{zx} and also the PSF width. The origins of this sensitivity variation is explained in the next section.

6 Discussions

The most common axial strain estimator is based on Eq. (23) where the axial displacements are estimated according to Eq. (3). The pre- and postcompression A-lines acquired by the same array element are crosscorrelated after segmented by overlapped

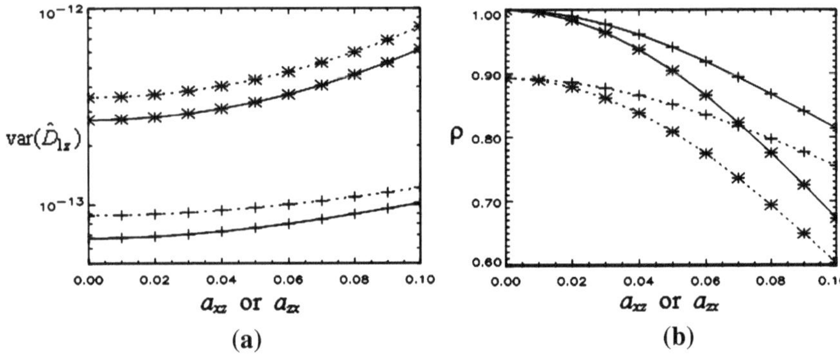

Fig. 3 Shows variations of **a** axial displacement estimation variance and **b** signal decorrelation with axial- or lateral-shear strains for $L_x = L_z/2$ (*solid lines*) and $L_x = 2L_z$ (*dotted lines*). The symbols represent; +: $a_{zz} = 0$ and $\times$: $a_{zz} = 0.015$. The other parameter values used in the calculations are $L_z = 300\,\mu\text{m}$, $k_0 L_z = 2\pi$, $L_z' = L_z$, $k_0' = k_0$, $Z_z = 20L_z$ and $SNR = 100$

and shifted window functions. When single A-lines used, the matrix **Z** describing the window function reduces to a scalar quantity $Z = Z_z^2$. Combining (21) and Eq. (26), the estimation performance can be measured from the variance formula

$$\mathrm{var}(\hat{a}_{zz}) = \mathrm{var}(\hat{D}_{1z}) \left\{ 2 \left(1 - \exp(-4\Delta Z_z^2/Z_z^2) \right) \right\}. \tag{31}$$

This expression is simple in the sense that the axial displacement variance and the window-shift dependent term in the curly brackets are separated, allowing better interpretation of how the window overlap contributes to the axial strain variance. The ratio $\Delta Z_z/Z_z$ is called the percentage window shift or overlap and is associated with the spatial resolution of axial strain estimates. It is usually kept small for improving the resolution and for these cases (i.e., $\Delta Z_z/Z_z < 25\,\%$), Eqs. (24–26) reduces to

$$\mathrm{var}(\hat{a}_{zz}) \approx 8 \frac{\mathrm{var}(\hat{D}_{1z})}{Z_z^2}. \tag{32}$$

Speckle decorrelation was defined as the attenuation of the correlation coefficient. In this paper, we presented similar results using Eq. (22), but for the rf-signals. Band-pass rf signals offer displacement estimates with higher precision due to the preserved phase information, but also decorrelate rapidly with deformation as compared to the base-band envelop signals. According to Eqs. (22) and (23), the displacement estimation variance increases with both of these parameters L_x and d_x, as illustrated in Fig. 1. This primarily occurs due to the signals being decorrelated by the lateral tissue motion and larger number of independent tissue scatterers contained in wider PSF. The last statement is correct provided that the amplifier gains are unchanged so that the SNR stays constant. Therefore, best estimation performance is achieved with pulses with small widths and when $d_x = 0$. This corresponds to the analysis of

A-lines coinciding with the axis of symmetry in the lateral motion, i.e., the center axis of the compressor. As the line of sight is moved away from this symmetry, the lateral motion increases according to $d_x = a_{xx}x$ for a uniform tissue, where x is the distance from the center axis, and thus significantly increase the variances for small width PSFs. This relationship indicates a tradeoff between the selection of PSF width and tolerable lateral motion. In order to compensate this effect, signal-processing strategies involving lateral search in the echo field can be implemented to employ highly coherent signal pairs received by two nearby transducer array elements in the motion estimation process.

The lateral motion can be estimated by tracking the changes in signals across the echo field. The error analysis for these estimates can be performed using the displacement covariance matrix derived above. Unlike axial signals, the signals in the lateral direction lack phase modulation and consequently the resulting estimates become less precise. This directional dependence of the errors induces anisotropy in the performance of the motion estimation.

Any type of deformation monotonically decorrelates the pre- and postcompression signals and subsequently reduces the accuracy of the motion estimates. As for the conditions that led to the calculations in Figs. 2 and 3, the sensitivity of the errors varied with the type of deformation and the width of the PSF. This sensitivity variation can be explained by analyzing the motion of tissue relative to the PSF profile after the deformation. Figure 4 graphically depicts two PSFs with different widths and a unit volume deforming in the presence of axial- and lateral shear strains. Both PSF profiles have the same axial length $2L_z$ and wavelength λ. In response to the strains, two hypothetical points (representing tissue scatterers) move along the directions indicated by the arrows. The strength of ultrasonic backscatter from each scatterer is determined by its position with respect to the PSF profile. Before and after the deformation, the scatterer on the left moves to different phase of the PSF but the scatterer on the right stays on the same phase. At their new positions, the scatterers experience different PSF amplitudes that change the amount of contribution to the backscattered echo. The difference in the amount of backscatter contribution before and after the deformation makes the pre- and post-compression signals decorrelate. For the scatterer on the left, the signals before and after the deformation completely loose phase coherence if the axial-shear strain satisfies $a_{zx} > \lambda/2L_x$. The scatterer on the right moves out of the beam when the lateral-shear strain is $a_{xz} > L_x/L_z$ and does not contribute to the post-compression signal at all. Under these conditions, the most severe decorrelation occurs. By generalizing this simple analysis, it can be stated that the axial-shear strains become more effective for wide pulses $L_x \gg L_z$ with short wavelength, and the lateral-shear strains become more important for long or narrow pulses $L_x \ll L_z$. Clearly, it is essential to investigate the communality of the deformation types in tissues so that appropriate PSF offering best compromise in image quality can be designed for different tissues (e.g, breast versus prostate tissue).

Signal companding was successfully shown to improve the performance of axial strain estimation. Signal warping is a generalized form of signal companding in 3D and produces a maximum-likelihood estimate for the motion. Therefore, it is

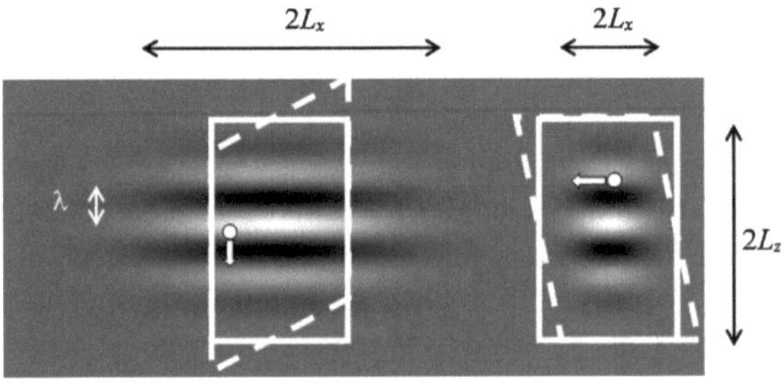

Fig. 4 Shows the interaction of PSF with the tissue deformation under axial- and lateral-shear strain conditions. On the *left* side, PSF with large width and axial-shear strain; and on the *right* side, PSF with small width and lateral-shear strain are illustrated. Notice that both PSF profiles have the same axial length $2L_z$ and wavelength λ. The applied strains move the two hypothetical scatters (points) along the directions indicated by the *arrows*. *Gray* scale values represent the amplitude of the PSF. Please refer to the main text for further explanations regarding the figures

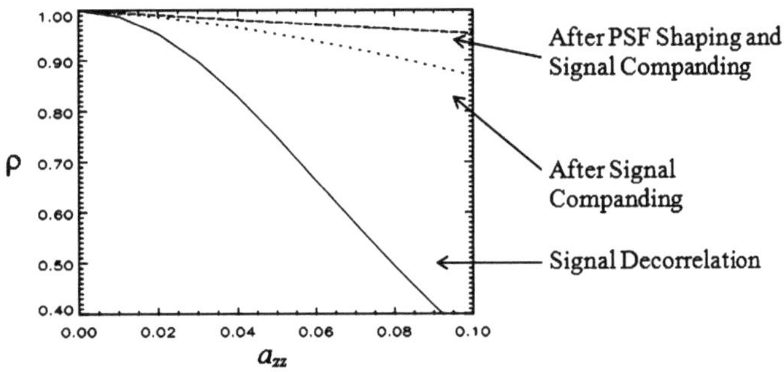

Fig. 5 Shows improvements in signal decorrelation (*solid line*) with signal companding by $(1 + a_{zz})$ alone (*dotted line*) and PSF compression followed by companding $L_z' = L_z/(1 + a_{zz})$, $k_0' = k_0(1 + a_{zz})$, (*dashed line*). The parameter values used in the calculations are $L_z = 300\,\mu\text{m}$, $k_0 L_z = 2\pi$, $Z_z = 20L_z$ and $SNR = 100$

advisable to apply these post-processing strategies to the pre- and post-compression signals. In addition, if possible, PSF shaping should be carried out during data acquisition to reduce the signal decorrelation. Figure 5 illustrates the improvements in the signal decorrelation with each of these attempts using 1D signal model. Although companding alone was effective, best restoration of the signal decorrelation was achieved with the acquisitions that employ PSF compression followed by signal companding as illustrated in the figure. These results together suggest better strategies for designing and developing new data acquisitions and instrumentation, and evaluating the elastography image formation algorithms, as appeared later in the literature [6, 7, 9, 10].

7 Conclusion

Given the experimental conditions, achieving accurate and precise motion estimates in elastography imaging is very challenging. Obtaining clinically useful elastograms requires careful adjustment of the experimental and signal processing parameters that interact in complex fashion. The analysis presented in this paper enhanced the basic understanding of some key elements and their effects on the performance of 3D motion estimation in elastography with the help of expressions derived for the correlation coefficient and displacement covariance matrix. It served as a guidance to determine the range of appropriate parameter settings that offer improved image quality for any elastography system and also to evaluate and design new systems with better performances. This paper did not consider phase aberration, curvature of motion and curvature of the PSF. Therefore, investigating their effects on motion estimation errors still remains for future studies.

References

1. Bae U, Dighe M, Dubinsky T, Minoshima S, Shamdasani V, Kim Y (2007) Ultrasound thyroid elastography using carotid artery pulsation: preliminary study. J Ultrasound Med 26(6): 797–805
2. Bilgen M (2000) Dynamics of errors in 3d motion estimation and implications for strain-tensor imaging in acoustic elastography. Phys Med Biol 45(6):1565–1578
3. Bilgen M, Insana MF (1998) Elastostatics of a spherical inclusion in homogeneous biological media. Phys Med Biol 43(1):1–20
4. Chakraborty A, Bamber JC, Dorward NL (2012) Slip elastography: a novel method for visualising and characterizing adherence between two surfaces in contact. Ultrasonics 52(3):364–376
5. Hall TJ, Zhu Y, Spalding CS (2003) In vivo real-time freehand palpation imaging. Ultrasound Med Biol 29(3):427–435
6. Ophir J, Garra B, Kallel F, Konofagou E, Krouskop T, Righetti R, Varghese T (2000) Elastography imaging. Ultrasound Med Biol 26:S23–S29
7. Ophir J, Srinivasan S, Righetti R, Thittai A (2011) Elastography: a decade of progress (2000–2010). Curr Med Imaging Rev 7(4):292–312
8. Parker KJ, Doyley MM, Rubens DJ (2011) Imaging the elastic properties of tissue: the 20 year perspective. Phys Med Biol 56(1):R1–R29
9. Rao M, Chen Q, Shi H, Varghese T, Madsen EL, Zagzebski JA, Wilson TA (2007) Normal and shear strain estimation using beam steering on linear-array transducers. Ultrasound Med Biol 33(1):57–66
10. Thittai AK, Yamal JM, Mobbs LM, Kraemer-Chant CM, Chekuri S, Garra BS, Ophir J (2011) Axial-shear strain elastography for breast lesion classification: further results from in vivo data. Ultrasound Med Biol 37(2):189–197
11. Wells PN, Liang HD (2011) Medical ultrasound: imaging of soft tissue strain and elasticity. J R Soc Interface 64(8):1521–1549

MIX
Papier aus verantwortungsvollen Quellen
Paper from responsible sources
FSC® C105338
FSC
www.fsc.org

If you have any concerns about our products,
you can contact us on
ProductSafety@springernature.com

In case Publisher is established outside the EU,
the EU authorized representative is:
Springer Nature Customer Service Center GmbH
Europaplatz 3, 69115 Heidelberg, Germany

Printed by Libri Plureos GmbH
in Hamburg, Germany